Lecture Notes in Computer Scie

Edited by G. Goos, J. Hartmanis, and J. van

Springer

Berlin
Heidelberg
New York
Barcelona
Hong Kong
London
Milan
Paris
Tokyo

Klaus Jansen   Marian Margraf
Monaldo Mastrolilli   José D. P. Rolim (Eds.)

Experimental and Efficient Algorithms

Second International Workshop, WEA 2003
Ascona, Switzerland, May 26-28, 2003
Proceedings

 Springer

Series Editors

Gerhard Goos, Karlsruhe University, Germany
Juris Hartmanis, Cornell University, NY, USA
Jan van Leeuwen, Utrecht University, The Netherlands

Volume Editors

Klaus Jansen
Marian Margraf
University of Kiel
Institute of Computer Science and Applied Mathematics
Olshausenstr. 40, 24098 Kiel, Germany
E-Mail:{kj / mma}@informatik.uni-kiel.de

Monaldo Mastrolilli
IDISIA - Instituto Dalle Molle di Studi sull'Intelligenza Artificiale
Galleria 2, 6928 Manno, Switzerland
E-Mail: monaldo@idsia.ch

José D. P. Rolim
Université de Genève 4, Switzerland
E-Mail: Jose.Rolim@cui.unige.ch

Cataloging-in-Publication Data applied for

A catalog record for this book is available from the Library of Congress

Bibliographic information published by Die Deutsche Bibliothek
Die Deutsche Bibliothek lists this publication in the Deutsche Nationalbibliographie;
detailed bibliographic data is available in the Internet at <http://dnb.ddb.de>.

CR Subject Classification (1998): F.2.1-2, E.1, G.1-2, I.3.5

ISSN 0302-9743
ISBN 3-540-40205-5 Springer-Verlag Berlin Heidelberg New York

Springer-Verlag Berlin Heidelberg New York
a member of BertelsmannSpringer Science+Business Media GmbH

http://www.springer.de

© Springer-Verlag Berlin Heidelberg 2003
Printed in Germany

Typesetting: Camera-ready by author, data conversion by DA-TeX Gerd Blumenstein
Printed on acid-free paper   SPIN: 10931127   06/3142   5 4 3 2 1 0

Preface

This volume contains the papers presented at the 2nd International Workshop on Experimental and Efficient Algorithms (WEA 2003), which took place at Monte Verità, Ascona, Switzerland, during May 26–28, 2003. WEA 2003 was concerned with the applications of efficient algorithms for combinatorial problems.

The volume contains 19 contributed papers, selected by the program committee from 40 submissions received in response to the call for papers. Invited talks by Kurt Mehlhorn, Roberto Solis-Oba and Dorothea Wagner were given at the workshop. We thank all of the authors who submitted papers, our invited speakers, the members of the program committee

J. Boyar, Odense	C. Ribeiro, Rio de Janeiro
A. Broder, IBM	J. Rolim, Geneva, Chair
A. Clementi, Rome	P. Sanders, MPII Saarbrüchen
A. Dandalis, Intel	I. Schiermeyer, Freiburg
A. Ferreira, Inria	M. Serna, Barcelona
A. Fiat, Tel Aviv	S. Skiena, New York
A. Goldberg, Microsoft	R. Solis-Oba, Waterloo
L. Gargano, Salerno	L. Trevisan, Berkeley
D. Johnson, AT&T	D. Trystam, Grenoble
S. Nikoletseas, Patras	E. Upfal, Brown
V. Prasanna, USC	P. Widmayer, Zurich

and the external reviewers: Ernst Althaus, Christoph Ambuehl, David Applegate, Daniel Brown, Annalisa De Bonis, Pierluigi Crescenzi, Pierre-Francois Dutot, Lionel Eyraud, Dimitris Fotakis, Antonio Frangioni, Ferran Hurtado, Miriam Di Ianni, Lefteris Kirousis, Spyros Kontogiannis, Balazs Kotnyek, Christos Makris, Jordi Petit, Adele Rescigno, Gianluca Rossi, Mitali Singh, Andrei Tchernykh, Costas Tsichlas, and Fatos Xhafa.

The best paper award was given for the paper "Algorithmic Techniques for Memory Energy Reduction" by Mitali Singh and Viktor K. Prasanna.

We gratefully acknowledge support from the Istituto Dalle Molle di Studi sull'Intelligenza Artificiale (IDSIA), the University of Lugano (USI), the University of Applied Sciences of Southern Switzerland (SUPSI), the University of Kiel (CAU), the Dipartimento dell'educazione, della cultura e dello sport del Canton Ticino, the city of Ascona, the city of Lugano, the Swiss National Science Foundation (project 20-63733.00/1), the Metaheuristics Network (grant HPRN-CT-1999-00106), ARACNE (grant HPRN-CT-199-00112), APPOL II (grant IST-2001-32007), and AntOptima.

March 2003

Klaus Jansen
Marian Margraf
Monaldo Mastrolilli
José D.P. Rolim

Table of Contents

Improving Linear Programming Approaches for the Steiner Tree Problem

Ernst Althaus[1], Tobias Polzin[1], and Siavash Vahdati Daneshmand[2]

[1] Max-Planck-Institut für Informatik
Stuhlsatzenhausweg 85, 66123 Saarbrücken, Germany
{althaus,polzin}@mpi-sb.mpg.de
[2] Theoretische Informatik, Universität Mannheim
68131 Mannheim, Germany
vahdati@informatik.uni-mannheim.de

Abstract. We present two theoretically interesting and empirically successful techniques for improving the linear programming approaches, namely graph transformation and local cuts, in the context of the Steiner problem. We show the impact of these techniques on the solution of the largest benchmark instances ever solved.

1 Introduction

In combinatorial optimization many algorithms are based (explicitly or implicitly) on linear programming approaches. A typical application of linear programming to optimization problems works as follows: First, the combinatorial problem is reformulated as an integer linear program. Then, some integrality constraints are relaxed and one of the numerous methods for solving (or approximating) a linear program is applied. For $\mathcal{NP}$-hard optimization problems, any linear relaxation of polynomial size (and any polynomial time solvable relaxation) is bound to have an integrality gap (unless $\mathcal{P} = \mathcal{NP}$). So the quality of the underlying relaxation can have a decisive impact on the performance of the overall algorithm. As a consequence, methods for generating tight lower bounds are significant contributions to elaborated algorithms for combinatorial optimization problems, see for example the long history of research for the Traveling Salesman Problem (TSP) focusing on linear programming [3, 4, 14, 15].

In this work, we improve the linear programming based techniques for the Steiner tree problem in networks, which is the problem of connecting a given subset of the vertices of a weighted graph at minimum cost. It is a classical $\mathcal{NP}$-hard problem [12] with many important applications in network design in general and VLSI design in particular. For background information on this problem, see [6, 11].

For the Steiner problem, linear programming approaches are particularly important, since the best known practical algorithms for optimal solutions, for heuristic Steiner trees, and for preprocessing techniques, which reduce the size of the problem instance without changing an optimal solution, all make frequent use

K. Jansen et al. (Eds.): WEA 2003, LNCS 2647, pp. 1–14, 2003.

of linear programming techniques [17, 18, 19, 20]. Typical situations where linear programming is used are the computation of lower bounds in the context of an exact algorithm, bound-based reduction techniques [17], and partitioning-based reduction techniques [18]. Especially for large and complex problem instances, very small differences in the integrality gap can cause an enormous additional computational effort in the context of an exact algorithm. Therefore, methods for improving the quality of the lower bounds are very important.

In Section 2, we give some definitions, including the directed cut relaxation, which is the basis for many linear programming approaches for the Steiner problem. Then, we will present two approaches for improving the lower bound provided by this relaxation:

- In Section 3, we introduce the "vertex splitting" technique: We identify locations in the network that contribute to the integrality gap and split up the decisive vertices in these locations. Thereby, we transform the problem instance into one that is equivalent with respect to the integral solution, but the solution of the relaxation may improve.

  This idea is inspired by the column replacement techniques that were introduced by Balas and Padberg [5] and generalized by Haus et. al. [10] and Gentile et. al. [9]. In these and other papers a general technique for solving integer programs is developed. However, these techniques are mainly viewed as primal algorithms, and extensions for combinatorial optimization problems are presented for the Stable Set problem only. Furthermore, these extensions are not yet part of a practical algorithm (the general integer programming techniques have been applied successfully). Thus, we are the first to apply this basic idea in a practical algorithm for a concrete combinatorial optimization problem.
- In Section 4, we show how to adopt the "local cuts" approach, introduced by Applegate, Bixby, Chvátal, and Cook [4] in the context of the TSP: Additional constraints are generated using projection, lifting and optimal solutions of subinstances of the problem. To apply this approach to the Steiner problem, we develop new shrinking operations and separation techniques.

In Section 5, we embed these two approaches into our successful algorithm for solving Steiner tree problems and present some experimental results. Like many other elaborated optimization packages our program consists of many parts (the source code has approximately 30000 instructions, not including the LP-solver code). Thus, this paper describes only a small part of the whole program. However, among other results, we will show that this part is decisive for the solution of the problem instance d15112, which is to our knowledge the largest benchmark Steiner tree instance ever solved. Furthermore, we believe that these new techniques are also interesting for other combinatorial optimization problems.

The other parts of the program package are described in a series of papers [17, 18, 19, 20]. Note that there is no overlapping between these papers and the work presented here. Some proofs in this paper had to be omitted due to the page constraint, they are given in [1].

2   Definitions

The Steiner problem in networks can be stated as follows (see [11] for details): Given an (undirected, connected) network $G = (V, E, c)$ (with vertices $V = \{v_1, \ldots, v_n\}$, edges E and edge weights $c_e > 0$ for all $e \in E$) and a set R, $\emptyset \neq R \subseteq V$, of *required vertices* (or *terminals*), find a minimum weight tree in G that spans R (a *Steiner minimal tree*). If we want to stress that v_i is a terminal, we will write z_i instead of v_i. We also look at a reformulation of this problem using the (bi-)directed version of the graph, because it yields stronger relaxations: Given $G = (V, E, c)$ and R, find a minimum weight arborescence in $\mathbf{G} = (V, A, c)$ ($A := \{[v_i, v_j], [v_j, v_i] \mid (v_i, v_j) \in E\}$, c defined accordingly) with a terminal (say z_1) as the root that spans $R^{z_1} := R \setminus \{z_1\}$.

A **cut** in $\mathbf{G} = (V, A, c)$ (or in $G = (V, E, c)$) is defined as a partition $C = \{\overline{W}, W\}$ of V ($\emptyset \subset W \subset V; V = W \dot\cup \overline{W}$). We use $\delta^-(W)$ to denote the set of arcs $[v_i, v_j] \in A$ with $v_i \in \overline{W}$ and $v_j \in W$. For simplicity, we write $\delta^-(v_i)$ instead of $\delta^-(\{v_i\})$. The sets $\delta^+(W)$ and, for the undirected version, $\delta(W)$ are defined similarly.

In the integer programming formulations we use (binary) variables $x_{[v_i, v_j]}$ for each arc $[v_i, v_j] \in A$, indicating whether this arc is in the solution ($x_{[v_i, v_j]} = 1$) or not ($x_{[v_i, v_j]} = 0$). For any $B \subseteq A$, $x(B)$ is short for $\sum_{a \in B} x_a$.

For every integer program P, LP denotes the linear relaxation of P, and $v(LP)$ denotes the value of an optimal solution for LP.

Other definitions can be found in [8, 11].

2.1   The Directed Cut Formulation

The directed cut formulation P_C was stated in [25]. An undirected version was already introduced in [2], but the directed variant yields a stronger relaxation.

$$c \cdot x \to \min,$$
$$x(\delta^-(W)) \geq 1 \quad (z_1 \notin W, R \cap W \neq \emptyset), \tag{1}$$
$$x \in \{0, 1\}^{|A|}. \tag{2}$$

The constraints (1) are called Steiner cut constraints. They guarantee that in any arc set corresponding to a feasible solution, there is a path from z_1 to any other terminal.

There is a group of constraints (see for example [13]) that can make LP_C stronger. We call them flow-balance constraints:

$$x(\delta^-(v_i)) \leq x(\delta^+(v_i)) \quad (v_i \in V \setminus R). \tag{3}$$

We denote the linear program that consists of LP_C and (3) by LP_{C+FB}. In [16] we gave a comprehensive overview on relaxations for the Steiner tree problem.

3   Graph Transformation: Vertex Splitting

In this section, we describe a new technique for effectively improving the lower bound corresponding to the directed cut relaxation by manipulating the underlying network.

We use the property that in an optimal directed Steiner tree, each vertex has in-degree at most 1. Implicitly, we realize a case distinction: If an arc $[v_i, v_j]$ is in an optimal Steiner tree, we know that other arcs in $\delta^-(v_j)$ cannot be in the tree. The only necessary operation to realize this case distinction for the Steiner problem is the splitting of a vertex. A vertex v_j is replaced by several vertices v_j^i, one for each arc $[v_i, v_j]$ entering v_j. Each new vertex v_j^i has only one incoming arc $[v_i, v_j^i]$, and essentially the same outgoing arcs as v_j. In Figure 1, the splitting of vertex v_j is depicted. The explanation of the figure also provides some intuition how splitting can be useful. In Section 3.3 we describe how we identify candidates for splitting.

The splitting operation is described formally by the pseudocode below. We maintain an array *orig* that points for each vertex in the transformed network to the vertex in the original network that it derives from. Initially, $orig[v_j] = v_j$ for all $v_j \in V$. With $P(v_i)$ we denote the longest common suffix of all paths from z_1 to v_i after every path is translated back to the original network. The intuition behind this definition is that if v_i is in an optimal Steiner arborescence, $P(v_i)$

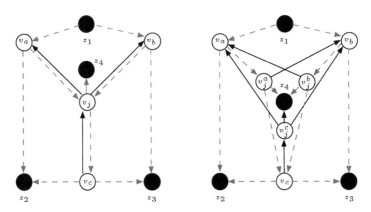

Fig. 1. Splitting of vertex v_j. The filled circles are terminals, z_1 is the root, all arcs have cost 1. An optimal Steiner arborescence has value 6 in each network. In the left network $v(LP_{C+FB})$ is 5.5 (set the x-values of the dashed arcs to 0.5 and of $[v_j, z_4]$ to 1), but 6 in the right network (again, set the x-values of the dashed arcs and of $[v_j^a, z_4]$ and $[v_j^b, z_4]$ to 0.5). The difference is that in the left network, there is a situation that is called "rejoining of flows": Flows from z_1 to z_2 and from z_1 to z_3 enter v_j on different arcs, but leave on the same arc, so they are accounted in the x variables only once. Before splitting, the x-value corresponding to the arc $[v_j, v_c]$ is 0.5, after splitting the corresponding x-values sum up to 1

must also be in the arborescence after it is translated into the original network. Note that the path $P(v_i)$ consists of vertices in the original network and may contain cycles; in this case, v_i cannot be part of an optimal arborescence. In Figure 1, $P(v_a)$ consists of v_a and $P(v_j^a)$ is the path of length 1 from v_a to v_j. To compute $P(v_i)$, one can reverse all arcs and use breadth-first-search. The main purpose of using $P(v_i)$ is to avoid inserting unnecessary arcs. This can improve the value of and the computation times for the lower bound. It is also necessary for the proof of termination in Section 3.2.

For the ease of presentation, we assume that the root terminal z_1 has no incoming arcs, and that all other terminals have no outgoing arcs. If this is not the case, we simply add copies of the terminals and connect them with appropriate zero cost arcs to the old terminals.

$SPLIT\text{-}VERTEX(G, v_j, orig)$:   (assuming $v_j \notin R$)
1   **forall** $[v_i, v_j] \in \delta^-(v_j)$:
2   **if** $P(v_i)$ *contains a cycle* **or** $orig[v_j]$ *in* $P(v_i)$:
3   **continue** *with next arc in* $\delta^-(v_j)$
4   *insert a new vertex* v_j^i *into* G, $orig[v_j^i] := orig[v_j]$
5   *insert an arc* $[v_i, v_j^i]$ *with cost* $c(v_i, v_j)$ *into* G
6   **forall** $[v_j, v_k] \in \delta^+(v_j)$:
7   **if** $orig[v_k]$ *not in* $P(v_i)$:
8   *insert an arc* $[v_j^i, v_k]$ *with cost* $c(v_j, v_k)$ *into* G
9   *delete* v_j
10   *delete all vertices that are not reachable from* z_1

3.1   Correctness

In this section, we prove that the transformation is valid, i.e., it does not change the value of an optimal Steiner arborescence.

Lemma 1. *Any optimal Steiner arborescence with root z_1 in the original network can be transformed into a feasible Steiner arborescence with root z_1 in the transformed network with the same cost and vice versa.*

Proof. We consider one splitting operation on vertex $v_j \in V \setminus R$, transforming a network G into G'. Repeating the argumentation extends the result to multiple splits. We use a condition (†) for a tree T denoting that for every v_k, v_l in T, it holds: $orig[v_k] = orig[v_l] \Leftrightarrow v_k = v_l$. Note that condition (†) holds for an optimal Steiner arborescence in the original network.

Let T be an optimal Steiner arborescence with root z_1 for G satisfying (†). If $v_j \notin T$, T is part of G' and we are done. If $v_j \in T$, there is exactly one arc $[v_i, v_j] \in T$. When $[v_i, v_j]$ is considered in the splitting, $P(v_i)$ is a subpath of the path from z_1 to v_i in T after it is translated to the original network. Together with (†) follows that neither $orig[v_j]$, nor $orig[v_k]$ for any $[v_j, v_k] \in T$ is in $P(v_i)$. Therefore, all arcs $[v_j, v_k] \in T$ can be replaced by arcs $[v_j^i, v_k]$ and the arc $[v_i, v_j]$ can be replaced by $[v_i, v_j^i]$. The transformed T is part of G', connects all terminals, has the same cost as T and satisfies condition (†).

Now, let T' be an optimal Steiner arborescence for G'. Obviously, T' can be transformed into a feasible solution T with no higher cost for G.

3.2 Termination

In this section, we show that iterating the splitting operation will terminate.[1]

Lemma 2. *For all non-terminals v_j, $P(v_j)$ is the common suffix of all paths $P(v_i)$ appended by $orig[v_j]$ for all $v_i, [v_i, v_j] \in \delta^-(v_j)$.*

Lemma 3. *For any two non-terminals v_s and $v_t, v_s \neq v_t$, $P(v_s)$ is not a suffix of $P(v_t)$.*

Lemma 4. *After splitting a vertex v_j with in-degree greater than 1, for any newly inserted vertex v_j^i it holds that $P(v_j^i)$ is longer than $P(v_j)$ was before the split.*

Lemma 5. *Repeated splitting of vertices with in-degree greater than 1 will stop with a network in which all non-terminals have in-degree 1. As a consequence, there is exactly one path from z_1 to v_i for all non-terminals v_i.*

3.3 Implementation Issues

Of course, for a practical application one does not want to split all vertices, which could blow up the network exponentially. In a cutting plane algorithm one first adds violated Steiner cut or flow-balance constraints. They can be found by min-cut computations [17], respectively with a summation of the incoming and outgoing arcs variables of non-terminals. If no such constraint can be found, we search for good candidates for the splitting procedure, i.e., vertices where more than one incoming arc and at least one outgoing arc have an x-value greater than zero. After splitting these vertices, the modified network will be used for the computation of new constraints, using the same algorithms as before. To represent this transformation in the linear program, we add new variables for the newly added arcs, and additional constraints that the x-values for all newly added arcs corresponding to an original arc $[v_i, v_j]$ must sum up to $x_{[v_i,v_j]}$. Using this procedure the constraints calculated for the original network can still be used.

4   Project, Separate, and Lift: Local Cuts

Let $S = (G, R) = (V, E, c, R)$ be an instance of the Steiner problem. Let $\mathcal{ST}(S)$ be the set of all incidence vectors of Steiner trees of S and $\mathcal{SG}(S) = \mathcal{ST}(S) + \mathbb{R}_+^{|E|}$.

---

[1] The proofs (also of Lemma 6 and Lemma 10) are given in [1].

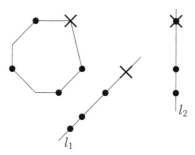

Fig. 2. The feasible integer solutions are marked as dots, the fractional solution to separate by the cross. If we project the solutions to the line l_1, we can obtain a valid violated inequality and lift it back to the original space. If we project to the line l_2, the fractional solution falls into the convex hull of the integer solutions and no such inequality can be found

We call the elements of $\mathcal{SG}(S)$ the Steiner graphs of S. We consider Steiner graphs, since Steiner graphs are invariant under the shrink operation (defined in Section 4.1). Note that the values $x_{(v_i,v_j)}$ are not restricted to be integral or bounded. It is obvious that if the objective function is non-negative, there exists a minimum Steiner graph that is a Steiner tree. Thus all vertices of the polyhedron conv($\mathcal{SG}(S)$) are Steiner trees. Furthermore, conv($\mathcal{SG}(S)$) is full dimensional if G is connected.

From a high level view, local cuts can be described as follows. Assume we want to separate x^* from conv($\mathcal{SG}(S)$). Using a linear mapping ϕ, we project the given point x^* into a small-dimensional vector $\phi(x^*)$ and solve the separation problem over conv($\phi(\mathcal{SG}(S))$). If we can find a violated inequality $a \cdot \tilde{x} \geq b$ that separates $\phi(x^*)$ from conv($\phi(\mathcal{SG}(S))$), we know that the linear inequality $a \cdot \phi(x) \geq b$ separates x^* from conv($\mathcal{SG}(S)$). The method is illustrated in Figure 2.

To make this method work, we have to choose ϕ such that

1. there is a good chance that $\phi(x^*) \notin$ conv($\phi(\mathcal{SG}(S))$) if $x^* \notin$ conv($\mathcal{SG}(S)$),
2. we can solve the separation problem over conv($\phi(\mathcal{SG}(S))$) efficiently and
3. the inequalities $a \cdot \phi(x) \geq b$ are strong.

We choose ϕ in such a way that for every solution $x \in \mathcal{SG}(S)$ of our Steiner problem instance S, the projected $\phi(x)$ is a Steiner graph of a small Steiner problem instance S^ϕ, i.e., conv($\phi(\mathcal{SG}(S))$) = conv($\mathcal{SG}(S^\phi)$) for an instance S^ϕ of the Steiner problem. Since our Steiner tree program package tends to be very efficient for solving small Steiner problem instances, we can handle the separation problem, as we will see in Section 4.2.

We use iterative shrinking to obtain the linear mappings. We review the well-known concept of shrinking in the next section. After that, we introduce our separation algorithm for small Steiner graph instances. So far, we always assumed that we are looking at the undirected version of the Steiner problem, since our separation algorithm is much faster for this variant. As seen above, the directed cut relaxation is stronger than the undirected variant. In Section 4.3, we discuss how we can use the directed formulation without solving directed Steiner graph instances in the separation algorithm.

4.1 Shrinking

We define our linear mappings as an iterative application of the following simple, well-known mapping, called shrinking. For the Steiner problem, shrinking was indroduced by Chopra and Rao [7].

Shrinking means to replace two vertices v_a and v_b by a new vertex $\langle v_a, v_b \rangle$ and replace edges (v_i, v_a) and (v_i, v_b) by an edge $(v_i, \langle v_a, v_b \rangle)$ with value $x^*_{(v_i, v_a)} + x^*_{(v_i, v_b)}$ (We assume $x^*_{(v_i, v_j)} = 0$ if $(v_i, v_j) \notin E$). The new vertex $\langle v_a, v_b \rangle$ is in the set of terminals R if v_a or v_b (or both) are in R. This informally defines the mapping ϕ and the instance S^ϕ. Note that for any incidence vector of a Steiner graph for the original problem, the new vector is the incidence vector of a Steiner graph in the reduced problem. Furthermore, for every Steiner graph $\tilde{x}$ in S^ϕ there is a Steiner graph $x \in S\mathcal{G}(S)$ such that $\phi(x) = \tilde{x}$. Thus $\mathrm{conv}(\phi(S\mathcal{G}(S))) = \mathrm{conv}(S\mathcal{G}(S^\phi))$.

Note that if we iteratively shrink a set of vertices $W \subset V$ into one vertex $\langle W \rangle$, the obtained linear mapping is independent of the order in which we apply the shrinks. We denote the unique linear mapping which shrinks a subset $W \subset V$ into one vertex by ϕ^W.

We have developed conditions on x^* under which we can prove that $\phi(x^*)$ is not in the convex hull of $S\mathcal{G}(S^\phi)$ if x^* is not in the convex hull of $S\mathcal{G}(S)$.

Lemma 6. *Let $x^* \geq 0$.*

1. *(edge of value 1): Let $x^*_{(v_a, v_b)} \geq 1$ and $W = \{v_a, v_b\}$. $x^* \in \mathrm{conv}(S\mathcal{G}(S)) \Leftrightarrow \phi^W(x^*) \in \mathrm{conv}(S\mathcal{G}(S^{\phi^W}))$.*
2. *(non-terminal of degree 2): Let v_a be in $V \setminus R$ and the vertices $(v_1, \ldots v_k)$ in V be ordered according to their $x^*_{(\cdot, v_a)}$ value (in decreasing order). Furthermore, let $W = \{v_a, v_1\}$. If $x^*_{(v_3, v_a)} = 0$, then $x^* \in \mathrm{conv}(S\mathcal{G}(S)) \Leftrightarrow \phi^W(x^*) \in \mathrm{conv}(S\mathcal{G}(S^{\phi^W}))$.*
3. *(cut of value 1): Let W be such that $x^*(\delta(W)) = 1$ and $\emptyset \neq R \cap W \neq R$. Let $\overline{W} = V \setminus W$. $x^* \in \mathrm{conv}(S\mathcal{G}(S)) \Leftrightarrow \phi^W(x^*) \in \mathrm{conv}(S\mathcal{G}(S^{\phi^W})) \wedge \phi^{\overline{W}}(x^*) \in \mathrm{conv}(S\mathcal{G}(S^{\phi^{\overline{W}}}))$.*
4. *(biconnected components): Let $U, W \subset V$ and $v_a \in V$ be such that $U \cup W = V$, $U \cap W = \{v_a\}$ and $x^*_{(v_k, v_l)} = 0$ for all $v_k \in U \setminus \{v_a\}$ and $v_l \in W \setminus \{v_a\}$. Furthermore, let $\emptyset \neq R \cap W \neq R$. $x^* \in \mathrm{conv}(S\mathcal{G}(S)) \Leftrightarrow \phi^U(x^*) \in \mathrm{conv}(S\mathcal{G}(S^{\phi^U})) \wedge \phi^W(x^*) \in \mathrm{conv}(S\mathcal{G}(S^{\phi^W}))$.*
5. *(triconnected components): Let $U, W \subset V$ and $v_a, v_b \in V$ be such that $U \cup W = V \setminus \{v_a\}$, $U \cap W = \{v_b\}$ and $x^*_{(v_k, v_l)} = 0$ for all $v_k \in U \setminus \{v_b\}$ and $v_l \in W \setminus \{v_b\}$. Let furthermore $x^*(\delta(v_a)) = 1$ and $v_a, v_b \in R$. $x^* \in \mathrm{conv}(S\mathcal{G}(S)) \Leftrightarrow \phi^U(x^*) \in \mathrm{conv}(S\mathcal{G}(S^{\phi^U})) \wedge \phi^W(x^*) \in \mathrm{conv}(S\mathcal{G}(S^{\phi^W}))$.*

Applying these "exact" shrinks does not project the solution of the current linear program into the projected convex hull of all integer solutions, i.e., if the solution of the current linear program has not reached the value of the integer optimum, we can find a valid, violated constraint in the shrunken graphs. Unfortunately, in many cases the graphs are still too large after applying these shrinks and we have to apply some "heuristic" shrinks afterwards.

In the implementation, we use a parameter *max-component-size*, which is initially 15. If the number of vertices in a graph after applying all "exact" shrinks is not higher than *max-component-size*, we start FIND-FACET (see Section 4.2), otherwise, we start a breadth-first-search from different starting positions, shrink everything except the first *max-component-size* vertices visited by the BFS, try the "exact" shrinks again and start FIND-FACET. If it turns out that we could not find a valid, violated constraint, we increase *max-component-size*. We also tried other "heuristic" shrinks by relaxing "exact" shrinks, e.g., accepting minimum Steiner cuts with value above 1, or edges that have an x-value close to 1. But we could not come up with a definitive conclusion which shrinks are best, and we believe that there is still room for improvement.

As we will see in the next section, our separation algorithm finds a facet of $\text{conv}(\mathcal{SG}(S^\phi))$. As shown in Theorem 4.1 of [7], the lifted inequality is then a facet of $\text{conv}(\mathcal{SG}(S))$.

4.2 Separation: Finding Facets

Assume we want to separate x^* from $\text{conv}(\mathcal{SG}(S))$. Note that we actually separate $\phi(x^*)$ from $\text{conv}(\mathcal{SG}(S^\phi))$, but this problem can be solved with the same algorithm.

As we will see, the separation problem can be formulated as a linear program with a row for every Steiner graph. Trying to solve this linear program using cutting planes, we have the problem that the number of Steiner graphs (contrary to the case of Steiner trees) is infinite and optimal Steiner graphs need not exist. Note that the same complication arises when applying local cuts to the Traveling Salesman Problem.

The solution for the separation problem is much simpler and more elegant for the Steiner tree case than for the Traveling Salesman case. The key is the following Lemma, a slight variation of Lemma 3.1.2 in [7].

Lemma 7. *All facets of* $\text{conv}(\mathcal{SG}(S))$ *different from* $x_{(v_a,v_b)} \geq 0$ *for an edge* $(v_a, v_b) \in E$ *can be written in the form* $a \cdot x \geq 1$ *with* $a \geq 0$.

Thus, if $x^* \notin \text{conv}(\mathcal{SG}(S))$, we can find an inequality of the form $a \cdot x \geq 1$, $a \geq 0$, that separates x^* from $\text{conv}(\mathcal{SG}(S))$. Note that if $a \geq 0$, there is a Steiner tree $t \in \mathcal{SG}(S)$ minimizing $a \cdot t$.

Thus an exact separation algorithm can be stated as follows (the name arises from the fact that the algorithm will find a facet of $\text{conv}(\mathcal{SG}(S))$, as we will see later).

FIND-FACET $(G = (V, E), R, x^*)$

1      $T :=$ *incidence vector of a Steiner tree for G, R*
2   **repeat:**
3   *solve LP:* $\min x^* \cdot \alpha,\ T\alpha \geq 1,\ \alpha \geq 0$ *(basic solution)*
4   **if** $x^* \cdot \alpha \geq 1$:   **return** *"$x^* \in \text{conv}(\mathcal{SG}(S))$"*
5   *find minimum Steiner tree t for $G = (V, E, \alpha), R$*
6   **if** $t \cdot \alpha < 1$:   *add t as a new row to matrix T*
7   **else:**   **return** $\alpha \cdot x \geq 1$

The algorithm terminates, since there are only a finite number of Steiner trees in $\mathcal{ST}(S)$ and as soon as the minimum Steiner tree t computed in Line 5 is already in T, we terminate because $\alpha \cdot t \geq 1$ is an inequality of the linear program solved in Line 3.

Lemma 8. *If FIND-FACET does not return an inequality, $x^* \in \text{conv}(\mathcal{SG}(S))$.*

Proof. Consider the dual of the linear program in Line 3: $\max \sum_i \lambda_i, T^T \lambda \leq x^*$, which has the optimal value $x^* \cdot \alpha \geq 1$. We divide λ by $x^* \cdot \alpha$, with the consequence that $\sum_i \lambda_i = 1$. Now, $T^T \lambda$ is a convex combination of Steiner trees and it still holds $T^T \lambda \leq x^*$.

Lemma 9. *If FIND-FACET returns an inequality $\alpha \cdot x \geq 1$, this inequality is a valid, separating, and facet-defining inequality.*

Proof. The value of the last computed minimum Steiner tree t is $t \cdot \alpha \geq 1$. Therefore, if $x \in \mathcal{SG}(S)$, the value can only be greater and it holds $x \cdot \alpha \geq t \cdot \alpha \geq 1$.
   As $x^* \cdot \alpha < 1$, the inequality is separating.
   From the basic solution of the linear program, we can extract $|E|$ linearly independent rows that are satisfied with equality. For each such row of the form $\alpha \cdot t \geq 1$, we add the tree t to a set S_λ and for each row $\alpha_e \geq 0$, we add the edge e to a set S_μ. Note that $|S_\lambda| + |S_\mu| = |E|$ and the incidence vectors corresponding to $S_\lambda \cup S_\mu$ are linearly independent.
   There is at least one tree t_j in S_λ. For each edge $e \in S_\mu$ we add to S_λ a new Steiner graph t_k that consists of t_j added by the edge e. Since $\alpha_e = 0$ we know that $\alpha \cdot t_k = 1$. Since the incidence vectors corresponding to $S_\lambda \cup S_\mu$ were linearly independent, replacing e with the t_k yields a new set of linearly independent vectors.
   Repeating this procedure yields $|E|$ linearly independent $t_i \in S_\lambda$ with $\alpha \cdot t_i = 1$. Thus, $\alpha \cdot x \geq 1$ is a facet.

As in [4], we can improve the running time of the algorithm by using the following fact. If we know some valid inequalities $a \cdot x \geq b$ with $a \cdot x^* = b$ then $x^* \in \text{conv}(\mathcal{SG}(S)) \Leftrightarrow x^* \in \text{conv}(\mathcal{SG}(S) \cap \{x \in \mathbb{R}^{|E|} \mid a \cdot x = b\})$. Thus we can temporarily remove all edges (v_i, v_j) with $x^*_{(v_i, v_j)} = 0$, since $x^*_{(v_i, v_j)} \geq 0$ is a valid inequality. Call the resulting instance S'. We use our algorithm to find a facet of $\text{conv}(\mathcal{SG}(S'))$. We can use sequential lifting to obtain a facet of $\text{conv}(\mathcal{SG}(S))$. For details see [4] and Theorem 4.2 of [7].

4.3   Directed versus Undirected Formulations

For computing the lower bounds, we focus on the directed cut formulation, because its relaxation is stronger than the undirected variant. However, in the local cut separation algorithm we want to solve undirected Steiner graph instances, since they can be solved much faster.

The solution is to use another linear mapping that maps arc-values of a bidirected Steiner graph instance $\boldsymbol{S} = (V, A, c, R)$ to edge-values of an undirected Steiner graph instance $S = (V, E, c', R)$.

We define S by $E = \{(v_i, v_j) \mid [v_i, v_j] \in A \}$ and $c'_{(v_i, v_j)} = c_{[v_i, v_j]} = c_{[v_j, v_i]}$. For a vector $x \in \mathbb{R}^{|A|}$ we define $\psi(x) \in \mathbb{R}^{|E|}$ by $\psi(x)_{(v_i, v_j)} = x_{[v_i, v_j]} + x_{[v_j, v_i]}$.

Lemma 10. $x^* \in \operatorname{conv}(\mathcal{SG}(\boldsymbol{S})) \Rightarrow \psi(x^*) \in \operatorname{conv}(\mathcal{SG}(S))$.

$\bar{x} \in \operatorname{conv}(\mathcal{SG}(S)) \Rightarrow \exists x^* \in \operatorname{conv}(\mathcal{SG}(\boldsymbol{S}))$ with $\psi(x^*) = \bar{x}$.

If $c \cdot x^*$ is smaller than the cost of an optimal Steiner arborescence, then $\psi(x^*) \notin \operatorname{conv}(\mathcal{SG}(S))$.

For lifting the undirected edges to directed arcs, one can use the computation of optimal Steiner arborescences. For the actual implementation, we used a faster lifting using a lower bound to the value of an optimal Steiner arborescence, provided by the fast algorithm DUAL-ASCENT [17, 25]. For producing facets for the directed Steiner problem, one could compute optimal Steiner arborescences in the FIND-FACET algorithm of Section 4.2.

5   Some Experimental Results

In this section, we present experimental results showing the impact of the methods described before. In this paper we confine ourselves to the presentation of some highlights, namely the largest benchmark instances ever solved (Table 1). Experiments on smaller instances show that vertex splitting can also significantly improve the solution time (some additional experimental results are presented in [1]). Note that in the TSP context, local cuts were helpful particularly for the solution of very large instances.

We have chosen the approach of applying these techniques together with the reduction methods [17], because this is the way they are actually used in our program package. Note that without the reductions, the impact of these techniques would be even more impressive, but then these instances could not be handled in reasonable time.

All results were obtained with a single-threaded run on a Sunfire 15000 with 900 MHz SPARC III+ CPUs, using the operating system SunOS 5.9. We used the GNU g++ 2.95.3 compiler with the -O4 flag and the LP-solver CPLEX version 8.0.

Table 1.   Results on large benchmark instances. In all cases, the lower bound reached the value of the integer optimum (and a tree with the same value was found). A dash means that the instance was already solved to optimality without local cuts. For the instance d15112, we used the program package GeoSteiner-3.1 [24] to translate the TSPLIB [21] instance into an instance of the Steiner problem in networks with rectilinear metric. No benchmark instance of this size has been solved before. The SteinLib [22] instances es10000 and fnl4461 were obtained in the same way. Warme et. al. solved the es10000 instance using the MSTH-approach [23] and local cuts. They needed months of cpu time. The instance fnl4461 was the largest previously unsolved geometric instance in Stein-Lib. The SteinLib instance lin37 originates from some VLSI-layout problem, is not geometric, and was not solved by other authors. Without lower bound improvement techniques, the solution of the instances would take much longer (or was not even possible in case of d15112). The number of vertex splits varied between 8 (lin37), 21 (es10000), 173 (fnl4461) and 321 (d15112). For d15112 only one additional local cut computation was necessary.

Instance	Orig. Size		Red.	Red. Size		LP_{C+FB}		+ vertex splitting		+ local cuts	
	$\|V\|$	$\|R\|$	time	$\|V\|$	$\|R\|$	val	time	val	time	val	time
d15112	51886	15112	5h	22666	7465	1553831.5	20.4h	1553995	21.9h	1553998	21.9h
es10000	27019	10000	988s	4061	1563	716141953.5	251s	716174280	284s	—	
fnl4461	17127	4461	995s	8483	2682	182330.8	5299s	182361	6353s	—	
lin37	38418	172	28h	2529	106	99554.5	1810s	99560	1860s	—	

6   Concluding Remarks

We presented two theoretically interesting and empirically successful approaches for improving lower bounds for the Steiner tree problem: vertex splitting and local cuts. Vertex splitting is a new technique and improves the lower bounds much faster than the local cut method, but the local cut method has the potential of producing tighter bounds. Vertex splitting, although inspired by a general approach (see Section 1), is not directly transferable to other problems, while local cuts are a more general paradigm. On the other hand, the application needs some effort, e.g., developing proofs for shrinks and implementation using exact arithmetic. A crucial point is the development of heuristic shrinks, where a lot of intuition comes into play and we believe that there is room for improvement. Although the local cut method was originally developed for the Traveling Salesman Problem, its application is much clearer for the Steiner tree problem.

Both methods are particularly successful if there are some local deficiencies in the linear programming solution. On constructed pathological instances the lower bounds are still improved significantly, but the progress is not fast enough to solve such instances efficiently.

Another interesting observation is that the power of the vertex splitting approach can be improved by looking at multiple roots simultaneously. In fact, we do not know any instance where repeated vertex splittings would not bring the lower bound to the integer optimum if multiple roots are used. It remains an open problem to find out if this is always the case.

References

[1] E. Althaus, T. Polzin, and S. Vahdati Daneshmand. Improving linear programming approaches for the steiner tree problem. Research Report MPI-I-2003-1-004, Max-Planck-Institut für Informatik, Stuhlsatzenhausweg 85, 66123 Saarbrücken, Germany, February 2003. 2, 6, 11

[2] Y. P. Aneja. An integer linear programming approach to the Steiner problem in graphs. *Networks*, 10:167–178, 1980. 3

[3] D. Applegate, R. Bixby, V. Chvátal, and W. Cook. Finding cuts in the TSP (A preliminary report). Technical report, Center for Discrete Mathematics and Theoretical Computer Science, Rutgers University, Piscataway, NJ, 1995. 1

[4] D. Applegate, R. Bixby, V. Chvátal, and W. Cook. TSP cuts which do not conform to the template paradigm. In Michael Jünger and Denis Naddef, editors, *Computational Combinatorial Optimization*, volume 2241 of *Lecture Notes in Computer Science*. Springer, 2001. 1, 2, 10

[5] E. Balas and M. Padberg. On the set-covering problem: II. An algorithm for set partitioning. *Operations Research*, 23:74–90, 1975. 2

[6] X. Cheng and D.-Z. Du, editors. *Steiner Trees in Industry*, volume 11 of *Combinatorial Optimization*. Kluwer Academic Publishers, Dordrecht, 2001. 1

[7] S. Chopra and M. R. Rao. The Steiner tree problem I: Formulations, compositions and extension of facets. *Mathematical Programming*, pages 209–229, 1994. 8, 9, 10

[8] T. H. Cormen, C. E. Leiserson, and R. L. Rivest. *Introduction to Algorithms*. MIT Press, 1990. 3

[9] C. Gentile, U.-U. Haus, M. Köppe, G. Rinaldi, and R. Weismantel. A primal approach to the stable set problem. In R. Möhring and R. Raman, editors, *Algorithms - ESA 2002*, volume 2461 of *Lecture Notes in Computer Science*, pages 525–537, Rom, Italy, 2002. Springer. 2

[10] U.-U. Haus, M. Köppe, and R. Weismantel. The integral basis method for integer programming. *Mathematical Methods of Operations Research*, 53(3):353–361, 2001. 2

[11] F. K. Hwang, D. S. Richards, and P. Winter. *The Steiner Tree Problem*, volume 53 of *Annals of Discrete Mathematics*. North-Holland, Amsterdam, 1992. 1, 3

[12] R. M. Karp. Reducibility among combinatorial problems. In R. E. Miller and J. W. Thatcher, editors, *Complexity of Computer Computations*, pages 85–103. Plenum Press, New York, 1972. 1

[13] T. Koch and A. Martin. Solving Steiner tree problems in graphs to optimality. *Networks*, 32:207–232, 1998. 3

[14] D. Naddef and S. Thienel. Efficient separation routines for the symmetric traveling salesman problem i: general tools and comb separation. *Mathematical Programming*, 92(2):237–255, 2002. 1

[15] D. Naddef and S. Thienel. Efficient separation routines for the symmetric traveling salesman problem ii: separating multi handle inequalities. *Mathematical Programming*, 92(2):257–283, 2002. 1

[16] T. Polzin and S. Vahdati Daneshmand. A comparison of Steiner tree relaxations. *Discrete Applied Mathematics*, 112:241–261, 2001. 3

[17] T. Polzin and S. Vahdati Daneshmand. Improved algorithms for the Steiner problem in networks. *Discrete Applied Mathematics*, 112:263–300, 2001. 2, 6, 11

[18] T. Polzin and S. Vahdati Daneshmand. Partitioning techniques for the Steiner problem. Research Report MPI-I-2001-1-006, Max-Planck-Institut für Informatik, Stuhlsatzenhausweg 85, 66123 Saarbrücken, Germany, 2001. 2

[19] T. Polzin and S. Vahdati Daneshmand. Extending reduction techniques for the steiner tree problem. In R. Möhring and R. Raman, editors, *Algorithms - ESA 2002*, volume 2461 of *Lecture Notes in Computer Science*, pages 795–807, Rom, Italy, 2002. Springer. 2

[20] T. Polzin and S. Vahdati Daneshmand. On Steiner trees and minimum spanning trees in hypergraphs. *Operations Research Letters*, 31, 2003. 2

[21] G. Reinelt. TSPLIB — a traveling salesman problem library. *ORSA Journal on Computing*, 3:376 – 384, 1991. 12

[22] SteinLib. http://elib.zib.de/steinlib, 1997. T. Koch, A. Martin, and S. Voß. 12

[23] D. M. Warme, P. Winter, and M. Zachariasen. Exact algorithms for plane Steiner tree problems: A computational study. In D-Z. Du, J. M. Smith, and J. H. Rubinstein, editors, *Advances in Steiner Trees*, pages 81–116. Kluwer Academic Publishers, 2000. 12

[24] D. M. Warme, P. Winter, and M. Zachariasen. GeoSteiner 3.1. http://www.diku.dk/geosteiner/, 2001. 12

[25] R. T. Wong. A dual ascent approach for Steiner tree problems on a directed graph. *Mathematical Programming*, 28:271–287, 1984. 3, 11

Algorithms and Experiments
on Colouring Squares of Planar Graphs

Maria I. Andreou, Sotiris E. Nikoletseas, and Paul G. Spirakis *

Computer Technology Institute (CTI) and Patras University, Greece
Riga Fereou 61, 26221 Patras, Greece
{mandreou,nikole,spirakis}@cti.gr
Fax: +30-2610-222086

Abstract. In this work we study the important problem of *colouring squares of planar graphs* (SQPG). We design and implement *two new algorithms* that colour in a different way SQPG. We call these algorithms *MDsatur* and *RC*. We have also implemented and experimentally evaluated the performance of most of the known approximation colouring algorithms for SQPG [14, 6, 4, 10]. We compare the quality of the colourings achieved by these algorithms, with the colourings obtained by our algorithms and with the results obtained from two well-known *greedy colouring heuristics*. The heuristics are mainly used for comparison reasons and unexpectedly give very good results. Our algorithm *MDsatur* outperforms the known algorithms as shown by the extensive experiments we have carried out.

The planar graph instances whose squares are used in our experiments are "non-extremal" graphs obtained by LEDA and hard colourable graph instances that we construct.

The most interesting conclusions of our experimental study are:
1) all colouring algorithms considered here have *almost optimal* performance on the squares of *"non-extremal"* planar graphs. 2) all known colouring algorithms especially designed for colouring SQPG, give *significantly better* results, even on hard to colour graphs, when the vertices of the input graph are *randomly named*. On the other hand, the performance of our algorithm, *MDsatur*, becomes worse in this case, however it still has the best performance compared to the others. *MDsatur* colours the tested graphs with $1.1\,OPT$ colours in most of the cases, even on hard instances, where OPT denotes the number of colours in an optimal colouring. 3) we *construct worst case instances* for the algorithm of Fotakis el al.[6], which show that *its theoretical analysis is tight*.

1 Introduction

Communication in wireless systems, such as radio networks and ad-hoc mobile networks, is accomplished through exploitation of a limited range of frequency

---

* This work has been partially supported by the EU IST/FET projects ALCOM-FT and CRESCCO.

K. Jansen et al. (Eds.): WEA 2003, LNCS 2647, pp. 15–32, 2003.

spectrum. One of the mechanisms utilised is to reuse frequencies where this does not result to unacceptable levels of signal interference. In graph theoretic terms, the interference between transmitters is usually modelled by the *interference graph* $G = (V, E)$, where V corresponds to the set of transmitters and E represents distance constrains (e.g. if two adjacent vertices in G get the same or nearby frequencies, then this causes unacceptable levels of interference).

In most real systems, the network topology has some special properties, e.g. G is a lattice network or G is a planar graph. Planar graphs are the object of study in this work, both because of their importance in real networks and also because of their independent combinatorial interest.

The *Frequency Assignment Problem* (FAP) is usually modelled by variations of the *vertex graph colouring* problem. The set of colours represents the available frequencies. In addition, in an acceptable assignment the colour of each vertex of G gets an integer value which has to satisfy certain inequalities compared to the colours of its nearby vertices in G (frequency-distance constraints). The FAP problem has been considered e.g. in [7, 8, 11].

Definition 1. (*k-colouring*) Given a graph $G = (V, E)$ let $F : V \rightarrow \{1, ..., \infty\}$ be a function, called *k-colouring* of G, such that for each $u, v \in V$ it is $|F_u - F_v| \geq x$ for $x = 0, 1, ..., k$ if $D(u, v) \leq k - x + 1$. $D(u, v)$ denotes the distance between u and v in G, i.e. the length of the shortest path joining them. The number of colours used by F is called *order* and is denoted by $\lambda = |F(V)|$. The range of them is denoted by $s = max_{v \in V} F(v) - min_{u \in V} F(u) + 1$ and is called *span*.

Definition 2. (Radiocolouring, RCP) When $k = 2$ then the function F is called a *radiocolouring* of G of order λ and span s. The *radiochromatic number* of G is the least order for which there exists a radiocolouring of G and is denoted by $X_{order}(G)$. The least span is denoted as $X_{span}(G)$, respectively.

A variation of RCP useful in practice is the following:

Definition 3. The *min order min span radiocolouring* on a graph G is a function that radiocolours G with the least number of distinct colours while the range of colours used is the least possible.

It can be easily seen, that the radiochromatic number of any graph G ($X_{order}(G)$) is equal to the chromatic number of its *square*, i.e. $\chi(G^2)$ [8].

Definition 4. The *square* of a graph G is denoted by G^2 and is a graph on the same vertex set as G with an edge between any pair of vertices of distance at most two in G. The graph G is a *square root* of graph G^2.

In this work we experimentally study the colouring problem on the square of any planar graph (i.e. the min order RCP) and the min order min span RCP on planar graphs. Ramanathan and Loyd proved that the first one of the above problems is NP-complete [15]. Fotakis et al. proved the NP-completeness of RCP on planar graphs [6]. Thus, efficient solutions for these problems necessarily rely on the choice of heuristics and approximation algorithms. To our knowledge, for the colouring of SQPG, there is an 1.66-approximation algorithm due to Molloy and Salavatipour [14] that currently gets the best approximation ratio. We are aware of another three colouring approximation algorithms designed for these graphs [10, 6, 4]. The first two of them have approximation ratio 2 and the third one has ratio 1.8.

Towards a global picture of the problem our work studies a variety of algorithms that have *distinct methodologies*. In this direction we have first implemented a new algorithm that we propose for colouring SQPG. We call this algorithm *MDsatur*, because it is a *non-trivial* modification of the well-known colouring algorithm *Dsatur* of Brélaz [5]. In our opinion, our approach is novel, because all known algorithms for this problem colour the uncoloured vertices of a given graph mainly based on the number of neighbours of each of these vertices either in graph G, or in the subgraph of graph G^2 that has already been coloured. On the other hand, our algorithm colours the uncoloured vertices mainly based on the number of distinct colours that have already been assigned to their neighbours.

We have also implemented the algorithm which is presented and theoretically analysed in the work of Fotakis et al. [6] and we call it $FNPS$. We also considered the algorithm of Agnarsson and Halldórsson presented in [4] and we call it AH. This algorithm actually concerns planar graphs of large maximum degree $\Delta \geq 749$ only. It is proved in [4] that the upper bound achieved by AH is tight.

To make our experimental study on colouring SQPG more complete we have also implemented two well known *greedy* colouring heuristics, which colour any graph. The first one is the *Randomised First Fit* (RFF). The second, which we call it *MaxIScover*, covers the given graph by maximal independent sets.

Finally, we also propose and implement a new min order min span radio-colouring algorithm, which we call *RadioColouring* (RC). RC min order min span radiocolours any planar graph G given a colouring on G^2. Based on its results we interestingly *conjecture* that $X_{span}(G) - X_{order}(G)$ is bounded from above by a small constant. So any efficient algorithm that solves the min order min span problem on planar graphs can be used as an efficient algorithm for the colouring problem on SQPG and vice-versa.

We use three sets of planar graph instances to evaluate the performance of the above algorithms. The first one, $S1$, has graphs which are obtained from the graphs generation library of LEDA [12]. These graphs are "*non-extremal*", because of the randomness in their generation. In the second set, $S2$, we create graphs that are modifications of the graph instances of the first set. We expected that the graphs of set S2 would be *harder* to colour graph instances for the colouring algorithms studied. The last set, $S3$, has hard colourable graph instances at least for our algorithm *MDsatur* and for algorithm *AH*.

From the experimental results we observe that each one of the colouring algorithms considered in this work colours the square of any planar graphs from set $S1$ with $\Delta + c$ colours, where c is a small constant (i.e. 4). It can be easily seen, that $\Delta + 1$ colours are necessary. The results of these algorithms are similar and on the graphs of set $S2$ as well. On the other hand, the results on the harder to colour graph instances of set $S3$ demonstrate the differences on the performance of each one of the algorithms considered. In particular, we remark that $MDsatur$ has the best *experimental performance* (EP) in all the cases. We measure the EP as the ratio of the number of colours the algorithm uses over the lower bound value on the chromatic number of the square of any graph G, that is $\Delta(G) + 1$. Our algorithm has EP less than 1.1 in most of the cases, even on hard to colour instances, while its theoretically claimed approximation ratio is 1.5 [2]. Heuristic $MaxIScover$ has the second best EP. Heuristic RFF also gives very good results. There are no theoretical results for these heuristics. Thus, we find their experimental results very interesting. The other algorithms give more or less expected results, much better than their theoretical upper bounds. For example, $FNPS$ has EP ranging from 1.1 to 1.4. However, we present in Section 3 a planar graph G and an order of its vertices such that if algorithm $FNPS$ selects the vertices according to this order, then it needs 2Δ colours to colour G^2. This proves that the theoretical upper bound of this algorithm is tight.

All our implementations are written in the C programming language. The programmes run on a Sun Enterprise Solaris 2.6 system with 4 processors at 300MHz and 256 MB RAM. The source codes and the planar graph instances used are available at: http://students.ceid.upatras.gr/~mandreou/.

2   Description of the Algorithms

The Algorithm MDsatur. We here present our algorithm $MDsatur$. As we have said in the introduction, this algorithm is based on a novel approach for the colouring problem on SQPG and is inspired by algorithm $Dsatur$ of Brélaz [5]. $Dsatur$ colours very well and optimally graphs whose chromatic number is strongly related to their clique number, i.e. random graphs and k-colourable random graphs [17, 16]. The chromatic number of SQPG is also strongly related to their clique number, because $\chi(G^2) \leq 1.66\Delta(G) + 24$ ([14]). Thus, $\chi(G^2) \leq 1.66\omega(G^2) + 24$ (where G is a planar graph and $\omega(G^2)$ is the size of the maximum clique in G^2).

Algorithm $Dsatur$ greedily colours a given graph G. At each point it colours a vertex that has the maximum value on its *degree of saturation* with the smallest allowable colour. The *degree of saturation*, Ds, of a vertex is the number of distinct colours already assigned to its neighbours.

Algorithm $MDsatur$ colours the vertices of a given graph G^2 based on a different to that of $Dsatur$ order, O, which is augmented by vertices that satisfy *more restrictions*. More precisely, let i be the current point of $MDsatur$ in its application on G^2. Then, the current uncoloured vertex which manages to become the vertex in position i of O, let it be v, satisfies the following requirements in turn: R1) has the maximum value on its degree of saturation in the current point of this algorithm (as $Dsatur$), R2) has the oldest coloured neighbour (i.e. sits in the leftmost position in O among the oldest neighbours of all the uncoloured vertices) and R3) is the 'closer' neighbour of vertex O_{i-1} (namely the last vertex coloured). We below provide a pseudo-code description of $MDsatur$.

Algorithm: *MDsatur*
Input: a graph $G = (V, E)$. *Output*: a colouring of G.
Begin

1. (a) Set each vertex of G as uncoloured, and
      set a dynamic order on them, called O, as empty.
   (b) Select any vertex of G.
      Add it in the beginning of O (in the position 1 of O),
      and colour it with colour 1.
2. **For** $i = 2$ **to** $n = |V|$ **do**
   (a) Set X to be the set of the current uncoloured vertices
      which have the maximum value on their *degree of saturation*
      (**Requirement R1**).
   (b) **For** each vertex $v \in X$ **do**
      Find the first neighbour of v coloured
      (i.e. its oldest coloured neighbour) (**Requirement R2**).
   (c) Select from X the vertex v which has the oldest coloured neighbour,
      denoted by ON, and also has the maximum number of common
      neighbours with the last coloured vertex (namely O_{i-1})
      (**Requirement R3**).
      If more than one of the vertices of X satisfy these requirements,
         then select at random one of them with equal probability.
      If none of the uncoloured neighbours of ON, which are in X, has
      common
         neighbours with vertex O_{i-1}, then select one of them at random.
      Set $O_i = v$.
   (d) Colour v with the smallest of its allowable colours, i.e. are not assigned
      to its neighbours.
      If none of the colours used is allowable for this vertex, insert a new
      colour.
3. Return the colour of each vertex.

End

The most '*closer*' neighbour of a vertex u is this one whose colouring is determined at most by the colours of the other neighbours of u. In most cases this neighbour of u, lies near to it in a planar embedding of G. Note that we achieve to well-approximate the nearby vertices in a planar embedding of G, without actually finding any such embedding. The effect of this is that our algorithm can colour any graph, so its results are more interesting.

At this point, let us just provide an intuitive explanation of why this algorithm gives better results than the original algorithm $Dsatur$ on SQPG. The main reason is that $MDsatur$ colours the given graph *more locally* based on the last coloured vertex. Also, the subgraph of the given graph that has already

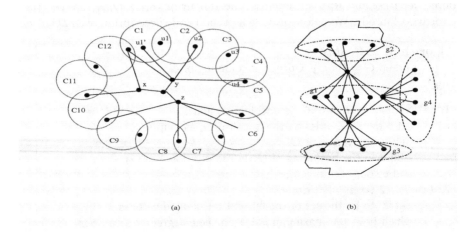

Fig. 1. The graphs that are used in order to explain why $MDsatur$ colours a given graph more locally and compact than $Dsatur$

been coloured at any point of the application of $MDsatur$ on this graph is *more compact* (dense) than the corresponding subgraph obtained from $Dsatur$. We explain how these properties of our algorithm are achieved using an example and we mention their consequences on the resulting colouring.

Let G be a planar graph and suppose that $MDsatur$ is applied on G^2. Also suppose that in an intermediate point of this application the following hold: Let $\{C1, C2, C3, C4, ...\}$ be a sequence of partially coloured cliques and $\{u1, u2, u3, u4, ...\}$, be a sequence of uncoloured vertices. Each vertex uj ($j = \{1, 2, 3, ...\}$), belongs in both cliques Cj and $C(j + 1)$ (see Fig. 1a). We assume that the values of the degree of saturation of all these vertices are the same. Then it would be possible to colour all these vertices sequentially before the colouring of other vertices which belong to the above cliques.

This fact, quite possibly yields an increase on the number of colours used, because it is possible to have vertices (like vertices x, y, z in the figure), such that their degree of saturation, when they will be coloured, is mainly computed based on the colours of non-adjacent vertices.

To avoid the above situation, we focus our effort on the guidance of algorithm $Dsatur$, so that to colour as compact as possibly a given graph. In this direction we add requirement R2 in our algorithm, $MDsatur$. This has to be satisfied by the next vertex to be coloured, in each point of this algorithm. The behaviour of our algorithm on the above example is explained in the following. Let as call ON_{u1} (ON_{u2}) the oldest coloured neighbour of vertex $u1$ ($u2$). If $u1$ is the next vertex to be coloured, then this means that vertex ON_{u1} was coloured before vertex ON_{u2} (due to req. R2). If $ON_{u1} \in C1$ (observe that $u1' \in C1$), then according to this algorithm the next vertex to be coloured cannot be vertex $u2$, because the oldest neighbour of $u1'$ is at least as old coloured as vertex ON_{u1}.

Thus the next vertex to be coloured is vertex $u1'$, which extends only the same clique as the predecessor vertex coloured (namely $u1$). So, the first of our goals is achieved on the colouring obtained by $MDsatur$. However, if $ON_{u1} \notin C1$, then unfortunately it is possible to colour vertex $u2$ before vertex $u1'$.

Finally, requirement R3 is used in order to achieve to colour in turn nearby vertices (in an embedding of graph G) of the same clique. Namely we try to locally colour these vertices. As we have said before, the most 'closer' neighbour of the last vertex coloured lies near to its location in a planar embedding of graph G. Fig. 1b is representative of this case. Let u be the next vertex to be coloured. Observe that the colouring of this vertex influences the value of the degree of saturation of each other vertex in the square of this graph. Suppose that all these vertices have the same value on their degree of saturation. By requirement R3 our algorithm colours after vertex u a vertex from set $g1$ because only vertices of this set (i.e. neighbours of u in the square of this graph) have the maximum number of common neighbours with the last vertex coloured (u).

Lemma 1. (Connectivity Property) Suppose that algorithm $MDsatur$ is applied on a connected graph G. If S is the subgraph of G that has already been coloured at any point of $MDsatur$ during its application on G, then S is a connected graph.

Proof. See full version [1].   □

Remark 1. Among the algorithms designed for colouring SQPG, [1, 6, 14, 2], the only one which has the Connectivity Property is algorithm $MDsatur$. This property of $MDsatur$ mainly differentiates it from the others and is one of the keypoints where the analysis of its performance is based on [2].

The FNPS Algorithm. $FNPS$ was first presented in [6]. It is a 2-approximation algorithm on SQPG.

Lemma 2. [10] Let G be a planar graph. Then there exists a vertex v in G with k neighbours, let them v_1, v_2, ..., v_k, with $d(v_1) \leq d(v_2) \leq ..., d(v_k)$ such that one of the following is true:
(i) $k \leq 2$; (ii) $k = 3$ with $d(v_1) \leq 11$; (iii) $k = 4$ with $d(v_1) \leq 7$ and $d(v_2) \leq 11$; (iii) $k = 5$ with $d(v_1) \leq 6$, $d(v_2) \leq 7$ and $d(v_3) \leq 11$, where $d(v)$ is the degree of vertex v.

Definition 5. Let G be a graph and $e = uv$ be one of its edges. Set as $G' = G/e$ the graph that is produced from G be deleting the vertex u from it and joining each one its neighbours to the vertex v.

FNPS Algorithm

Input: a planar graph G and its square G^2. *Output*: a colouring of G^2.

Begin

1. Sort the vertices of G by their degree.
2. **If** $\Delta \leq 12$ **then** follows Procedure 1 below:

   Procedure 1: Every planar graph G has at least one vertex of degree ≤ 5. Now, inductively assume that any proper (in vertices) subgraph of G^2 can be coloured by 66 colours. Consider a vertex v in G with $degree(v) \leq 5$. Delete v from G to get G'. Now recursively colour G'^2 with 66 colours. The number of colours that v has to avoid is at most $5\Delta + 5$. Thus, there is one free colour for v.
3. **If** $\Delta \geq 13$ **then**

   (a) Find a vertex v and a neighbour v_1 of it, as described in Lemma 2, and set $e = vv_1$.

   (b) Form $G' = G/e$ and modify the sorted list of vertices according to their new degrees.

   (c) $F(G') = FNPS(G', G'^2)$ that is the colouring which is produced by the recursive call of this function on G' and G'^2.

   (d) Extend $F(G')$ to a valid colouring of G^2 as follows:

   Colour v with one of the proper colours used in the colouring F of G'^2. If no colour used is allowable for vertex v insert a new one.

End

The Algorithm *AH*. We note that algorithm *AH* colours the square of a planar graph G if $\Delta(G) \geq 749$.

AH Algorithm

Input: a graph G^2 and the maximum degree of G, Δ.

Output: a colouring of G^2.

Begin

1. $inductiveness = 1.8$;
2. **While** there are uncoloured vertices **do**

   (a) find any uncoloured vertex with degree less than $inductiveness * \Delta$ in the current subgraph of G^2 that has already been coloured i.e. vertex v

   (b) colour v with the smallest allowable colour

   (c) delete v from G
3. Return the colour of each vertex.

End

The Heuristic RFF. Let G be a graph and O a random order on its vertices. RFF assigns to each vertex the first colour that is not assigned to any one of its previous neighbours (i.e. are before the vertex v in order O).

The Heuristic MaxIScover. This heuristic finds a cover of the given graph G by maximal independent sets (MISs). Namely, it uses an algorithm that finds a MIS in G. Then, it colours the vertices of this set with a new colour and it deletes them from the current graph G. This process is done recursively on until the graph G becomes empty. Here we use heuristic $MaxIS$ [3] to find a maximal independent set of the current graph G. It is clear that instead of this algorithm any algorithm which finds MISs in SQPG can be used.

The RadioColouring Algorithm. We propose an algorithm that finds a min order min span radiocolouring. Since our experiments seem to suggest that in planar graphs the difference between $X_{span}(G)$ and $X_{order}(G)$ is a small constant (e.g. 4), the algorithm below can be also used to evaluate the chromatic number of the square of a planar graph.

Algorithm RadioColouring (RC)

Input   : a graph G and a colouring of its square, let it be $Col(G^2)$
Output : a min order min span radiocolouring of G, let it be $rad(G)$.
Begin

1. Initialise the radiocolouring of G to be the colouring of its square (i.e. $rad(G) := Col(G^2)$).
2. Sort the vertices of G in decreasing order by their degrees. Let O be the resulting permutation.
3. Take into account the vertices of G based on their rank in O. Radiocolour each one of them which has a wrong radiocolour, let it be v, as follows: $rad[v] :=$ smallest proper radiocolour, based on the radiocolours of the neighbours of v in G^2, which have bigger degree than its degree (i.e. are to the left of vertex v in O).
4. Return the radiocolour of each vertex.

End

---

3   Hard Instances

In this section we present the planar graphs whose squares are hard to colour by most of the algorithms considered. Let G be any such graph. Agnarsson and Halldórsson in [4] characterise the graphs G^2 where their algorithm's performance bounds are tight. These graphs are worst case graph instances for AH. Andreou and Spirakis in [2] claim which the hard graph instance for our algorithm $MDsatur$ are. Here we identify a worst case instance for algorithm $FNPS$ and based on this we prove that its theoretical analysis is tight.

Based on Agnarsson et al, Andreou et al. [2, 4] we present here the potential structures of G. We also present an intuitive explanation of why the corresponding graphs G^2 are hard to colour graphs.

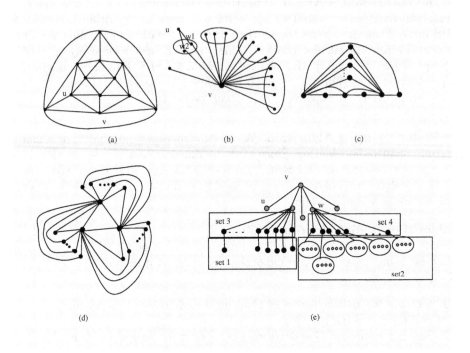

Fig. 2. a) an example of a 5-regular planar graph, the icosahedron. b) the neighbours of vertex v of icosahedron in the hard to colour graph obtained from the icosahedron when $k1 = 1, k2 = 2, k3 = 3, k4 = 4, k5 = 5$. c) a planar square root of a clique when there is a cycle of size six in its exterior in a planar embedding. d) a square root of a clique of size more than $\Delta+3$. e) the graph that leads to the proof that the theoretical analysis on the performance of algorithm $FNPS$ is tight

3.1 Hard Graph Instances for Algorithms AH and $MDsatur$

Let G' be a planar r-regular graph. The graph G is obtained from G' by replacing each of its edges, i.e. e, with a set of vertices and joining them with the endpoints of the corresponding edge. We remark that $r \leq 5$ in the case of planar graphs [1]. An example of a 5-regular planar graph is shown in Fig. 2a.

Observe that each vertex in G^2 may belong in two cliques of maximum size. This happens when the degree of each vertex of G' becomes $\Delta(G)$.

About Algorithm AH. If G' is a 5-regular planar graph, then the graph G obtained from this graph as described above is one of the graphs, which makes the performance of algorithm AH tight. This is true, because if all the edges of graph G' are replaced by sets of vertices of the same size, i.e. $\Delta/5$, then each vertex has 1.8Δ neighbours in graph G^2. We recall that at each point of the application of this algorithm on G^2 it chooses a vertex of degree at most 1.8Δ in the current subgraph of G^2 that has already been coloured and it colours it with the first allowable colour. Let v be a vertex of G and H be the subgraph of G on vertex v and its neighbours (see Fig. 2a). Observe that H^2 is a clique in G^2 of size $\Delta + 1$. Thus, if all neighbours of each vertex of H in $G^2 - H$ have already been coloured using a distinct colours from $\{1, 2, 3, ..., 0.8\Delta\}$, then AH has to assign a distinct new colour to each of them. Thus, AH uses 1.8Δ colours to colour G^2.

About Algorithm *MDsatur*. The graph G that is obtained from a planar r-regular graph G' as described above also has a square that is a hard to colour graph by our algorithm $MDsatur$. Andreou and Spirakis claim this in [2]. The basic tools used in that work is the Connectivity Property of $MDsatur$ and the fact that no planar graph has as subgraph the graph $K_{3,3}$ [9]. Here we briefly present the basic steps of the proof of the assertion following [2]. Let $MDsatur$ apply on a graphs G^2.

1. Let H^2 be a currently uncoloured subgraph of G^2, then $MDsatur$ will colour its vertices with new colours iff the coloured neighbours of each of them have all the colours currently used. For example if the graph H is the subgraph of G on the vertex v with its neighbour (like in Fig. 2b), then in order to colour graph H^2 with as many new colours as possible, it is implied that the vertices of graph $G^2 - H^2$ have to be already coloured. In [2] it is proved that at most half of the vertices of H can have neighbours coloured with Δ distinct colours before their colouring.

2. In that work it is also proved that the structure of graph H that leads to significantly hard to colour instances for $MDsatur$ is when H is the star graph (observe that in this case H^2 is a clique). Otherwise, if H^2 is a clique and H is not a star graph, then some of its vertices lie inside a cycle C (in an embedding of H in the plane) and by the planarity of G it is impossible to have neighbours vertices of $G - H$ that lie outside cycle C (see Fig. 2c). Thus, because of the connectivity property of $MDsatur$ it is proved in [2] that the graphs obtained under these conditions are not harder to colour, than the graphs obtained when H is a star graph. It is also proved there, that the insertion of new vertices between any pair of vertices similar to $(w1, w2)$ in Fig.2b does not affect the performance of this algorithm, even if these new vertices are Δ. This is correct, because by the connectivity property it is impossible to colour the new vertices after the colouring of at least six vertices of H. This is true, because each group of new vertices (one for each pair of vertices $(w1, w2)$) are lie inside distinct bounded faces. Hence, their colouring affects only the degree of saturation of vertices of H.

3. If H is the star graph, it is also proved in [2], that each set of at least four vertices of H cannot belong to more than two cliques of size at least eight with at least three distinct vertices each in G^2 (because of $K_{3,3}$). Based on this it is proved in [2] that when H is a star graph, G^2 is very hard to colour by $MDsatur$. By the argument used at the end of the above step we conclude

that even if vertices of H belong to more than two cliques of maximum size in H^2 this does not affect the performance of our algorithm.

4. Finally, if the graph H^2 is a clique of size greater than $\Delta + 3$, then H has the structure of the graph shown in Fig. 2d as it is claimed in [2]. So again by the connectivity property it is proved in that work that the graph G^2 obtained under this condition is not hard to be coloured graph for $MDsatur$.

3.2   Worst Case Instance for Algorithm *FNPS*

In this subsection we prove that the graph G properly obtained from the graph presented in Fig. 2e shows that the theoretical analysis of $FNPS$ is tight.

Lemma 3. *The approximation ratio of algorithm FNPS is tight.*

Proof. Let G the graph obtained by the graph shown in Fig. 2e as follows: replace each supervertex of *set 2* by $\Delta - 1$ new vertices and joining each of the new vertices with the neighbour of the corresponding supervertex. To prove this Lemma it is enough to show that $FNPS$ colours vertex x with colour 2Δ.

Suppose that algorithm $FNPS$ firstly selects the vertex v (see in the pseudo-code step 3) to be a vertex from *set 1*. Observe that the degree of vertex v in this case is two in the current graph G. Hence it is not needed to find a vertex that will satisfy the conditions of vertex $v1$ (see also the pseudo-code). Subsequently, $FNPS$ sequentially selects all the vertices of this set and all the vertices from *set 2* and then it deletes them from G. Observe that all these vertices had degree two in G, so their selection was allowable. Subsequently $FNPS$ can choose sequentially all the vertices of *set 3* and all the vertices of *set 4*. This is an allowable action, because the vertices of *set 2* have already been deleted from G. So the vertices of *set 3* had degree one at their selection time. Finally we assume that $FNPS$ selects in turn the vertices u, w and x.

$FNPS$ is a recursive algorithm. Let G' be the current graph G in iteration i and let v be the vertex that is selected by $FNPS$ at this iteration. Then, from the way that $FNPS$ colours the vertices of G we conclude that it colours vertex v after the colouring of the graph $G - \{v\}$. Thus, $FNPS$ colours the vertices in the order that they are deleted from G. Hence, the vertices of *set 1* are first coloured by colour 1. The vertices of *set 2* which have replaced the same super vertex of the graph of Fig. 2e have the colours $\{1, 2, ..., \Delta - 1\}$. To the vertices of *set 3* are assigned the colours $\{2, ..., \Delta - 1\}$. Hence, the vertices of *set 4* have the colours $\{\Delta, \Delta + 1, \Delta + 2, ..., 2\Delta - 2\}$. Finally vertex u receives colour 1, vertex w receives colour $2\Delta - 1$ and vertex x the colour 2Δ. $\square$

4   Graph Generation

The planar graph instances, which give the squares used as input to the colouring algorithms, that are experimentally evaluated in this work, are split in three sets. In the first one, we call it S1, are graphs created using the graphs generation library of LEDA [12]. More specific, we used the following LEDA procedures:

Class C1: random_planar_graph(G, n, m = 1.5*n); which produces a random maximal planar graph G with $n - 1$ vertices. Then it adds a new vertex into a random selected face F and it joins this vertex with all the vertices of face F. An appropriate number of edges of graph G are deleted randomly in order to have only m edges, in the resulting graph. We mention here that $m \leq 3n - 6$ in the case of planar graphs [9]. The graphs of Class C3 are maximal planar graphs and they have almost the maximum number of allowable edges. So, in this case we prefer an intermediate value on m and we choose $m = 1.5n$

Class C2: triangulated_planar_graph(G, n); which produces planar graphs without any cycle of size at least four.

Class C3: maximal_planar_graph(G, n); which produces a maximal planar graph with n vertices.

The planar graphs produced by the procedures of both classes C2 and C3 are initially generated the procedure of class C1, i.e. are random-planar graphs, and then these procedures appropriately add edges to these graphs to become triangulated planar and maximal planar, respectively.

In the second set, we call it S2, we take each of the graphs obtained from LEDA in classes C1, C2, C3 and we create a new graph modifying the original graph in order to get harder to colour graph instances for the considered algorithms. Their hardness has the characteristics of the hard graph instances which are discussed in Section 3. Hence, if G is a planar graph from set S1 and k is a positive integer, then the planar graph G' obtained from G by replacing each one of its edges, let it be e, with a set of new vertices of size k and joining each one of the new vertices with the endpoints of edge e.

In the last set, S3, we create graphs obtained from the icosahedron ($20dron$) (see Fig. 2(a)) by replacing each of its edges with a set of new vertices. The size of these sets has one of the following values: $K = \{k1, k2, ..., k5\}$. We remark here that the icosahedron is a 5-regular planar graph with 30 edges. We also mention that the edges incident to each of its vertices have to be replaced by sets of vertices for each of the sizes $k1$ to $k5$. With this condition we achieve the degree of each vertex of the icosahedron to become Δ in the resulting graph (so each vertex of G belongs in two cliques of maximum size in G^2. This is a hard case as it is explained in Section 3).

To specify the size of the set of vertices which replaces each edge of the $20dron$ we implement an exhaustive algorithm which edge-colours the $20dron$ with five colours and returns all possible edge colourings of this graph [1]. This algorithm returns 780 different edge colourings. Having an edge colouring we properly replace each edge of colour i with a set of new vertices of size ki ($1 \leq i \leq 5$).

The size n of the planar graph instances of set S1 is 1024. In most past experimental works the size of the input graphs does not exceed 1024 [13]. We have done experiments on smaller (e.g. n = 256) and bigger graphs (e.g. n = 2048) and we observed that the asymptotic behaviours of the algorithms considered remain the same. So, in this sense, our experiments can be considered large enough and representative.

For graphs of set S2 we performed tests on the squares of the graphs obtained from the graphs of set S1 with size $n = 256$ and $k = 6$. In most cases the value of n', that is the number of vertices of the resulting graph, is $3n * 6 \geq 4000$. As shown in Tables 1 and 2 the graphs of set S2, which are expected to be harder to colour instances than the corresponding graphs of set S1, do not affect the behaviour of the algorithms w.r.t. their behaviour on the squares of the graphs of set S1. Thus, we conclude that our experiments are representative.

We also performed some special experiments on the graphs of set S3. If $k1 = k2 = k3 = k4 = k5$, then we denote by k this unique value. We produce graphs with $k = 20$ and $k = 100$. $20dron$ has only 30 edges, so it can be handled even with such a large value on k. From our results we conclude that the number of colours used by each algorithm considered is proportional to the increase of the value of k, while their performance remains the same. Thus, the results on these values of k are representative.

In the case where the value of each ki $(1 \leq i \leq 5)$ is distinct, we have done experiments for the case where $k1 = x, k2 = 2x, k3 = 3x, k4 = 4x, k5 = 5x$, for $x = 5, 15$. These results have greater interest on our algorithm, $MDsatur$ in contrast to the results of the previous case, which are more interesting for the rest of the algorithm and especially for $FNPS$.

5   Experimental Results

We split this section in two parts. In the first one, we present the experimental results of the colouring algorithms RFF, $MaxIScover$, $MDsatur$ and $FNPS$, which colour squares of planar graphs. In the second part, we present the result of the radiocolouring algorithm RC, when it has as input a planar graph G and a colouring of G^2 obtained from each one of the above algorithms. In the following tables are depicted: a) the average number of colours used from each of the above algorithms on the squares of 10 planar graphs from each of the classes $C1, C2, C3$. b) the average maximum degrees of the corresponding planar graphs. We need this information in order to be able to make some comparisons between a good estimation of the optimal colouring (of each of these graphs, i.e. $\Delta + 1$) and the solutions produced from each one of the above algorithms. c) the value of k or the value of x depends on whether ki's have the same or different values $(1 \leq i \leq 5)$.

5.1   Colouring the Square of a Planar Graph

The results of the application of the colouring algorithms, which are considered here, on SQPG are depicted in Tables 1-3. In Table 1 are shown the results on the squares of the graphs of set S1, in Table 2 the results on the squares of the graphs of set S2. In Table 3 we considered the squares of the graphs of set S3 and in Table 4 are shown the ER of each of the algorithm considered on these graphs. The most important observation on the results depicted in Table 1 is that each one of the considered algorithms uses $\Delta + constant$ colours to colour

planar graph class	max degree	Greedy	FindMIS	MDsatur	FNPS
random_planar_graph	60,4	63	61	60,4	61
triangulated_planar_graph	24,1	26	28	24,5	25
maximal_planar_graph	120,1	121	120,5	120,5	121

Table1: 1

planar graph class	max degree	Greedy	FindMIS	MDsatur	FNPS
random_planar_graph	188,5	188,5	188,5	188,5	189
triangulated_planar_graph	99,4	102	101,2	99,8	100,5
maximal_planar_graph	295	295,5	295	295	296

Table1: 2

	max degree	Greedy min % max%	FindMIS min % max %	MDsatur min % max%	FNPS min% max%
k=20	100	115 ~ 100%	105 – 115 ~ 100%	102,104 90% 120 10%	135- 140 ~100%
k=100	500	540 – 560 ~ 100%	510 – 550 ~ 100%	502–505~90% 600 ~10%	690–705 ~100%
x = 5	75	85-96 80% 100~20%	80-85 ~100%	77-82 50% 90-94 0.06%	101 ~100%
x =15	225	247 – 260 ~ 100%	240-260 ~100%	227-247 50% 270-281 0.06%	300 ~100%

Table1: 3

	max deg.	Greedy min % max%	FindMIS min % max %	MDsatur min % max%	FNPS min% max%
k=20	100	1.15 ~ 100%	1.05–1.15 ~ 100%	1.02-1.04 90% 1.20~10%	1.35- 1.40 ~100%
k=100	500	1.1 ~ 100%	1.02–1.10 ~ 100%	1.004–1.01~90% 1.20~10%	1.38–1.41 ~100%
x = 5	75	1.13-1.28 80%1.3 20%	1.06,1.13%~100%	1.01-1.09 50%1.2-1.25 0.06%	1.35 ~ 100%
x =15	225	1.09 – 1.15 ~ 100%	1.08-1.15 ~ 100%	1.02-1.1 50% 1.2-1.25 0 .06%	1.33 ~ 100%

Table1: 4

Fig. 3.   Table 1: In this table are depicted the results, on average, of the colouring algorithms: RFF, $MaxIScover$, $MDsatur$ and $FNPS$. on 10 planar graphs from set S1 with 1024 vertices each. In Table 2 are shown the results, on average, of the above colouring algorithms on 10 planar graphs from set S2 with 256 vertices each and $k = 6$. In Table 3 are depicted the results, on average, of these algorithms on the icosahedron when $k = 20, 100$ and $x = 5, 15$. Finally in Table 4 are displayed the experimental ratios (ER) of each of the above algorithm on the results depicted in Table 3

the square of any planar graph which belongs in set S1. Actually this constant term is very small, i.e. 4. This very good result yields the first indication that it is maybe possible to get colouring algorithms that colour the square of each planar graph with less than 1.66Δ colours (that currently is the best known result). In order to further support this assumption we get results on harder to colour squares of planar graphs. I.e. the squares of the graphs of sets S2, S3.

In Table 2 are depicted the results on the squares of the graphs of set S2. From these results we conclude that the colouring algorithms we consider still colour very well the squares of the graph instances of set S2. There is no significant difference in the number of colours used from each one of the algorithms.

As we mention in the introduction the graphs obtained from LEDA are expected to be "non-extremal" graphs, because of the randomness in the way they are constructed (see Section 4). From the results we conclude that there does not exist a clique of size greater than $\Delta + 3$ in these graphs, because all the algorithms considered colour them with $\Delta + constant$ colours. Because of this fact the results obtained on these graphs are reasonable (EP about 1), because as we can see from Tables 3, 4 almost all the algorithms considered have very good EP even on hard colourable graph instances.

The last experiments have been done on the squares of graphs obtained from an $20dron$. I.e. the squares of the graphs of set S3. Our algorithm $MDsatur$ in the case where $k = 20$ uses 1.02Δ to 1.04Δ colours to colour the 90% from the 780 graphs tested. We recall here that we create one graph for each edge colouring of $20dron$. In only 10% of these squares it uses 1.20Δ colours. When $k = 100$ the behaviour of our algorithm is improved. It uses at most 1.004Δ to 1.01Δ colours to colour 90% of the graphs tested. This is reasonable, because $MDsatur$ actually uses a constant number of colours with value bigger than $\Delta + 1$, so as greater is the value of Δ so smaller is the ratio among them (i.e. $\frac{\Delta+c}{\Delta+1}$).

Algorithm $FNPS$ uses in almost all cases 1.35Δ to 1.40Δ colours in order to colour the 780 graphs tested in the case where $k = 20$. In the case where $k = 100$ it uses slightly more colours, i.e. 1.38Δ to 1.41Δ.

When ki's have different values we examine the cases where $x = 5$ and $x = 15$. In these cases observe that the performance of our algorithm gets worse, but it is still better from the performance of the other algorithms. Algorithm $MDsatur$ gives similar results as in the case where all ki's have the same value in only 50% of the tested graphs (i.e. it uses 1.01Δ to 1.09Δ colours). We observe that it uses at most 1.25Δ colours to colour 0.06% of the graphs tested. The rest of these graphs are coloured with 1.1Δ to 1.19Δ colours by $MDsatur$. The colouring of each of these graphs is uniformly distributed in this range. On the other hand, the performance of $FNPS$ becomes better in this case.

The results presented in Tables 3 and 4 show the worst behaviour of algorithms $MDsatur$ and $FNPS$ on the squares of the graphs of set S3. For our algorithm these results are obtained when we named the vertices of G randomly. On the other hand, when we label these vertices sequentially (namely nearby vertices in an embedding of G receive consecutive integers as index), then it colours with $\Delta + c$ colours almost all (about 100%) of the graphs of set S3. The behaviour of algorithm $FNPS$ is exactly opposite. $FNPS$ colours almost all the squares of the graphs of set S3 with at most 1.2Δ colours, when the vertices are randomly named, in contrast to 1.4Δ in the other case.

Finally, on the squares of the graphs of set S3 both greedy heuristics considered (RFF,$MaxIScover$) colour the square of each such planar graph with ER at most 1.15 in almost all the cases. The first of this heuristics, RFF, uses the same number of colours, that is $1.15\Delta(G)$, in almost each of the 780 graphs tested (for each value of k). The second heuristic $MaxIScover$ uses 1.02Δ to 1.15Δ colours uniformly spread in the above range on the squares of these graphs.

For the case where the values of ki's are not the same, the performance of RFF gets worse. Especially, when $x = 5$. In this case it uses 1.3Δ colours in 20% of the graphs tested. Observe that its performance is better when $x = 15$. The heuristic $MaxIScover$ has almost the same results as in the above case.

At this point we discuss the experimental performance of algorithm AH. This algorithm colours the square of a planar graph with maximum degree at least 749. Because of this condition it is obvious why we do not present results from this algorithm. On the other hand, in order to study its behaviour and not the exact number of colours that it uses, we did experiments increasing the value of the variable $inductiveness$ (see the pseudo code). From these results we can conclude that it has similar behaviour as algorithm $FNPS$. Namely, when it selects the next vertex to be coloured at random (beyond the vertices which satisfy the proper condition i.e. have less than 1.8Δ neighbours in the already coloured subgraph of G^2), then it colours even the modified icosahedron with $k = \Delta/5$ with 1.2Δ colours in almost all cases. Observe that exactly this graph gives the tightness on its performance as we explain in Section 3.

5.2 RadioColouring a Planar Graph

In this paragraph we notice that our algorithm $RadioColouring$ min order min span radiocolours each of the planar graphs, G, used in the experiments shown in Tables 1-3 using at most four new extra colours in addition to the colours used by any of the algorithms: RFF, $MaxIScover$, $MDsatur$ and $FNPS$, to colour G^2.

The similarity in its behaviour on each of the set of graphs used and on each of the colourings of their squares allow as to make the following conjecture:
$X_{span}(G) - X_{order}(G) \leq c$, where c is a small constant.

References

[1] Maria Andreou, Sotiris Nikoletseas and Paul Spirakis: Algorithms and Experiments on Colouring Squares of Planar Graphs, TR2003/03 Computer Technology Institute, Greece, 2003. http://students.ceid.upatras.gr/~mandreou/. 21, 24, 27

[2] Maria Andreou and Paul Spirakis: Efficient Colouring of Squares of Planar Graphs, TR2002/11/01 Computer Technology Institute, Greece, 2002. 18, 21, 23, 25, 26

[3] Maria Andreou and Paul Spirakis: Planar Graphs, Hellenic Conference on Informatics, EPY, 2002. 23

[4] Geir Agnarsson, Magnus M.Hallorsson: Coloring Powers of Planar Graphs, ACM Symposium on Discrete Algorithms (SODA). 15, 17, 21, 23

[5] D.Brélaz: New methods to color the vertices of a graph, Communications of the ACM 22, 1979, pp. 251-256. 17, 18

[6] D. A. Fotakis, S. E. Nikoletseas, V. G. Papadopoulou and P. G. Spirakis: NP-completeness Results and Efficient Approximations for Radiocoloring in Planar Graphs, In the Proceedings of the 25th International Symposium on Mathematical Foundations of Computer Science (MFCS), Editors Mogens Nielsen, Branislav Rovan, LNCS 1893, pp 363-372, 2000. 15, 16, 17, 21

[7] D. Fotakis, G. Pantziou, G. Pentaris and P. Spirakis: Frequency Assignment in Mobile and Radio Networks. Networks in Distributed Computing, DIMACS Series in Discrete Mathematics and Theoretical Computer Science, 45, American Mathematical Society (1999) 73-90. 16

[8] D. Fotakis and P. Spirakis: Assignment of Reusable and Non-Reusable Frequencies, International Conference on Combinatorial and Global Optimization (1998). 16

[9] Harary: Graph Theory, Addison-Wesley, 1972. 25, 27

[10] J. Van D. Heuvel and S. McGuiness: Colouring the Square of a Planar Graph, CDAM Research Report Series, July (1999). 15, 17, 21

[11] Katsela I. and M. Nagshineh: Channel assignment schemes for cellular mobile telecommunication system, IEEE Personal Communication Complexity, 1070, 1996. 16

[12] K. Mehlhorn and S. Naher: The LEDA Platform of Combinatorial and Geometric Computing, Cambridge University Press, 1999. 17, 26

[13] D. Johnson, C. Aragon, L. Mcgeoch, C. Schevon: Optimization by simulated annealing: an experimental evaluation; Part II, Graph Coloring and Number Partitioning, Operating Research, Vol. 39, No. 3, 1991. 27

[14] M. Molloy and M. R. Salavatipour: Frequency Channel Assignment on Planar Networks , To appear in: Proceedings of 10th European Symposium on Algorithms, ESA 2002. The journal version is : A Bound on the Chromatic Number of the Square of a Planar Graph", submitted. ESA 2002. 15, 17, 18, 21

[15] S. Ramanathan, E. R. Loyd: The complexity of distance2-coloring, 4th International Conference of Computing and information, (1992) 71-74. 16

[16] J. S. Turner: Almost all k-colorable graphs are easy to color: Journal of Algorithms, 9, pp. 217-222, 1988. 18

[17] G. Tinhofer, E. Mayr, H. Noltemeier, M. M. Syslo (eds) in cooreration with R. Albrecht: Computational Graph Theory, Springer-Verlag/Wien, 1990 chapter "heuristics for graph colouring". 18

Experimental Analysis of Online Algorithms for the Bicriteria Scheduling Problem*

Vittorio Bilò, Michele Flammini, and Roberto Giovannelli

Dipartimento di Informatica, Università di L'Aquila
Via Vetoio loc. Coppito, I-67100 L'Aquila, Italy
{bilo,flammini}@di.univaq.it

Abstract. In this paper we experimentally evaluate the performances of some natural online algorithms for the bicriteria version of the classical Graham's scheduling problem. In such a setting, jobs are characterized by a processing time and a memory size. Every job must be scheduled on one of the m processors so as to minimize the time makespan and the maximum memory occupation per processor simultaneously. We consider four fundamental classes of algorithms obtained by combining known single-criterion algorithms according to different strategies. The performances of such algorithms have been evaluated according to real world sequences of jobs and to sequences generated by fundamental probability distributions. As a conclusion of our investigation, three particular algorithms have been identified that seem to perform significantly better than the others. One has been presented in [4] and is the direct bicriteria extension of the basic Graham's greedy algorithm, while the other ones are given by two different combinations of the Graham's algorithm and the Albers' algorithm proposed in [1].

1 Introduction

In the last years a considerable research activity has been devoted to multicriteria optimization problems, in which a feasible solution must be simultaneously evaluated with respect to different criteria and/or cost functions. For instance, one can ask for the determination of a spanning tree of a graph whose global cost is low with respect to two different weightings of the edges [17, 22], or such that the diameter is low with respect to first weighting and has a low global cost with respect to the second one [17].

Many scheduling problems have been investigated under a multicriteria optimization point of view. As an example, in [24] the authors considered the problem of minimizing the makespan on one machine among the solutions with minimum maximum lateness, in [23] scheduling algorithms have been presented for optimizing the average completion time given an upper bound or budget on

* Work supported by the IST Programme of the EU under contract number IST-1999-14186 (ALCOM-FT), by the EU RTN project ARACNE, by the Italian project REAL-WINE, partially funded by the Italian Ministry of Education, University and Research, and by the Italian CNR project CNRG003EF8 (AL-WINE).

K. Jansen et al. (Eds.): WEA 2003, LNCS 2647, pp. 33–46, 2003.

the possible makespan, and in [5, 25] the authors dealt with the simultaneous minimization of the makespan and total weighted completion time. Other results with respect to various multicriteria objectives can be found in [9, 12, 13, 14, 18, 20, 21, 25, 27]. A survey on the multicriteria scheduling literature up to '93 is given in [19].

In the bicriteria extension of the classical Graham's problem jobs are characterized by a processing time and a memory size and have to be scheduled so as to simultaneously minimize the time makespan and the maximum memory occupation per processor. Such a problem has been first considered in [4], where among the others an online algorithm has been presented whose two competitive ratios are both always less than 3. However, no results are given concerning its practical behavior.

In this paper we propose four natural classes of online algorithms for the bicriteria scheduling problem that include the one of [4]. They are obtained by combining according to different strategies the single-criterion algorithms of Graham [10], Bartal et al.[3], Karger et al.[16] and Albers [1], that in the order achieved competitive ratios 2, 1.985, 1.945 and 1.923.

We evaluate the performances of the presented bicriteria algorithms according to real world sequences of jobs and to sequences generated by fundamental probability distributions. Our investigation shows that the results heavily depend on the characteristic of the particular sequence. However, three algorithms seem equivalent and to outperform all the others in all the different cases. One corresponds to [4], while the other ones are obtained by two different combinations of the Graham's and Albers's algorithms.

The paper is organized as follows. In the next section we introduce the basic notation and the various bicriteria algorithms. In Section 3 we give a detailed description of the experiments. In Section 4 we discuss the obtained results and finally, in Section 5, we give some conclusive remarks.

2   Definitions and Notation

In this section we introduce the basic notation and the the bicriteria algorithms used in the experimental analysis.

We denote by m the number of machines and by $\sigma = < p_1, ..., p_n >$ the input sequence. Each job p_j, $1 \leq j \leq n$, is characterized by a pair of costs (t_j, s_j), where t_j represents the time required to process job p_j and s_j its memory occupation.

A scheduling algorithm A is online if it assigns each job $p_j \in \sigma$ to one of the m machines without knowing any information about the future jobs. We denote as $T_j^i(A)$ the completion time of machine i after job p_j is scheduled according to A and as T_j the sum of the processing times of the first j jobs, that is, $T_j = \sum_{h=1}^{j} t_h$, or analogously $T_j = \sum_{i=1}^{m} T_j^i(A)$. Similarly, let $S_j^i(A)$ be the memory occupation on machine i after job p_j is scheduled according to A and S_j be the total memory occupation of the first j jobs. The time and the memory costs of algorithm A are, respectively, $t(A, \sigma) = \max_{i=1}^{m} T_n^i(A)$ and

$s(A, \sigma) = \max_{i=1}^{m} S_n^i(A)$. For the sake of simplicity, when the algorithm A is clear from the context, we will drop A from the notation.

Let $t^*(\sigma)$ be the minimum makespan required to process σ, independently from the memory. Analogously, let $s^*(\sigma)$ be the minimum memory occupation per machine required by σ ignoring the time.

Definition 1. *An algorithm A is said c-competitive if, for all possible sequences σ, $t(\sigma) \leq c \cdot t^*(\sigma)$ and $s(\sigma) \leq c \cdot s^*(\sigma)$.*

Let us briefly describe the single-criterion online algorithms used in our bi-criteria extensions. Let ω_j be the cost of the job p_j, M_j^i be the machine with the i-th smallest load l_j^i after the first j jobs have been scheduled, $1 \leq i \leq m$, and let A_j^i be the average load on the i smallest loaded machines after the first j jobs have been scheduled. Then the single-criterion algorithms are defined as follows:

Graham's algorithm: Assign job p_j to the least loaded machine, i.e. M_j^1.
Bartal's algorithm: Set $k = [0.445m]$ and $\epsilon = \frac{1}{70}$. Assign job p_j to machine M_{j-1}^{k+1} if $l_{j-1}^{k+1} + \omega_j \leq (2 - \epsilon)A_{j-1}^k$ (basic choice), otherwise assign p_j to the machine with the smallest load (alternative choice).
Karger's algorithm: Set $\alpha = 1.945$. Assign job p_j to the highest loaded machine M_{j-1}^k (that is with the highest k) such that $l_{j-1}^k + \omega_j \leq \alpha \cdot A_{j-1}^{k-1}$ (basic choice). If there is not such a machine, assign p_j to the machine with the smallest load (alternative choice).
Albers's algorithm: Set $c = 1.923$, $k = \lfloor \frac{m}{2} \rfloor$ and $r = 0.29m$. Set $\alpha = \frac{(c-1)k - \frac{r}{2}}{(c-1)(m-k)}$. Let L_l be the sum of the loads on machines $M_j^1, ..., M_j^k$ if p_j is scheduled on the smallest loaded machine. Similarly, let L_h be the sum of the loads obtained considering the machines $M_j^{k+1}, ..., M_j^m$ if p_j is scheduled on the smallest loaded machine. Let δ_j^m be the makespan if p_j is scheduled on the machine with the $(k+1)$-st smallest load. Assign p_j to the smallest loaded machine if at least one of the following conditions holds: (a) $L_l \leq \alpha \cdot L_h$; (b) $\delta_j^m > c \cdot \frac{L_l + L_h}{m}$ (basic choice). Otherwise, assign p_j to the machine with the $(k+1)$-st smallest load (alternative choice).

Notice that these last three algorithms maintain a list of machines ordered by non decreasing load. In most of their bicriteria extensions, a total order has still to be maintained and is obtained by means of a single cost measure given by a proper linear combination between the time and memory loads of each machine. To this aim, we observe that a simple combination that at each step j assigns to each processor i a combined load $C_j^i = T_j^i + S_j^i$ might be extremely compromising for one cost measure when the other is characterized by substantially bigger values. In particular, this is true when the respective optima are very different. However, equal optima can be obtained by scaling up one of the two costs measures, thus guaranteeing a substantial equivalence. More precisely, if σ_j is the subsequence of the first j jobs of σ, it suffices to define $C_j^i = T_j^i + \lambda_j S_j^i$

with $\lambda_j = \frac{t^*(\sigma_j)}{s^*(\sigma_j)}$. Unfortunately, computing $t^*(\sigma_j)$ and $s^*(\sigma_j)$ is an intractable problem. However, the following lower bounds on $t^*(\sigma_j)$ and $s^*(\sigma_j)$ provide approximations that rapidly converge to $t^*(\sigma_j)$ and $s^*(\sigma_j)$ as j increases. Let $\bar{t}_1, \ldots, \bar{t}_j$ be the processing times of the first j jobs listed in non increasing order, that is in such a way that $\bar{t}_1 \geq \ldots \geq \bar{t}_j$. Then $t^*(\sigma_j) \geq Lt(\sigma_j) = max\{T_j/m, \bar{t}_1, 2\cdot\bar{t}_{m+1}, 3\cdot\bar{t}_{2m+1}, \ldots, h\cdot\bar{t}_{(h-1)m+1}\}$, where h is the largest integer such that $(h-1)m + 1 \leq j$. The lower bound $Ls(\sigma_j)$ on $s^*(\sigma_j)$ can be defined accordingly, so that $\lambda_j = \frac{Lt(\sigma_j)}{Ls(\sigma_j)}$ can be easily calculated at step j.

Let us now introduce the different families of algorithms.

2.1   Bicriteria Graham Based Algorithms

All the algorithms belonging to this family select at each step the subset of the k machines with the lowest time load, where k is a parameter at most equal to $\lceil\frac{m}{2}\rceil$ whose value has been determined during the experimental analysis so as to minimize the maximum competitive ratio. The particular machine among the k selected ones chosen to schedule the current job is then determined according to the memory loads by applying one of the above single-criterion algorithms, thus obtaining corresponding bicriteria algorithms that we call respectively Graham-Graham, Graham-Bartal, Graham-Karger and Graham-Albers. The algorithm Graham-Graham corresponds to the one of [4].

The idea behind this family is that of improving the performance of Graham-Graham by exploiting the better behavior of the other single-criterion algorithms when applied to real world cases (see [2]).

2.2   Weak Bicriteria Algorithms

These algorithms are obtained by reducing each bicriteria instance to a single-criterion one. More precisely, this is accomplished by assigning each job $p_j = (t_j, s_j)$ a cost $w_j = t_j + \lambda_j s_j$, with λ_j determined as above, and then by applying the above single-criterion algorithms assuming at each step j a combined load C^i_{j-1} for each machine i, $1 \leq i \leq m$. We call Weak-Graham, Weak-Bartal, Weak-Karger and Weak-Albers the corresponding algorithms.

2.3   Half Bicriteria Algorithms

This family extends all the single-criterion algorithms, except Graham's, by considering at each step j the total order induced by the combined loads C^i_j, $1 \leq i \leq m$. Then, if the basic choice can be performed both on the time and memory measures, the job is scheduled on such a machine, otherwise on the unique one given by the alternative choice. We denote as Half-Bartal, Half-Karger and Half-Albers the corresponding algorithms.

2.4   Strong Bicriteria Algorithms

As in the previous case, this family extends all the single-criterion algorithms, except Graham's and the basic choice is made by considering the order induced by the combined loads. However, the alternative choice is performed according to the bicriteria algorithm Graham-Graham, which in some sense chooses the smallest loaded machine in a bicriteria way. We call Strong-Bartal, Strong-Karger and Strong-Albers the corresponding algorithms.

The main reason leading to the definition of this family is that, as observed during our experiments, the basic choice is always taken only the 10% of the steps. Thus, strong bicriteria algorithms are very close to Graham-Graham and might improve when basic choices correspond to better selections.

Particular attention deserves Strong-Albers. In such an algorithm, the $(k+1)$-st smallest loaded machine $(k = \lfloor \frac{m}{2} \rfloor)$ corresponding to the alternative choice, for each single measure is rather close to the one chosen by Graham-Graham. In fact, Graham-Graham selects a machine whose time and memory loads are always below the average. Such a machine tends to be not far from the $(k+1)$-st position in both the two orderings that can be obtained according to the two measures.

3   Description of the Experiments

The experiments performed to investigate the behavior of the various bicriteria algorithms belong to two different classes, according to whether the sequences of the jobs come from real systems or are generated according to probability distributions.

In the first case, the traces used in the tests are taken from two different systems. One sequence consists of traces obtained from the log files of the San Diego Supercomputer Center (SDSC) [8]. The second sequence is relative to the Computer CM-5 installed in the Los Alamos National Lab (LANL) and is part of the NPACI job traces repository [11].

The following tables summarize the main characteristics of the two traces respectively for the time and memory measures. They will be useful for the interpretation of our results.

System	Year	Numb. of jobs	Time Avg	Min Time	Max Time	Variation
SDSC	1998-2000	67,667	84,079	1	7,345,105	4
LANL	1994-1996	201,387	1,000	1	25,200	2

System	Year	Numb. of jobs	Mem. Avg	Min Mem.	Max Mem.	Variation
SDSC	1998-2000	67,667	15,034	1	360,000	1.3
LANL	1994-1996	201,387	3,677	1	29,124	1.1

The sequences of the second class have been generated according the following probability distributions: uniform, exponential, Erlang, hyperexponential and Bounded-Pareto distribution [15, 26]. Each distribution is characterized by corresponding parameters that, following the suggestions in [6, 7], have been properly set in order to get realistic sequences and to cover a great range of values. The most common distributions used to model the processing times of the jobs in computing systems are the exponential, hyperexponential and Erlang distributions. However, for the sake of completeness, we have considered also the uniform distribution. In addition, we included the Bounded-Pareto distribution, which seems to be close to real world instances [2].

Similar considerations hold also for the memory sizes of the jobs, although in this case values tend to be more uniform and with a lower variation with respect to the processing times. As a consequence, the hyperexponential and Bounded-Pareto distributions are less realistic.

Since we are randomly generating both processing times and memory sizes, we also consider different combinations of the above distributions. In such a setting we restricted to the most reasonable and mostly used combinations, that is hyperexponential and Bounded-Pareto for the time, and uniform, exponential and Erlang for the memory (again see [6, 7] for technical motivations).

The algorithms have been tested on sequences of 10, 000 jobs and their current maximum competitive ratio has been computed after having scheduled each job, even if the competitive ratios have been evaluated exploiting the lower bounds $Lt(\sigma_j)$ and $Ls(\sigma_j)$ as approximations of $t^*(\sigma_j)$ and $s^*(\sigma_j)$. The number of machines has been set to 10, 50, 100, 200 and 500.

4   Analysis of the Results

In this section we present the experimental results of our tests and give a detailed analysis of the performances of the various bicriteria algorithms defined in section 2.

Due to space limitations, in the following figures we give only the experiments for a number of machines $m = 10$ and $m = 500$. In every case, the results for $m = 50$ are close to the ones for $m = 10$, while there is a substantial equivalence for $m = 100$, $m = 200$ and $m = 500$. Moreover, for the sake of clarity, we depict only the curves representing the maximum competitive ratio of the best algorithm in each class, plus the one of Graham-Graham.

4.1   Results for Real World Jobs

Let us first discuss the results obtained for a small number of machines, that is for $m = 10$ and $m = 50$.

Figure 1a shows the (maximum) competitive ratio of each algorithm applied on the first trace when $m = 10$. For all the algorithms it is possible to note a greater fluctuation of the competitive ratios for the first 2, 000 jobs. In this interval, the oscillations are between the values 1.2 and 1.8. The only exception (even though it is not in the figure) is the Weak-Karger algorithm, for which the competitive ratio rapidly grows to reach the value 3. The bad behavior of this algorithm confirms the results in [2]. After the first 2, 000 jobs, the competitiveness tends to become more stable and the best results are obtained in the order by Strong-Albers, Weak-Graham, Graham-Graham, Graham-Bartal, Graham-Albers and Graham-Karger, with Strong-Albers, Weak-Graham and Graham-Graham very close to each other. For all these algorithms the competitive ratio is bounded between 1.1 and 1.2.

Figure 2a shows the results for the second trace. The general trend is similar to the one of the first trace, even if the competitive ratios are significantly better as they tend to 1. It is also possible to note a higher stability of the competitive ratios and thus a faster convergence. The best performances are achieved in the order by Graham-Albers, Graham-Graham, Strong-Albers, Weak-Graham, Graham-Bartal and Graham-Karger.

For both the two traces, the best performance of the Graham based algorithms is achieved when $k = \lceil \frac{m}{2} \rceil$.

Let us now consider the results for a high number of machines, that is for $m = 100$, $m = 200$ and $m = 500$. The evolution of the competitive ratios is better in the first trace, due to a longer transient phase in the second one (see Figures 1b and 2b). However, the situation seems to change right after $10,000$ jobs. At the end of the traces the competitive ratios are still greater than 1.4, while for a small number of machines they approached 1.1. For the first trace the best algorithms are Strong-Albers and Graham-Graham, while for the second one Graham-Albers, Graham-Bartal, Graham-Karger, Strong-Bartal, Strong-Albers and Graham-Graham.

The best performance of the Graham based algorithms is achieved when k is close to $\frac{m}{4}$.

Notice that all the executions correspond to the prefix of the first $10,000$ jobs of each trace. However, our experiments show that the results are completely analogous considering any subsequence of $10,000$ jobs.

In order to understand our experimental results, it is necessary to recall first the classification presented in [2] concerning the effects of each incoming new job p_j on the competitive ratio of an online single-criterion algorithm. In particular, three different effects have been identified:

1. If the cost of p_j is small with respect to the average load of the machines, the job cannot have significant effects on the competitive ratio of an online algorithm. This happens mainly at the end of the traces.
2. If the cost of p_j is of the same order of the average load of the machines, the competitive ratio grows. This is a consequence of the fact that any online algorithm tries to keep the load of the machines quite balanced in order to avoid a too high competitive ratio. Thus, the machine receiving p_j in general reaches a load significantly higher than the average one. The optimal algorithm instead will reserve a low loaded machine for this job, thus causing a significant increase on the competitive ratio which justifies the fluctuations observed in the tests when the number of jobs j is small.
3. If the cost of p_j is very big with respect to the average load of the machines, the job will dominate both the online and the optimal solution. As a consequence, the competitive ratio approaches the value 1.

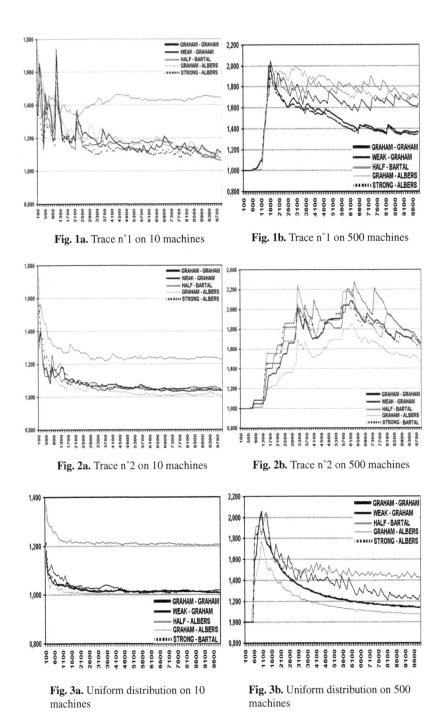

Fig. 1a. Trace n°1 on 10 machines

Fig. 1b. Trace n°1 on 500 machines

Fig. 2a. Trace n°2 on 10 machines

Fig. 2b. Trace n°2 on 500 machines

Fig. 3a. Uniform distribution on 10 machines

Fig. 3b. Uniform distribution on 500 machines

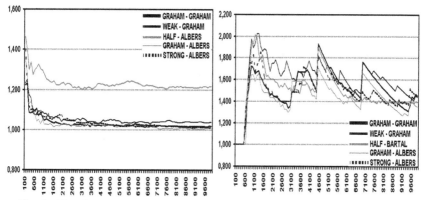

Fig. 4a. Time hyp. and memory unif. distr. on 10 machines

Fig. 4b. Time hyp. and memory unif. distr. on 500 machines

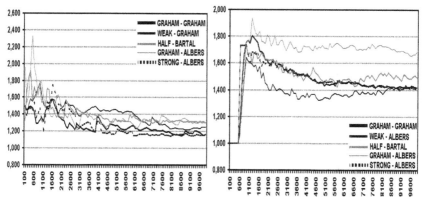

Fig. 5a. Time B.Pareto and memory unif. distr. on 10 machines

Fig. 5b. Time B.Pareto and memory unif. distr. on 500 machines

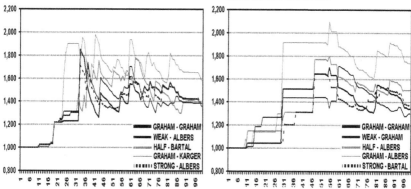

Fig. 6a. Trace n°1: fluctuation for the first 100 jobs on 10 machines

Fig. 6b. Trace n°2: fluctuation for the first 100 jobs on 10 machines

At this point, it is easy to understand that the situation in which a sequence of small jobs (effect 1) is followed by a job causing an effect of type 2 is the worst for any Graham based algorithm. The other kinds of algorithms instead, in general tend to keep some of the machines lightly loaded in order to reserve a portion of the resources for jobs causing an effect of type 2.

The probability that one of the three effects occurs and the magnitude of its influence on the competitive ratio heavily depend on the characteristic of the trace and of the tested algorithm. Obviously, if the variability of the costs of the jobs is low, effects of type 2 and 3 are very unlikely to occur. From the tables of the previous section it can be seen that the coefficient of variation of trace 2 is lower than the one of trace 1 and this explains why on this trace the competitive ratios of the algorithms are more stable and the Graham based ones perform better. This is especially evident for a big number of machines, like when $m = 500$.

After having analyzed all the different experiments, including the ones not depicted in our figures, we can state that the algorithms that have shown the best practical behavior are Graham-Graham, Graham-Albers and Strong-Albers, with the last two ones performing slightly better. This appears more evident on the first trace where the variability of the jobs is greater. More in general we can state that both the Weak-bicriteria and the Half-bicriteria algorithms are dominated by the Graham-Graham algorithm, the best performances are achieved by the Graham based and Strong-bicriteria classes and finally the Karger's algorithm in all of its extensions has not a good behavior.

4.2   Results on Probabilistic Sequences

Let us now discuss the results obtained for the sequences of jobs generated according to probability distributions.

Figure 3a shows the trend of the competitive ratios when $m = 10$ and the jobs are generated using the uniform distribution. Except for the first 500 jobs, there is an almost total lack of oscillations. The best performance in the order is obtained by Graham-Albers, Strong-Albers, Graham-Bartal, Graham-Graham, Weak-Graham and Graham-Karger. For these algorithms the competitive ratios approach 1 already after the first $1,000$ jobs. For the other distributions the results are essentially the same, even if for the hyperexponential and Bounded-Pareto distributions a general increase in the number of jobs is needed before the competitive ratios become stable.

Figure 3b shows the results for $m = 500$ again when jobs are generated using the uniform distribution. As we expected, the ratios follow the same trend of the previous case with a natural increase in the magnitude of the oscillations and a slower convergence, this time toward 1.1. While the results for the exponential and Erlang distributions are essentially the same, after $10,000$ jobs the hyperexponential and Bounded-Pareto distributions are still not stable and the ratios tend to be significantly higher, that is respectively around 1.4 and 1.8.

Figure 4a, 4b, 5a and 5b concern the more interesting case of different distributions for 10 and 500 machines. The depicted combinations are the time hyperexponential and Bounded Pareto distributions versus the memory uniform one. When using the Erlang or exponential distributions for the memory instead of the uniform one, the trends are almost coincident. For the time hyperexponential distribution the best algorithms in the order are Graham-Albers, Strong-Karger, Graham-Graham and Strong-Albers, with ratios converging to 1 for $m = 10$ and 1.4 for $m = 500$. For the Bounded-Pareto distribution, the best algorithms are Graham-Albers, Graham-Karger, Graham-Bartal and Graham-Graham, with ratios converging to 1.2 for $m = 10$ and 1.4 for $m = 500$. It is worth noting the good behavior of Weak-Albers for the test in Figure 5b, especially on the first half of the sequence.

In all the above cases, the best Graham based algorithms are obtained for $k = \lceil \frac{m}{2} \rceil$ for a small number of machines, with k tending to $\frac{m}{4}$ as m increases.

The key point for the interpretation of these results is again the coefficient of variation of the sequences. It is well known that the uniform, the exponential and the Erlang distributions have a coefficient not greater than 1, thus generating sequences of jobs for which the competitive ratios of our algorithms are quite stable. The fluctuations become relevant only for a large number of machines. On the other hand, by using the hyperexponential or the Bounded-Pareto distributions, it is possible to generate sequences of job whose characteristics are similar to those of real world traces, that is with a coefficient greater than 1, thus obtaining bigger fluctuations and a much slower convergence. However, the algorithms with the best performances remain the same identified for real world traces, thus confirming their good behavior.

4.3   Results for $j = 100$ and $m = 10$

In this subsection we focus on short sequences in order to better understand the behavior of the competitive ratios during the transient phases in which they have big oscillations. Here we can also perform a tighter analysis since for a relatively small number of jobs it is possible to efficiently compute the optimal offline solutions.

Figure 6a and 6b show the results achieved considering sequences of 100 jobs on only 10 machines for the two real world traces. As to the performances of the algorithms, besides Graham-Graham, Graham-Albers and Strong-Albers algorithms, it is possible to observe a good behavior of Weak-Albers, Half-Bartal, Strong-Bartal and Half-Albers. This suggests that using the technique of reducing to a single-criterion instance and running the Bartal's or Albers's algorithm works during the transient phases. However, in general, apart from the Karger based algorithms, all the others are more or less equivalent. Often some algorithms dominate the others at the beginning of the trace, but the situation changes toward the end.

Similar results hold also for the sequences generated according to probability distributions.

5   Conclusions

As a result of our experimental analysis, it is possibile to derive the following conclusions.

The performances of the algorithms depends heavily on the characteristics of the sequences. Graham-Albers, Strong-Albers and Graham-Graham seem to be the best algorithms, with the first two ones performing only slightly better in a few cases.

As the number of the machine increases, the transient phases tend to be longer. In fact, the lower average load at each step makes the competitive ratios more sensible to effects of types 2 and 3 (see Section 4), thus giving rise to bigger oscillations and slower convergence. As it can be observed, the competitive ratios become stable when the ratio $\frac{n}{m}$ becomes greater than 100.

In general, the Graham based algorithms and the strong bicriteria ones (except Graham-Karger and Strong-Karger) have always a good behavior and outperform the other families. The worse performance of the extensions of the Karger's algorithm is not surprising, as it aims to maintain some processors more loaded than the others.

The performance of the Graham based algorithms is better when the parameter k tends to $\frac{m}{4}$ as m increases. In fact, from a worst case point of view, setting $k = \lceil \frac{m}{2} \rceil$ guarantees that the k lightest time loaded machines have all a load at most $2T_j/m$, that is at most twice the average. However, the selection of the single machine among these k ones is done ignoring the time load, thus not letting unlikely the choice of a machine with time load close to $2T_j/m$. The situation is better for the memory measure. In fact, while we are guaranteed about the existence of a machine among the k ones with memory load at most $2S_j/m$, as m increases the probability of the existence of machines with memory load much less than $2S_j/m$ increases. Therefore, a selection considering the memory loads in general yields a much better outcome with respect to the worst case. Decreasing k in general causes a better choice for the time and a worse one for the memory. This confirms why experimentally it results that the outcome becomes balanced for values of k decreasing as m increases.

We notice also that different families of algorithms are possible. However, all the ones that exploit single-criterion reductions for making selections in general have a worse performance. Weak-bicriteria algorithms are the worst ones and their behavior is satisfactory only during the transient phases.

Surprisingly, in many cases the maximum competitive ratio tends to values close to 1. It would be interesting to experimentally evaluate the performance of these algorithms for more than $d > 2$ criteria and to estimate the increase of the ratios as a function of d.

Clearly our experiments are not exhaustive, as they considered the cases that have been suggested as more realistic in the scientific literature. It would be nice to determine and test other realistic scenarios.

Acknowledgments

We would like to thank Dr. Dror Feitelson for his help during the retrieval of the real world traces.

References

[1] Susanne Albers. Better bounds for online scheduling. *SIAM Journal on Computing*, 29(2):459–473, 1999. 33, 34

[2] Susanne Albers and Bianca Schröder. An experimental study of online scheduling algorithms. In *4th International Workshop on Algorithm Engineering (WAE)*, volume 1982 of *Lecture Notes in Computer Science*, pages 11–22. Springer-Verlag, 2000. 36, 38, 39

[3] Yair Bartal, Amos Fiat, Howard Karloff, and Rakesh Vohra. New algorithms for an ancient scheduling problem. In *24th ACM Symposium on Theory of Computing (STOC)*, pages 51–58, 1992. 34

[4] Vittorio Bilò and Michele Flammini. Time versus memory tradeoffs for multiprocessor scheduling. *Manuscript.* 33, 34, 36

[5] S. Chakrabarti, C. Phillips, A. S. Schulz, D. B. Shmoys, C. Stein, and J. Wein. Improved approximation algorithms for minsum criteria. In *23rd International Colloquium on Automata, Languages and Programming (ICALP)*, volume 1099 of *Lecture Notes in Computer Science*, pages 646–657. Springer-Verlag, 1996. 34

[6] D. G. Feitelson and editors L. Rudolph. Job scheduling stategies for parallel processing. In *9th IEEE International Parallel Processing Symposium (IPPS)*, volume 949 of *Lecture Notes in Computer Science*. Springer-Verlag, 1995. 37, 38

[7] D. G. Feitelson and editors L. Rudolph. Job scheduling stategies for parallel processing. In *10th IEEE International Parallel Processing Symposium (IPPS)*, volume 1162 of *Lecture Notes in Computer Science*. Springer-Verlag, 1996. 37, 38

[8] Dror Feitelson. Parallel workloads archive, http://www.cs.huji.ac.il/labs/parallel/workload. 37

[9] M. R. Garey, R. E. Tarjan, and G. T. Wilfong. One-processor scheduling with symmetric earliness and tardiness penalties. *Mathematics of Operations Research*, 13:330–348, 1988. 34

[10] R. L. Graham. Bounds for certain multiprocessing anomalies. *Bell System Technical Journal*, 45:1563–1581, 1966. 34

[11] Victor Hazlewood. Npaci joblog repository, http://joblog.npaci.edu. 37

[12] J. A. Hoogeveen. Minimizing maximum promptness and maximum lateness on a single machine. *Mathematics of Operations Research*, 21:100–114, 1996. 34

[13] J. A. Hoogeveen. Single machine scheduling to minimize a function of two or three maximum cost criteria. *Journal of Algorithms*, 21(2):415–433, 1996. 34

[14] J. A. Hoogeveen and S. L. Van de Velde. Minimizing total completion time and maximum cost simultaneously is solvable in polynomial time. *Operations Research Letters*, 17:205–208, 1995. 34

[15] R. Jain. *The Art of Computer Systems Performance Analysis*. Wiley, 1991. 37

[16] David R. Karger, Steven J. Phillips, and Eric Torng. A better algorithm for an ancient scheduling problem. In *5th ACM-SIAM Symposium on Discrete Algorithms (SODA)*, 1994. 34

[17] Madhav V. Marathe, R. Ravi, Ravi Sundaram, S. S. Ravi, Daniel J. Rosenkrantz, and Harry B. Hunt III. Bicriteria network design problems. In *22nd International Colloquium on Automata, Languages and Programming (ICALP)*, volume 944 of *Lecture Notes in Computer Science*, pages 487–498. Springer-Verlag, 1995. 33

[18] S. T. McCormick and M. L. Pinedo. Scheduling n indipendent jobs on m uniform machines with both flow time and makespan objectives: A parametric approach. *ORSA Journal of Computing*, 7:63–77, 1992. 34

[19] A. Nagar, J. Haddock, and S. Heragu. Multiple and bicriteria scheduling: a literature survey. *European Journal of Operations Research*, 81:88–104, 1995. 34

[20] R. T. Nelson, R. K. Sarin, and R. L. Daniels. Scheduling with multiple performance measures: the one-machine case. *Management Science*, 32:464–479, 1986. 34

[21] A. Rasala, C. Stein, E. Torng, and P. Uthaisombut. Existence theorems, lower bounds and algorithms for scheduling to meet two objectives. In *Proceedings of the 13th Annual ACM-SIAM Symposium on Discrete Algorithms (SODA)*, pages 723–731. ACM Press, 2002. 34

[22] R. Ravi and Michel X. Goemans. The constrained minimum spanning tree problem. In *5th Scandinavian Workshop on Algorithm Theory (SWAT)*, volume 1097 of *Lecture Notes in Computer Science*, pages 66–75. Springer-Verlag, 1996. 33

[23] B. D. Shmoys and E. Tardos. An approximation algorithm for the generalized assignment problem. *Mathematical Programming A*, 62:461–474, 1993. 33

[24] W. E. Smith. Various optimizers for single-stage production. *Naval Research Logistics Quarterly*, 3:59–66, 1956. 33

[25] C. Stein and J. Wein. On the existence of scheduling that are near-optimal for both makespan and total weighted completion time. *Operations Research Letters*, 21:115–122, 1997. 34

[26] D. von Seggen. *CRC Standard Curves and Surfaces*. CRC Press, 1993. 37

[27] L. N. Van Wassenhove and F. Gelders. Solving a bicriterion scheduling problem. *European Journal of Operations Research*, 4:42–48, 1980. 34

Fast-Search: A New Efficient Variant of the Boyer-Moore String Matching Algorithm

Domenico Cantone and Simone Faro

Università di Catania, Dipartimento di Matematica e Informatica
Viale Andrea Doria 6, I-95125 Catania, Italy
{cantone,faro}@dmi.unict.it

Abstract. We present a new variant of the Boyer-Moore string matching algorithm which, though not linear, is very fast in practice.
We compare our algorithm with the Horspool, Quick Search, Tuned Boyer-Moore, and Reverse Factor algorithms, which are among the fastest string matching algorithms for practical uses. It turns out that our algorithm achieve very good results in terms of both time efficiency and number of character inspections, especially in the cases in which the patterns are very short.

Keywords: string matching, experimental algorithms, text processing.

1 Introduction

Given a text T and a pattern P over some alphabet Σ, the *string matching problem* consists in finding *all* occurrences of the pattern P in the text T. It is a very extensively studied problem in computer science, mainly due to its direct applications to several areas such as text processing, information retrieval, and computational biology.

A very comprehensive description of the existing string matching algorithms and a fairly complete bibliography can be found respectively at the following URLs

- http://www-igm.univ-mlv.fr/~lecroq/string/
- http:// liinwww.ira.uka.de/bibliography/Theory/tq.html.

We first introduce the notation and terminology used in the paper. We denote the empty string by ε. A string P of length m is represented as an array $P[0..m-1]$. Thus, $P[i]$ will denote the $(i+1)$-st character of P, for $i = 0, \ldots, m-1$. For $0 \le i \le j < \text{length}(P)$, we denote by $P[i..j]$ the substring of P contained between the $(i+1)$-st and the $(j+1)$-st characters of P. Moreover, for any i and j, we put

$$P[i..j] = \begin{cases} \varepsilon & \text{if } i > j \\ P[\max(i,0), \min(j, \text{length}(P)-1)] & \text{otherwise .} \end{cases}$$

If P and P' are two strings, we write $P' \sqsupseteq P$ to indicate that P' is a suffix of P, i.e., $P' = P[i..\text{length}(P)-1]$, for some $0 \le i \le \text{length}(P)$. Similarly we

K. Jansen et al. (Eds.): WEA 2003, LNCS 2647, pp. 47–58, 2003.

write $P' \sqsubset P$ to indicate that P' is a prefix of P, i.e., $P' = P[0..i-1]$, for some $0 \leq i \leq \text{length}(P)$.

Let T be a text of length n and let P be a pattern of length m. If the character $P[0]$ is aligned with the character $T[s]$ of the text, so that the character $P[i]$ is aligned with the character $T[s+i]$, for $i = 0, \ldots, m-1$, we say that the pattern P has *shift* s in T. In this case the substring $T[s..s+m-1]$ is called the *current window* of the text. If $T[s..s+m-1] = P$, we say that the shift s is *valid*.

Most string matching algorithms have the following general structure:

Generic_String_Matcher(T, P)
   Precompute_Globals(P)
   $n = \text{length}(T)$
   $m = \text{length}(P)$
   $s = 0$
   while $s \leq n - m$ do
      $s = s + \textit{Shift\_Increment}(s, P, T)$

---

where

- the procedure *Precompute_Globals*(P) computes useful mappings, in the form of tables, which may be later accessed by the function *Shift_Increment*(s, P, T);
- the function *Shift_Increment*(s, P, T) checks whether s is a valid shift and computes a *positive* shift increment.

Observe that for the correctness of procedure *Generic_String_Matcher*, it is plainly necessary that the shift increment Δs computed by *Shift_Increment*(s, P, T) is *safe*, namely no valid shift can belong to the interval $\{s+1, \ldots, s + \Delta s - 1\}$.

In the case of the naive string matching algorithm, for instance, the procedure *Precompute_Globals* is just dropped and the function *Shift_Increment*(s, P, T) always returns a unitary shift increment, after checking whether the current shift is valid. The latter can be instantiated as follows:

Naive_Shift_Increment(s, P, T)
   for $i = 0$ to $\text{length}(P) - 1$ do
      if $P[i] \neq T[s+i]$ then
         return 1
   print(s)
   return 1

Therefore, in the worst case, the naive algorithm requires $\mathcal{O}(mn)$ character comparisons.

Information gathered during the execution of the *Shift_Increment*(s, P, T) function, in combination with the knowledge of P as suitably extracted by procedure *Precompute_Globals*(P), can yield shift increments larger than 1 and ultimately lead to more efficient algorithms. Consider for instance the case in which

Shift_Increment(s, P, T) processes P from right to left and finds immediately a mismatch between $P[m-1]$ and $T[s+m-1]$, where additionally the character $T[s+m-1]$ does not occur in P in any other position; then the shift can safely be incremented by m. In the best case, when the above case occurs repeatedly, it can be verified that the text T does not contain any occurrence of P in sublinear time $\mathcal{O}(n/m)$.

1.1   The Boyer-Moore Algorithm

The Boyer-Moore algorithm (cf. [BM77]) is a progenitor of several algorithmic variants which aim at efficiently computing shift increments close to optimal. Specifically, the Boyer-Moore algorithm can be characterized by the following function *BM_Shift_Increment*(s, P, T) which, as in the previous example, scans the pattern P from right to left. *BM_Shift_Increment*(s, P, T) computes the shift increment as the maximum value suggested by the *good suffix rule* and the *bad character rule* below, via the functions gs_P and bc_P respectively, provided that both of them are applicable.

$$
\begin{array}{|l|}
\hline
BM\_Shift\_Increment(s, P, T) \\
\quad \text{for } i = \text{length}(P) - 1 \text{ downto } 0 \text{ do} \\
\qquad \text{if } P[i] \neq T[s+i] \text{ then} \\
\qquad\quad \text{return } \max(gs_P(i), i - bc_P(T[s+i])) \\
\quad print(s) \\
\quad \text{return } gs_P(0) \\
\hline
\end{array}
$$

If a mismatch occurs at position i of the pattern P, while it is scanned from right to left, the good suffix rule suggests to align the substring $T[s+i+1 \ldots s+m-1] = P[i+1 \ldots m-1]$ with its rightmost occurrence in P preceded by a character different by $P[i]$. If such an occurrence does not exist, the good suffix rule suggests a shift increment which allows to match the longest suffix of $T[s+i+1 \ldots s+m-1]$ with a prefix of P.

More formally, if the first mismatch occurs at position i of the pattern P, the good suffix rule states that the shift can be safely incremented by $gs_P(i+1)$ positions, where

$$
gs_P(j) =_{\text{Def}} \min\{0 < k \leq m \mid P[j-k..m-k-1] \sqsupset P
$$
$$
\text{and } (k \leq j-1 \rightarrow P[j-1] \neq P[j-1-k])\} ,
$$

for $j = 0, 1, \ldots, m$. (The situation in which an occurrence of the pattern P is found can be regarded as a mismatch at position -1.)

The bad character rule states that if $c = T[s+i] \neq P[i]$ is the first mismatching character, while scanning P and T from right to left with shift s, then P can be safely shifted in such a way that its rightmost occurrence of c, if present, is aligned with position $(s+i)$ of T. In the case in which c does not occur in P, then P can be safely shifted just past position $(s+i)$ of T. More formally, the shift increment suggested by the bad character rule is given by the expression $(i - bc_P(T[s + i]))$, where

$$bc_P(c) =_{\text{Def}} \max(\{0 \le k < m | P[k] = c\} \cup \{-1\}) \ ,$$

for $c \in \Sigma$, and where we recall that Σ is the alphabet of the pattern P and text T. Notice that there are situations in which the shift increment given by the bad character rule can be negative.

It turns out that the functions gs_P and bc_P can be computed during the preprocessing phase in time $\mathcal{O}(m)$ and $\mathcal{O}(m+|\Sigma|)$, respectively, and that the overall worst-case running time of the Boyer-Moore algorithm, as described above, is linear (cf. [GO80]).

For the sake of completeness, we notice that originally the Boyer-Moore algorithm made use of a good suffix rule based on the following simpler function

$$gs'_P(j) =_{\text{Def}} \min\{0 < k \le m | P[j - k..m - k - 1] \sqsupseteq P\} \ ,$$

for $j = 0, 1, \ldots, m$, which led to a non-linear worst-case running time.

Several variants of the Boyer-Moore algorithm have been proposed over the years. In particular, we mention Horspool, Quick Search, Tuned Boyer-Moore, and the Reverse Factor algorithms, which are among the fastest variants in practice (cf. [Hor80], [Sun90], [HS91], and [CCG+94], respectively). In Sect. 3, we will compare them with our proposed variant of the Boyer-Moore algorithm.

1.2   The Horspool Algorithm

Horspool suggested a simplification to the original Boyer-Moore algorithm, defining a new variant which, though quadratic, performed better in practical cases (cf. [Hor80]). He just dropped the good suffix rule and based the calculation of the shift increments only on the following variation of the bad character rule. Specifically, he observed that when the first mismatch between the window $T[s..s + m - 1]$ and the pattern P occurs at position $0 \le i < m$ and the rightmost occurrence of the character $T[s + i]$ in P is at position $j > i$, then the bad character rule would shift the pattern backwards. Thus, he proposed to compute the shift advancement in such a way that the rightmost character $T[s + m - 1]$ is aligned with its rightmost occurrence on $P[0..m - 2]$, if present (notice that the character $P[m-1]$ has been left out); otherwise the pattern is advanced just past the window. This corresponds to advance the shift by $hbc_P(T[s + m - 1])$ positions, where

$$hbc_P(c) =_{\text{Def}} \min(\{1 \le k < m | P[m - 1 - k] = c\} \cup \{m\}) \ .$$

It turns out that the resulting algorithm performs well in practice and can be immediately translated into programming code (see Baeza-Yates and Régnier [BYR92] for a simple implementation in the **C** programming language).

1.3   The Quick-Search Algorithm

The Quick-Search algorithm, presented in [Sun90], uses a modification of the original heuristics of the Boyer-Moore algorithm, much along the same lines of the Horspool algorithm. Specifically, it is based on the following observation: when a mismatch character is encountered, the pattern is always shifted to the right by at least one character, but never by more than m characters. Thus, the character $T[s + m]$ is always involved in testing for the next alignment. So, one can apply the bad-character rule to $T[s + m]$, rather than to the mismatching character, obtaining larger shift advancements. This corresponds to advance the shift by $qbc_P(T[s + m])$ positions, where

$$qbc_P(c) =_{\mathrm{Def}} \min(\{0 \leq k < m | P[m - 1] = c\} \cup \{m + 1\}) .$$

Experimental tests have shown that that the Quick-Search algorithm is very fast especially for short patterns (cf. [Lec00]).

1.4   The Tuned Boyer-Moore Algorithm

The Tuned Boyer-Moore algorithm (cf. [HS91]) can be seen as an efficient implementation of the Horspool algorithm. Again, let P be a pattern of length m. Each iteration of the Tuned Boyer-Moore algorithm can be divided into two phases: *last character localization* and *matching phase*. The first phase searches for a match of $P[m - 1]$, by applying rounds of three blind shifts (based on the classical bad character rule) until needed. The matching phase tries then to match the rest of the pattern $P[0..m - 2]$ with the corresponding characters of the text, proceeding from right to left. At the end of the matching phase, the shift advancement is computed according to the Horspool bad character rule. Moreover, in order to compute the last shifts correctly, the algorithm in the first place adds m copies of $P[m - 1]$ at the end of the text, as a sentinel.

The fact that the blind shifts require no checks is at the heart of the very good practical behavior of the Tuned Boyer-Moore, despite its quadratic worst-case time complexity (cf. [Lec00]).

1.5   The Reverse Factor Algorithm

Unlike the variants of the Boyer-Moore algorithm summarized above, the Reverse Factor algorithm computes shifts which match prefixes of the pattern, rather than suffixes. This is accomplished by means of the smallest suffix automaton of the reverse of the pattern, while scanning the text and pattern from right to left (for a complete description see [CCG+94]).

The Reverse Factor algorithm has a quadratic worst-case time complexity, but it is very fast in practice (cf. [Lec00]). Moreover, it has been shown that on the average it inspects $\mathcal{O}(n \log(m)/m)$ text characters, reaching the best bound shown by Yao in 1979 (cf. [Yao79]).

2   Fast-Search: A New Efficient Variant of the Boyer-Moore Algorithm

We present now a new efficient variant of the Boyer-Moore algorithm, called Fast-Search, which will use the *Fast_Search_Shift_Increment* procedure to be given below as shift increment function. As before, let P be a pattern of length m and let T be a text of length n over a finite alphabet Σ; also, let $0 \le s \le m - n$ be a shift. The main observation upon which our Fast-Search algorithm is based is the following:

> the Horspool bad character rule leads to larger shift increments than the good suffix rule if and only if a mismatch occurs immediately, while comparing the pattern P with the window $T[s..s + m - 1]$, namely when $P[m - 1] \ne T[s + m - 1]$.

The above observation, which will be proved later in Sect. 2.1, suggests at once that the following shift increment rule should lead to a faster algorithm than the Horspool one:

> to compute the shift increment use the Horspool bad character rule, if a mismatch occurs during the first character comparison; otherwise use the good suffix rule.

This translates into the following pseudo-code:

```
Fast_Search_Shift_Increment(s, P, T)
    m = length(P)
    for i = m - 1 downto 0 do
        if P[i] ≠ T[s + i] then
            if i = m - 1 then
                return hbcₚ(T[s + m − 1])
            else
                return gsₚ(i)
    print(s)
    return gsₚ(0)
```

Notice that $hbc_P(a) = bc_P(a)$, whenever $a \ne P[m - 1]$, so that the term $hbc_P(T[s + m - 1])$ can be substituted by $bc_P(T[s + m - 1])$ in the above procedure, as will be done in the efficient implementation of the Fast-Search algorithm to be given in Sect. 2.2.

Experimental data which will be presented in Sect. 3 confirm that the Fast-Search algorithm is faster than the Horspool algorithm. In fact, we will see that, though not linear, Fast-Search compares well with the fastest string matching algorithms, especially in the case of short patterns. We also notice that the functions hbc_P and gs_P can be precomputed in time $\mathcal{O}(m)$ and $\mathcal{O}(m + |\Sigma|)$, respectively, by *Precompute_Globals(P)*.

2.1   The Horspool Bad Character Rule versus the Good Suffix Rule

We will show in Proposition 1 that the Horspool bad character rule wins against the good suffix rule only when a mismatch is found during the first character comparison. To this purpose we first prove the following technical lemma.

Lemma 1. *Let P be a pattern of length $m \geq 1$ over an alphabet Σ. Then the following inequalities hold:*

(a) $gs_P(m) \leq hbc_P(c)$, for $c \in \Sigma \setminus \{P[m-1]\}$;
(b) $gs_P(j) \geq hbc_P(P[m-1])$, for $j = 0, 1, \ldots, m-1$.

Proof. Concerning (a), let $c \in \Sigma \setminus \{P[m-1]\}$ and let $\bar{k} = hbc_P(c)$. If $\bar{k} = m$, then $gs_P(m) \leq hbc_P(c)$ follows at once. On the other hand, if $\bar{k} < m$, then we have $P[m-1-\bar{k}] = c$, so that $P[m-1-\bar{k}] \neq P[m-1]$, since by assumption $P[m-1] \neq c$ holds. Therefore $gs_P(m) \leq \bar{k} = hbc_P(c)$, proving (a).

Next, let $0 \leq j < m$ and let $0 < k \leq m$ be such that

- $P[j-k..m-k-1] \sqsupset P$, and
- $P[j-1] \neq P[j-1-k]$, provided that $k \leq j-1$,

so that $gs_P(j) \leq k$. If $k < m$, then $P[m-k-1] = P[m-1]$, and therefore $hbc_P(P[m-1]) \leq k$. On the other hand, if $k = m$, then we plainly have $hbc_P(P[m-1]) \leq k$. Thus, in any case, $gs_P(j) \geq hbc_P(P[m-1])$, proving (b). □

Then we have:

Proposition 1. *Let P and T be two nonempty strings over an alphabet Σ and let $m = |P|$. Let us also assume that we are comparing P with the window $T[s..s+m-1]$ of T with shift s, scanning P from right to left. Then*

(a) if the first mismatch occurs at position $(m-1)$ of the pattern P, then
$$gs_P(m) \leq hbc_P(T[s+m-1]);$$
(b) if the first mismatch occurs at position $0 \leq i < m-1$ of the pattern P, then
$$gs_P(i+1) \geq hbc_P(T[s+m-1]);$$
(c) if no mismatch occurs, then
$$gs_P(0) \geq hbc_P(T[s+m-1]).$$

Proof. Let us first assume that $P[m-1] \neq T[s+m-1]$, i.e., the first mismatch occurs at position $(m-1)$ of the pattern P, while comparing P with $T[s..s+m-1]$ from right to left. Then by Lemma 1(a) we have $gs_P(m) \leq hbc_P(T[s+m-1])$, yielding (a).

On the other hand, if $P[m-1] = T[s+m-1]$, i.e., the first mismatch occurs at position $0 \leq i < m-1$ or no mismatch occurs, then Lemma 1(b) implies immediately (b) and (c). □

```
Fast-Search(P, T)
1.      n = length(T)
2.      m = length(P)
3.      T' = T.P
4.      bc_P = precompute-bad-character(P)
5.      gs_P = precompute-good-suffix(P)
7.      s = 0
8.      while bc_P(T'[s + m − 1]) > 0 do s = s + bc_P(T'[s + m − 1])
9.      while s ≤ n − m do
10.         j = m − 2
11.         while j ≥ 0 and P[j] = T'[s + j] do j = j − 1
12.         if j < 0 then print(s)
13.         s = s + gs_P(j + 1)
14.         while bc_P(T'[s + m − 1]) > 0 do s = s + bc_P(T'[s + m − 1])
```

Fig. 1. The Fast-Search algorithm

2.2   An Efficient Implementation

A more effective implementation of the Fast-Search algorithm can be obtained much along the same lines of the Tuned Boyer-Moore algorithm. The main idea consists in iterating the bad character rule until the last character $P[m − 1]$ of the pattern is matched correctly against the text, and then applying the good suffix rule, at the end of the matching phase. More precisely, starting from a shift position s, if we denote by j_i the total shift advancement after the i-th iteration of the bad character rule, then we have the following recurrence:

$$j_i = j_{i-1} + bc_P(T[s + j_{i-1} + m − 1]) \ .$$

Therefore, starting from a given shift s, the bad character rule is applied k times in row, where $k = \min\{i \mid T[s + j_i + m − 1] = P[m − 1]\}$, with a resulting shift advancement of j_k. At this point it is known that $T[s + j_k + m − 1] = P[m − 1]$, so that the subsequent matching phase can start with the $(m − 2)$-nd character of the pattern.

As in the case of the Tuned Boyer-Moore algorithm, the Fast-Search algorithm benefits from the introduction of an external sentinel, which allows to compute correctly the last shifts with no extra checks. For this purpose, we have chosen to add a copy of the pattern P at the end of the text T, obtaining a new text $T' = T.P$. Plainly, all the valid shifts of P in T are the valid shifts s of P in T' such that $s ≤ n − m$, where, as usual, n and m denote respectively the lengths of T and P.

The code of the Fast-Search algorithm is presented in Fig. 1.

3   Experimental Results

In this section we present experimental data which allow to compare the running times and number of character inspections of the following string matching algorithms in various conditions: Fast-Search (FS), Horspool (HOR), Quick-Search (QS), Tuned Boyer-Moore (TBM), and Reverse Factor(RF).

All five algorithms have been implemented in the **C** programming language and were used to search for the same strings in large fixed text buffers on a PC with AMD Athlon processor of 1.19GHz. In particular, the algorithms have been tested on three Randσ problems, for $\sigma = 2, 8, 20$, and on a natural language text buffer.

A Randσ problem consisted in searching a set of 200 random patterns over an alphabet Σ of size σ, for each assigned value of the pattern length, in a 20Mb random text over the same alphabet Σ. We have performed our tests with patterns of length $2, 4, 6, 8, 10, 20, 40, 80$, and 160.

The tests on a natural language text buffer have been performed on a 3.13Mb file obtained from the WinEdt spelling dictionary by discarding non-alphabetic characters. All words in the text buffer have been searched for.

In the following tables, running times are expressed in hundredths of seconds. Concerning the number of character inspections, these have been obtained by taking the average of the total number of times a text character is accessed, either to perform a comparison with a pattern character, or to perform a shift, or to compute a transition in an automaton, and dividing it by the total number of characters in the text buffer.

Experimental results show that the Fast-Search algorithm obtains the best runtime performances in most cases and, sporadically, it is second only to the Tuned Boyer-Moore algorithm.

Concerning the number of text character inspections, it turns out that the Fast-Search algorithm is quite close to the Reverse Factor algorithm, which generally shows the best behaviour. We notice, though, that in the case of very short patterns the Fast-Search algorithm reaches the lowest number of character accesses.

4   Conclusion

We have presented a new efficient variant of the Boyer-Moore string matching algorithm, named Fast-Search, based on the classical bad character and good suffix rules to compute shift advancements, as other variations of the Boyer-Moore algorithm.

Table 1. Running times for a Rand2 problem

$\sigma = 2$	2	4	6	8	10	20	40	80	160
HOR	46.05	44.75	44.77	45.12	44.83	42.10	41.23	40.83	42.13
QS	38.13	40.59	42.11	41.27	41.13	38.97	38.09	37.04	37.54
TBM	**36.27**	36.26	38.42	38.87	38.69	37.75	37.81	37.36	38.44
RF	268.38	197.88	149.83	120.14	100.02	60.37	37.91	28.40	22.63
FS	38.38	**32.96**	**30.19**	**27.35**	**25.40**	**21.04**	**18.90**	**18.16**	**17.39**

Table 2. Number of text character inspections for a Rand2 problem

$\sigma = 2$	2	4	6	8	10	20	40	80	160
HOR	1.83	1.72	1.66	1.66	1.64	1.59	1.64	1.61	1.68
QS	1.54	1.65	1.69	1.64	1.63	1.64	1.67	1.60	1.63
TBM	1.23	1.35	1.42	1.45	1.45	1.42	1.46	2.43	2.49
RF	1.43	1.06	**.78**	**.62**	**.51**	**.29**	**.16**	**.09**	**.05**
FS	**1.00**	**.92**	.80	.70	.63	.45	.34	.26	.22

Table 3. Running times for a Rand8 problem

$\sigma = 8$	2	4	6	8	10	20	40	80	160
HOR	30.22	21.99	21.85	18.62	18.04	17.27	17.24	17.11	17.38
QS	22.41	20.43	19.48	17.63	17.41	16.93	16.86	16.82	16.94
TBM	23.14	**19.51**	18.95	17.34	17.07	16.79	16.78	16.73	16.97
RF	120.5	74.29	63.99	48.61	42.84	29.16	22.23	19.71	16.48
FS	**22.06**	**19.51**	**18.77**	**17.11**	**16.96**	**16.65**	**16.64**	**16.54**	**16.47**

Table 4. Number of text character inspections for a Rand8 problem

$\sigma = 8$	2	4	6	8	10	20	40	80	160
HOR	1.191	.680	.507	.422	.374	.294	.282	.275	.281
QS	.842	.575	.456	.393	.358	.291	.282	.278	.285
TBM	.663	.386	.291	.245	.218	.174	.168	.164	.167
RF	.674	.381	.278	.225	**.191**	**.112**	**.063**	**.360**	**.020**
FS	**.600**	**.348**	**.260**	**.217**	.193	.150	.137	.126	.120

Table 5. Running times for a Rand20 problem

$\sigma = 20$	2	4	6	8	10	20	40	80	160
HOR	24.51	18.56	17.03	16.39	16.01	15.19	14.78	14.84	14.98
QS	19.16	17.16	16.19	15.77	15.51	14.93	14.70	14.67	14.69
TBM	19.12	16.68	15.80	15.48	**15.25**	14.79	14.64	14.57	14.79
RF	96.16	56.63	43.32	36.69	32.29	23.43	19.46	17.83	14.62
FS	**19.11**	**16.67**	**15.78**	**15.43**	15.26	**14.74**	**14.58**	**14.55**	**14.51**

Table 6. Number of text character inspections for a Rand20 problem

$\sigma = 20$	2	4	6	8	10	20	40	80	160
HOR	1.075	.566	.395	.311	.259	.161	.119	.106	.103
QS	.735	.463	.346	.282	.241	.156	.118	.107	.103
TBM	.563	.297	.208	.164	.137	.086	.064	.057	.055
RF	.565	.302	.214	.171	.143	.084	**.049**	**.027**	**.014**
FS	**.538**	**.284**	**.198**	**.156**	**.131**	**.082**	.060	.054	.051

Table 7. Running times for a natural language problem

NL	2	4	6	8	10	20	40	80	160
HOR	3.56	2.71	2.48	2.39	2.32	2.18	2.17	2.15	2.01
QS	2.77	2.48	2.38	2.33	2.23	2.19	2.16	2.14	1.99
TBM	**2.81**	2.47	2.32	**2.27**	2.23	2.21	2.15	2.19	**1.91**
RF	14.44	8.69	6.67	5.69	4.97	3.47	2.84	2.77	5.41
FS	2.85	**2.39**	**2.27**	**2.27**	**2.20**	**2.15**	**2.13**	**2.12**	1.93

Table 8. Number of text character inspections for a natural language problem

NL	2	4	6	8	10	20	40	80	160
HOR	1.094	.590	.418	.337	.282	.172	.111	.077	.059
QS	.759	.489	.375	.309	.261	.175	.125	.086	.069
TBM	.584	.318	.226	.182	.153	.096	.062	.044	.034
RF	.588	.321	.231	.185	.153	**.084**	**.045**	**.024**	**.013**
FS	**.550**	**.299**	**.211**	**.171**	**.143**	.087	.055	.038	.028

Rather than computing the shift advancement as the larger of the values suggested by the bad character and good suffix rules, our algorithm applies repeatedly the bad character rule until the last character of the pattern is matched correctly, and then, at the end of each matching phase, it executes one application of the good suffix rule.

It turns out that, though quadratic in the worst-case, the Fast-Search algorithm is very fast in practice and compares well with other fast variants of the Boyer-Moore algorithm, as the Horspool, Quick Search, Tuned Boyer-Moore, and Reverse Factor algorithms, in terms of both running time and number of character inspections.

References

[BM77]   R. S. Boyer and J. S. Moore. A fast string searching algorithm. *Commun. ACM*, 20(10):762–772, 1977. 49

[BYR92] R. A. Baeza-Yates and M. Régnier. Average running time of the Boyer-Moore-Horspool algorithm. *Theor. Comput. Sci.*, 92(1):19–31, 1992. 50

[CCG+94] M. Crochemore, A. Czumaj, L. Gąsieniec, S. Jarominek, T. Lecroq, W. Plandowski, and W. Rytter. Speeding up two string matching algorithms. *Algorithmica*, 12(4/5):247–267, 1994. 50, 51

[GO80]   L. J. Guibas and A. M. Odiyzko. A new proof of the linearity of the Boyer-Moore string searching algorithm. *SIAM J. Comput.*, 9(4):672–682, 1980. 50

[Hor80] R. N. Horspool. Practical fast searching in strings. *Softw. Pract. Exp.*, 10(6):501–506, 1980. 50

[HS91]   A. Hume and D. M. Sunday. Fast string searching. *Softw. Pract. Exp.*, 21(11):1221–1248, 1991. 50, 51

[Lec00]   T. Lecroq. New experimental results on exact string-matching. Rapport LIFAR 2000.03, Université de Rouen, France, 2000. 51

[Sun90]   D. M. Sunday. A very fast substring search algorithm. *Commun. ACM*, 33(8):132–142, 1990. 50, 51

[Yao79]   A. C. Yao. The complexity of pattern matching for a random string. *SIAM J. Comput.*, 8(3):368–387, 1979. 51

An On-Line Algorithm for
the Rectangle Packing Problem with Rejection

Massimiliano Caramia[1], Stefano Giordani[2,3], and Antonio Iovanella[2]

[1] Istituto per le Applicazioni del Calcolo "M. Picone"
Viale del Policlinico, 137 - 00161 Rome, Italy
`caramia@iac.rm.cnr.it`
[2] Dipartimento di Informatica, Sistemi e Produzione
University of Rome "Tor Vergata"
Via del Politecnico, 1 - 00133 Rome, Italy
`{giordani,iovanella}@disp.uniroma2.it`
[3] Centro Interdipartimentale "Vito Volterra"
University of Rome "Tor Vergata"
Via Columbia 2, I-00133 Rome, Italy

Abstract. In this paper an on-line algorithm for the Rectangle Packing Problem is presented. The method is designed to be able to accept or reject incoming boxes to maximize efficiency. We provide a wide computational analysis showing the behavior of the proposed algorithm as well as a comparison with existing off-line heuristics.

1 Introduction

Given a set J of two-dimensional rectangular-shaped boxes, where each box $j \in J$ is characterized by its width w_j and its height h_j, we consider the problem of orthogonally packing a subsets of the boxes, without overlapping, into a single bounding rectangular area A of width W and height H, maximizing the ratio ρ between the area occupied by the boxes and the total available area A, i.e., minimizing the wasted area of A. Clearly, ρ is in between 0 and 1. It is assumed that the boxes cannot be guillotined, and have fixed orientation, i.e., that cannot be rotated; moreover, we assume that all input data are positive integers, with $w_j \leq W$ and $h_j \leq H$, for every $j \in J$.

The problem is known as "*Rectangle (or Two-Dimensional) Packing Problem*" (see, for example, [10], [18]), and has been shown to be $\mathcal{NP}$-complete ([13]). It is a special case of the Two-Dimensional Cutting Stock (or Knapsack) Problem, where each box j has an associated profit $p_j > 0$ and the problem is to select a subset of the boxes to be packed in a single finite rectangular bin maximizing the total selected profit ([15]); clearly, in our case $p_j = w_j \cdot h_j$.

The Rectangle Packing Problem appears in many practical applications in the cutting and packing (transportation and warehousing) industry, e.g., in cutting wood or foam rubber into smaller pieces, and in placing goods on a shelf; other important applications are newspaper paging, VLSI floor planning, and also GRID computing.

K. Jansen et al. (Eds.): WEA 2003, LNCS 2647, pp. 59–69, 2003.

A natural application occurs also in multiprocessor task scheduling problems, since scheduling tasks with shared resources involves two dimensions (the resource and the time). If we consider the width as the processing time and the height as the resources requirement (say processors), we may represent a multiprocessor task by a box. In particular, the Rectangle Packing Problem models a multiprocessor task scheduling problem where a set of H processors arranged in a linear configuration as in an array ([4]), and a set J of tasks are given, with each task $j \in J$ requiring h_j physically adjacent processors for a certain time w_j, and the objective is to schedule as many tasks as possible without interruption within a deadline W, maximizing the processor total busy time.

In the literature two versions of the problem have been investigated: the off-line and the on-line, respectively. While in the off-line version all the problem data are known in advance, in the on-line version where e.g. the over list paradigm is considered ([7]), the boxes (with their dimensions) arrive from a list without any knowledge on further boxes; in particular, the boxes along with their dimensions are known one by one, and when a new box is presented it is to be decided if it can be placed into a free rectangular (with equal dimensions) sub-area of A, i.e., it can be accepted or rejected. The on-line problem is to accept (place) or reject the incoming boxes, maximizing the ratio between the occupied area and the total available area A.

Most of the contributions in the literature are devoted to the off-line problem that is solved using several approaches based on optimal algorithms. The basic formulation issues and solution procedures for the two-dimensional cutting stock problems were presented in [8]. Optimal algorithms for the orthogonal two-dimensional cutting were proposed in [1] and in [8] but such techniques may be non practical for large instances. Anyhow, for a complete presentation on this problem the reader is referred to the survey by Lodi *et al.* in [15]. Various approximation schemes have been proposed e.g. in [9]. Heuristic approaches have been considered in [10], [11] and in [18], where rejection is also concerned.

The on-line case is investigated in many variants mostly deriving from the bin packing problem. The problem of packing one-dimensional items in $k \geq 1$ or fewer active bins, where each bin becomes active when it receives its first item, is solved with several techniques based on combining the so called HARMONIC Algorithm ([12]) and the FIRST and BEST FIT rules; the best approximation ratio tends to 1.69103 as k tends to infinity. For the more general case of the d-dimensional packing problem, a worst ratio equal to $h^d \approx 1.69^d$ was demonstrated ([5]) and for $d = 2$ the best value for the lower bound currently known is 1.907 ([3]). Recently, for the on-line strip packing version, an algorithm with modifiable boxes was developed in [16], where a 4-competitive algorithm and a 1.73 lower bound on the value of the competitive ratio were presented. Such results were developed using similar ideas from [17]. For a survey on on-line algorithms for the Packing Problem and many others variants, the reader is referred to the chapter by Csiric and Woeginger in [7]. To the best of our knowledge, the on-line version of the problem we consider has not been investigated, and no on-line algorithms with guaranteed competitive ratio are known so far.

In this paper, we consider the on-line version of the Rectangle Packing Problem in the paradigm of boxes arriving over list, and provide an on-line algorithm. Computational results are given to show the performance of the proposed algorithm, and a comparison with algorithms in [18] is provided. The remainder of the paper is organized as follows. In Section 2, we sketch the on-line algorithm, and in Section 3 computational results are provided. Section 4 concludes the paper with some final remarks.

2   The On-Line Algorithm

Without loss of generality, we consider the boxes indexed according to the order in which they are presented. We recall that given a list $\mathcal{L} = 1, 2, \ldots, n$ of such boxes, an algorithm $\mathcal{A}$ considering $\mathcal{L}$ is said to be on-line if:

- $\mathcal{A}$ considers boxes in the order given by the list $\mathcal{L}$;
- $\mathcal{A}$ considers each box i without knowledge of any box j, with $j > i$;
- $\mathcal{A}$ never reconsiders a box already considered.

The algorithm operates in $n = |J|$ iterations. During j-th iteration, in which a sub-area $A^{(j-1)} \subseteq A$ is available (free), box j is considered to be accepted or rejected, and a new (possibly empty) free sub-area A^j of A is defined. Let us consider in detail iteration j-th. Let $A^{(j-1)}$ be the non-assigned (free) area of A, and $\{A_1^{(j-1)}, \ldots, A_j^{(j-1)}\}$ a given partition of $A^{(j-1)}$, that is, $A_p^{(j-1)} \cap A_q^{(j-1)} = \emptyset$, for $p \neq q \in \{1, \ldots, j\}$ and $\cup_{k=1}^{j} A_k^{(j-1)} = A^{(j-1)}$, where each $A_k^{(j-1)}$ is a free rectangular area of A, of width $W_k^{(j-1)}$, height $H_k^{(j-1)}$, and size equal to $W_k^{(j-1)} \cdot H_k^{(j-1)}$. See for example Figure 1. Clearly, at the beginning, we have $A_1^{(0)} = A^{(0)} = A$.

The box j is accepted if there is a free rectangular area $A_k^{(j-1)} \in \{A_1^{(j-1)}, \ldots, A_j^{(j-1)}\}$ that may satisfy the requirement of j, that is both $W_k^{(j-1)} \geq w_j$ and $H_k^{(j-1)} \geq h_j$, otherwise j is rejected. In particular, if j is accepted let $A_k^{(j-1)}$ be the smaller (in terms of size) free rectangular area satisfying the requirement of j; ties are broken by selecting the smallest area $A_k^{(j-1)}$ with the minimum index k. When j is accepted (see Figure 2), a sub-area X_j (of height h_j and width w_j) in the north-west corner of $A_k^{(j-1)}$ is assigned to j, leaving two free rectangular sub-areas, namely $A_k^{(j)}$ and $A_{j+1}^{(j)}$, of $A_k^{(j-1)}$. In particular, let $A_k^{(j)} = A_k^{(j-1)} \setminus M_k^{(j)}$ and $A_{j+1}^{(j)} = M_k^{(j)} \setminus X_j$, where $M_k^{(j)}$ is the rectangular sub-area of $A_k^{(j-1)}$, with size $m_k^{(j)} = \min\{w_j \cdot H_k^{(j-1)}, h_j \cdot W_k^{(j-1)}\}$, being located in the west side of $A_k^{(j-1)}$ if $m_k^{(j)} = w_j \cdot H_k^{(j-1)}$, and in the north side otherwise; with this choice we have that the size of $A_k^{(j)}$ is not less than the size of $A_{j+1}^{(j)}$, and the considered free rectangular area $A_k^{(j-1)}$ is reduced by a minimal amount. If j is rejected, we consider $A_k^{(j)} = A_k^{(j-1)}$ and $A_{j+1}^{(j)} = \emptyset$. Finally, let $A_h^{(j)} = A_h^{(j-1)}$ for each $h \in \{1, \ldots, j\} \setminus \{k\}$. The value of the solution found by the on-line algorithm is:

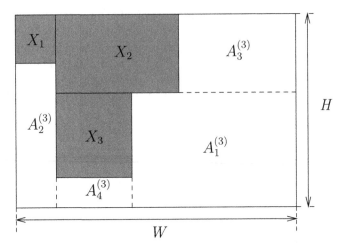

Fig. 1. Assigned and free rectangular areas at the beginning of iteration (4)

$$\rho = \frac{W \cdot H - \sum\limits_{s=1}^{n+1} (W_s^{(n)} \cdot H_s^{(n)})}{\min\{W \cdot H, \sum\limits_{j=1}^{n} (w_j \cdot h_j)\}} \tag{1}$$

3   Computational Results

3.1   Setup of the Experiments

The proposed algorithm has been tested both on random instances and on benchmarks. Referring to randomly generated instances, we experimented with $n = 10, 20, 50$ and 100 rectangular boxes getting a total number of 4 classes of instances. For each class we have considered different test cases according to different choices of two parameters, say w_{max} and h_{max}, being the maximum width and maximum height, respectively, for the boxes, and different values for the width W and the height H of the bounding rectangular area A; in particular, we have considered $h_{max} = 5$, 10, 15 and $w_{max} = 5, 10, 15$, and 4 different rectangular bounding areas with the following values for the width W and the height H: $(W, H) = \{(15, 20), (20, 30), (25, 50), (15, 50)\}$, for a total number of 36 instances for each class. For each one of the 144 different test cases, we randomly generated ten instances where the widths and heights of the boxes are uniformly distributed in the intervals $[1, w_{max}]$ and $[1, h_{max}]$, respectively.

As previously said, the proposed algorithm was also tested on some benchmark instances publicly available on the web[1]. The site contains some rectangle

---

[1] http://www.ms.ic.ac.uk/jeb/pub/strip1.txt

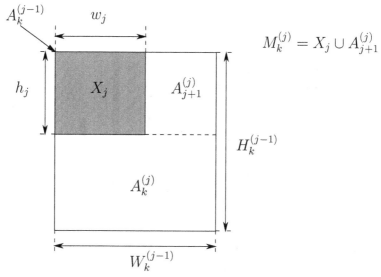

$$M_k^{(j)} = X_j \cup A_{j+1}^{(j)}$$

Fig. 2. Accepting box j

packing data set that have been introduced by Hopper and Turton [10]. As we work in an on-line scenario we decided to consider boxes in the same order as they appear in the data set. The results obtained by our algorithm on such instances are compared with the results achieved on the same instance set by an off-line heuristic algorithm with rejection proposed by Wu *et al.* ([18]), in which a *"quasi-human"* rule based on an ancient Chinese philosophical approach has been used. Basically, the rule, called *"Less Flexibility First"*, is based on the principle that empty corners inside the bounding area should be filled first, then the boundary lines of the free area, and then other empty areas. In other words it states a priority order in selecting the empty space inside the bounding area to be filled. Under this rule, the rectangle with less flexibility should be packed earlier. In [18], Wu *et al.* implemented such principle in two heuristics, called H1 and H2, which differ by the fact that H2 is implemented in a more strictly manner, i.e., in each packing step only the box with *"the least flexibility"* will be packed. H1 and H2 are ones of the most effective off-line procedures for the rectangle packing problem.

Our algorithm and the instance generator have been implemented in the C language, compiled with the GNU CC 2.8.0 with the -o3 option and tested on a PC Pentium 600 MHz with Linux OS.

3.2 Experimental Results and Analysis

In Figures 3–6 we summarize average results of the efficiency index ρ. It can be noted that the lowest efficiency values are achieved by our algorithm on instances with a small number of boxes whose dimensions are very close to those of the

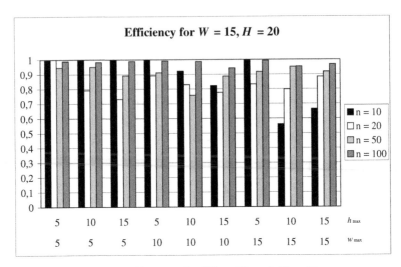

Fig. 3. Efficiency for $W = 15$ and $H = 20$

bounding area. As soon as the boxes are of small size (i.e., with small width and/or small height) the efficiency reach a value almost one. This is due to the chance the algorithm has to accept and place "small" boxes in the bounding area.

For the sake of completeness, we also list in Tables 1 and 2 the complete results for the scenarios ($W = 15$, $H = 20$, $n = 10$) and ($W = 15$, $H = 20$, $n = 20$), respectively, which are the cases where we obtained the worst results. For these instances we analyzed also the number of rejected, i.e., non-accepted, boxes. In Tables 1 and 2, the first two columns are the maximum width and height of a box, respectively, the third column shows the number of the average rejected requests, and the last three columns are the minimum value of ρ, say $\rho_{\min}$, the average values of ρ, say ρ_{ave}, and the maximum value of ρ, say $\rho_{\max}$, respectively, computed over ten different instances.

The worst results are obtained for the cases ($W = 15$, $H = 20$, $n = 10$, $w_{\max} = 15$, $h_{\max} = 10$) and ($W = 15$, $H = 20$, $n = 10$, $w_{\max} = 15$, $h_{\max} = 15$), both in terms of number of rejected boxes and efficiency. This is explainable by the fact that in these cases we have to consider boxes of large sizes requiring almost the whole bounding area. A similar situation occurs for the cases ($W = 15$, $H = 20$, $n = 20$, $w_{\max} = 5$, $h_{\max} = 15$) and ($W = 15$, $H = 20$, $n = 20$, $w_{\max} = 10$, $h_{\max} = 5$), even though the efficiency is higher with respect to the previous two cases with $n = 10$; this can be justified because with $n \geq 20$ we have more chances to efficiently use the available bounding area. As one can expect, the worst cases to deal with are those ones with $w_{\max}$ and/or $h_{\max}$ values close to W and H, respectively; indeed, in these cases we could have a very small chance to place a box whose size is very close to the size of the whole bounding area, especially if small requests have been accepted before implying a reduction

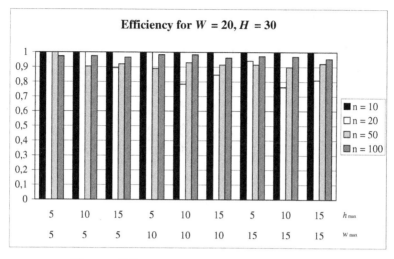

Fig. 4. Efficiency for $W = 20$ and $H = 30$

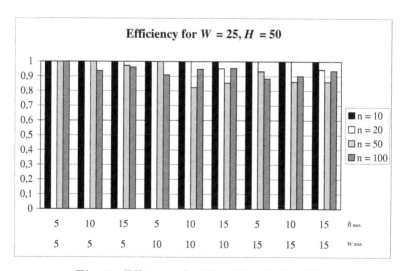

Fig. 5. Efficiency for $W = 25$ and $H = 50$

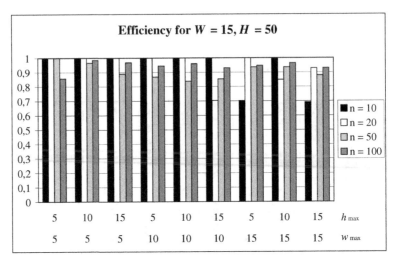

Fig. 6. Efficiency for $W = 15$ and $H = 50$

on the size of the available area. Nevertheless, the algorithm seems to perform well providing almost always solutions with efficiency value ρ greater than 0.8.

In Table 3, we list the results of our on-line algorithm compared with the results of the off-line heuristics presented by Wu *et al* in [18] and tested on the same instance set. As discussed in Section 3.1 Wu *et al.* presented two heuristic algorithms, namely H1 and H2 in which two variants of the *"Less Flexibility First"* principle are implemented. The data set is formed by 21 instances with a number of boxes to be packed that ranges from 16 to 197 and bounding rectangular area size that ranges from 400 to 38, 400. In Table 3 we list in the columns, respectively, the instance ID, the number n of boxes in that instance, the width of the bounding area, the height of the bounding area, the values of ρ, ρ_{H1} and

Table 1. Results for $W = 15$, $H = 20$ and $n = 10$

w_{max}	h_{max}	*Rejected*	ρ_{min}	ρ_{ave}	ρ_{max}
5	5	0.0	1.000	1.000	1.000
5	10	0.0	1.000	1.000	1.000
5	15	1.0	0.911	1.000	1.000
10	5	0.0	1.000	1.000	1.000
10	10	1.5	0.623	0.923	1.000
10	15	3.0	0.731	0.822	0.913
15	5	0.5	0.935	1.000	1.000
15	10	3.4	0.547	0.561	0.570
15	15	6.4	0.637	0.665	0.780

Table 2. Results for $W = 15$, $H = 20$ and $n = 20$

w_{max}	h_{max}	Rejected	ρ_{min}	ρ_{ave}	ρ_{max}
5	5	0.0	1.000	1.000	1.000
5	10	2.7	0.680	0.790	1.000
5	15	5.8	0.687	0.731	0.767
10	5	2.3	0.833	0.891	0.937
10	10	6.0	0.783	0.829	0.937
10	15	9.2	0.763	0.776	0.807
15	5	7.2	0.797	0.832	0.840
15	10	12.6	0.643	0.797	0.940
15	15	14.5	0.863	0.884	0.947

ρ_{H2} (being the efficiency of our algorithm and algorithms H1 and H2, respectively), the number of rejected boxes by our algorithm, the number of rejected rectangles by algorithm H1 and the number of rejected rectangles by algorithm H2.

Before passing to the analysis of the results, we remark that H1 and H2 are off-line algorithms, contrarily to our on-line algorithm. If we consider the values of the efficiency measure ρ, H1 and H2 of course provide in general better results, even if our algorithm provide the best possible value in 4 out of 21 instances; on average, the value of ρ over the 21 benchmark instances provided by our on-line algorithm is $\rho_{ave} = 0.868$, while for the off-line heuristics H1 and H2 are 0.973 and 0.954, respectively. Nevertheless, H1 and H2 are not efficient at all, requiring $O(n^5 \log n)$ and $O(n^4 \log n)$ computing time, respectively, and the CPU time spent by these algorithms is very large (e.g., for instances with $n = 97$ H1 and H2 take about one hour and more than 10 minutes, respectively); note that instances with $n = 196$ and 197 are not even considered. On the contrary, our algorithm is much more efficient, processing n boxes in $O(n^2)$ time, and the CPU time spent is negligible for all the test problems.

Moreover, if we compare the number of rejected boxes, several times our algorithm shows the same or even better performance with respect to H1 and H2. As it can be inferred by the comparison, our algorithm:

- is better than H1 and H2 in 4 instances, i.e., instances *prob1*, *prob4*, *prob10* and *prob19*, where the best possible result is achieved;
- gives the same number of rejected boxes as H1 in instances *prob13*, *prob16* (in which it is also superior to H2) and *prob17* (in which it equals H2, too);
- shows a worst behavior than H1 but better or equal than H2 in instances *prob2*, *prob5*, *prob8*, *prob12*;
- has a worst behavior than H1 and H2 in instances *prob3*, *prob6*, *prob7*, *prob9*, *prob11*, *prob14*, *prob15*, *prob18*.

Note that a comparison with H1 is not possible for *prob12* because the values *Rejected* and ρ are not reported in [18].

Table 3. Performance comparison of our on-line algorithm with off-line algorithms on benchmarks

Instance ID	n	W	H	Area	ρ	ρ_{H1}	ρ_{H2}	Rejected	H1	H2
prob1	16	20	20	400	1.000	0.980	0.925	0	2	2
prob2	17	20	20	400	0.744	0.980	0.925	2	1	2
prob3	16	20	20	400	0.870	0.975	0.937	2	1	1
prob4	25	40	15	600	1.000	0.993	0.951	0	1	3
prob5	25	40	15	600	0.827	1.000	0.958	1	0	1
prob6	25	40	15	600	0.795	1.000	0.991	3	0	1
prob7	28	60	30	1,800	0.846	0.993	0.977	3	1	1
prob8	29	60	30	1,800	0.764	0.991	0.941	3	1	3
prob9	28	60	30	1,800	0.798	0.992	0.960	2	1	1
prob10	49	60	60	3,600	1.000	0.990	0.976	0	1	2
prob11	49	60	60	3,600	0.849	0.997	0.948	4	1	3
prob12	49	60	60	3,600	0.736	-	0.987	3	-	3
prob13	73	60	90	5,400	0.898	0.997	0.976	1	1	4
prob14	73	60	90	5,400	0.800	0.999	0.978	6	1	2
prob15	73	60	90	5,400	0.874	0.991	0.988	4	2	3
prob16	97	80	120	9,600	0.922	0.997	0.966	1	1	4
prob17	97	80	120	9,600	0.840	0.962	0.966	3	3	3
prob18	97	80	120	9,600	0.899	0.994	0.986	5	2	3
prob19	196	160	240	38,400	1.000	-	-	0	-	-
prob20	197	160	240	38,400	0.884	-	-	5	-	-
prob21	196	160	240	38,400	0.896	-	-	9	-	-

It is worth noticing that, while instances prob19, prob20 and prob21 are not processed by H1 and H2, our algorithm provide the best possible solution for instance prob19. Moreover, our algorithm achieves four times $\rho = 1$, i.e., a complete covering of the bounding area, while the worst performance ratio achieved is $\rho = 0.736$ (instance prob12).

4 Conclusions

In this paper, we presented an on-line algorithm for the Rectangle Packing Problem able to accept or reject incoming boxes to maximize efficiency. We performed a wide computational analysis showing the behavior of the proposed algorithm. Performance results show that the on-line algorithm provides good solutions in almost all the tested cases. Moreover, a comparison with existing off-line heuristics confirms the promising performance of our algorithm.

Acknowledgements

The authors wish to dedicate this paper to the memory of their friend Mario Lucertini passed away one year ago.

References

[1] Beasley, J. E.: An exact two-dimensional non-guillotine cutting tree search procedures. Operation Research **33** (1985) 49-64 60

[2] Borodin, A., El-Yaniv, R.: Online computation and competitive analysis. Cambridge University Press (1998)

[3] Blitz, D. Van Vliet, A., Woeginger, G. J.: Lower bounds on the asymptotic worst case ratio of the on-line bin packing problem. Unpublished manuscript (1996) 60

[4] Confessore, G., Dell'Olmo, P., Giordani, S.: Complexity and approximation results for scheduling multiprocessor tasks on a ring. Discrete Applied Mathematics, (to appear) 60

[5] Csirik, J., Van Vliet, A.: An on-line algorithm for multidimensional bin packing. Operations Research Letters 13 (1993) 149-158 60

[6] Drozdowski, M.: Scheduling multiprocessor task. An overview. European Journal of Operations Research **94** (1996) 215-230

[7] Fiat, A. (ed.), Woeginger, G. J. (ed): Online Algorithms: The State of the Art. Lecture Notes in Computer Science, Vol. 1442. Springer-Verlag, Berlin Heidelberg (1998) 60

[8] Haessler, R. W., Sweeney, P. E.: Cutting stock problems and solution procedures. European Journal of Operational Research **54** (1991) 141-150 60

[9] Hochbaum, D. S., Maass, W.: Approximation schemes for covering and packing problems in image processing and VLSI. Journal of the Association for Computing Machinery **32 (1)** (1985) 130-136 60

[10] Hopper, E., Turton, B. C. H.: An empirical investigation of meta-heuristic and heuristic algorithms for a 2D packing problem. European Journal of Operational Research **128** 1 (2001) 34-57 59, 60, 63

[11] Hopper, E., Turton, B. C. H.: A Review of the Application of Meta-Heuristic Algorithms to 2D Strip Packing Problems. Artificial Intelligence Review **16** 4 (2001) 257-300 60

[12] Lee, C. C., Lee, D. T.: A simple on-line bin packing algorithm, Journal of ACM **32** (1985) 562-572 60

[13] Leung, J., Tam, T., Wong, C. S., Young, G., Chin, F.: Packing squares into square. Journal of Parallel and Distributed Computing **10** (1990) 271-275 59

[14] Lodi, A., Martello, S., Vigo, D.: Recent advances on two-dimensional bin packing problems. Discrete Applied Mathematics **123** (2002) 379-396

[15] Lodi, A., Martello, S., Monaci, M.: Two-dimensional packing problems: A survey. European Journal of Operational Research **141** (2002) 241-252 59, 60

[16] Imreh, C.: Online strip packing with modifiable boxes. Operation Research Letters **29** (2001), 79-85 60

[17] Shmoys, D., Wien, J., Williamson, D. P.: Scheduling parallel machines on-line. SIAM Journal of Computing **24** (1995) 1313-1331 60

[18] Wu, Y. L., Huang, W., Lau, S. C., Wong, C. K., Young, G. H.: An effective quasi-human based heuristic for solving the rectangle packing problem. European Journal of Operational Research **141** (2002) 341-358 59, 60, 61, 63, 66, 67

New Lower and Upper Bounds
for Graph Treewidth

François Clautiaux[1], Jacques Carlier[1], Aziz Moukrim[1], and Stéphane Nègre[2]

[1] HeuDiaSyC UMR CNRS 6599 UTC, BP 20529 60205, Compiègne
{francois.clautiaux,jacques.carlier,aziz.moukrim}@hds.utc.fr
[2] Institut Supérieur de Sciences Et Techniques
48 rue Raspail, 02100 St Quentin
stephane.negre@insset.u-picardie.fr

Abstract. The notion of tree-decomposition has very strong theoretical interest related to NP-Hard problems. Indeed, several studies show that it can be used to solve many basic optimization problems in polynomial time when the treewidth is bounded. So, given an arbitrary graph, its decomposition and its treewidth have to be determined, but computing the treewidth of a graph is NP-Hard. Hence, several papers present heuristics with computational experiments, but for many instances of graphs, the heuristic results are far from the best lower bounds.
The aim of this paper is to propose new lower and upper bounds for the treewidth. We tested them on the well known DIMACS benchmark for graph coloring, so we can compare our results to the best bounds of the literature. We improve the best lower bounds dramatically, and our heuristic method computes good bounds within a very small computing time.

1 Introduction

The notion of tree-decomposition has been introduced by Robertson and Seymour in the context of their research on *graph minors* [15]. It is based on the decomposition of the representative graph $G = (V, E)$ into i separating vertex subsets, called *separators*, connected in a tree. The maximum size of a separator minus one in an optimal tree-decomposition is called *treewidth*. With this method, the exponential factor of the complexity depends only on the treewidth $tw(G)$ of the graph G and not on its number of vertices n. It uses the property that several states of a subgraph can be summarized by the state of a separator. So, it can be used to tackle problems on large size graphs with a bounded treewidth, using dynamic programming methods.

There are numerous applications of this method to classical optimization problems [13, 6], to probabilistic networks [10], to graph coloration [14], to the frequency assignment problem [11], etc.

Computing the treewidth of a graph is NP-Hard [2]. Several exact methods study the decision problem $tw(G) \leq k$, but only for very small values of k (see [2] for $k = 1, 2, 3$ and [3] for an arbitrary k). A way to tackle the problem

K. Jansen et al. (Eds.): WEA 2003, LNCS 2647, pp. 70–80, 2003.

is to compute good solutions using heuristic methods. A recent paper proposes new bounds and computational experiments, but the gap between the lower and upper bounds remains large for several instances [12].

In this paper, we propose a method to improve the lower bound which uses properties stated by Hans Bodlaender [4] and a new heuristic. We developed these methods and tested them on the DIMACS benchmark for graph coloring [1], so we can compare our results to previous ones. We improve dramatically the values of the best lower bounds of the literature and the results of our heuristic are close in average to those obtained by the most recent ones, within a far smaller computing time.

In section 2 we explain our notation and give the definition of treewidth. Section 3 is devoted to the lower bound. In section 4, we recall the notion of triangulated graphs, triangulation and elimination orderings, and we present a new heuristic developed to compute upper bounds for the treewidth. The computational experiments are presented in section 5 before concluding remarks and ideas for future works.

2   Definitions and Notation

Let $G = (V, E)$ be an undirected graph with vertex set V and edge set $E \subset V \times V$. Let $n = |V|$ and $m = |E|$. Let v be a vertex, u is a *neighbor* of v in G if $[v, u] \in E$. The set of neighbors of v is called the *neighborhood* of v and it is denoted $N(v)$. Let $deg(v) = |N(v)|$ be the *degree* of v. A set of vertices Q is called a *clique* if there is an edge between each pair of distinct vertices of Q. For $U \subset V$, the *subgraph induced* by U is the graph $G[U] = (U, E[U])$ with $E[U] = (U \times U) \cap E$. The maximum cardinality of a clique in G is denoted $\omega(G)$ and the minimum coloring number of G, $\chi(G)$.

For distinct vertices $v, u \in V$, a *chain* is a sequence of distinct vertices $[v = v_1, v_2, \ldots, v_j = u]$ such that $\forall i < j, [v_i, v_{i+1}] \in E$. A *cycle* is a sequence of distinct vertices $[v = v_1, v_2, \ldots, v_j = v]$ such that $\forall i < j, [v_i, v_{i+1}] \in E$. A *chord* is an edge between two non-consecutive vertices in a cycle. A graph is *connected* if $\forall v, u \in V$, there is a chain between v and u. A *tree* $T = (V, E)$ is a connected graph with no cycle.

Definition 2.1. *(See [15]) A tree-decomposition D_T of $G = (V, E)$ is a pair (X, T), where $T = (I, F)$ is a tree with node set I and edge set F, and $X = \{X_i : i \in I\}$ is a family of subsets of V, one for each node of T, and:*
$$D_T = (\{X_i / i \in I\}, T = (I, F)) \text{ such that}$$

1. $\cup_{i \in I} X_i = V$
2. $\forall [v, u] \in E$, there is a X_i, $i \in I$ with $v \in X_i$ and $u \in X_i$
3. $\forall i, j, k \in I$, if j is on the path from i to k in T, then $X_i \cap X_k \subseteq X_j$

The *width* of a tree-decomposition is $max_{i \in I} |X_i| - 1$. The *treewidth* of a graph G, denoted $tw(G)$, is the minimum width over all possible tree-decompositions of G.

A tree-decomposition $D_T = (T, X)$ of G is *minimal* if removing any vertex v from a subset X_i makes D_T to violate one of the three properties of the tree-decomposition.

3   Lower Bounds

Even in the most recent papers, the lower bounds for the treewidth are far smaller than the heuristic results for a large number of graphs. The aim of this section is to present an iterative method which improves the value of the best lower bounds. It exploits properties stated by Hans Bodlaender [4].

3.1   Maximum Clique Bound

Proposition 3.1. *(See [5]) Suppose $(\{X_i/i \in I\}, T = (I, F))$ is a tree-decomposition of $G = (V, E)$. If $W \subseteq V$ forms a clique in G, then there exists an $i \in I$ with $W \subseteq X_i$.*

Proposition 3.1 induces that $\omega(G) - 1$ is a lower bound for $tw(G)$. The computation of this bound is NP-hard and thus it is unknown for a large number of graphs. Moreover, the *MMD* bound [12] introduced below is always greater or equal to $\omega(G) - 1$.

3.2   Maximum Minimum Degree Bound (MMD) [12]

Consider (T, X) an optimal tree-decomposition for G. Let $i \in I$ be a leaf of T with predecessor j. As the decomposition is minimal (Cf. 2), there exists a vertex $v \in X_i \setminus X_j$ (otherwise the node i contains no new information and can be deleted). Property 2 in the definition of the tree-decomposition induces that the neighbor set of v is included in the node X_i. We have $tw(G) \geq |N(v)|$.

So, a lower bound can be computed by the following method [12]: At each step of the algorithm, the vertex of G with lower degree is deleted and its degree recorded (Cf. algorithm 3.1). The algorithm stops when the vertex set V is empty. It returns the maximum degree recorded. This bound is always greater or equal to $\omega(G) - 1$ and even $\chi(G) - 1$. Moreover, it is very fast to compute.

3.3   Improving the MMD Bound

Properties stated in [4] can be used to find better lower bounds for the treewidth. We use the concept of *improved graph* defined by Hans Bodlaender in his paper [4].

Algorithm 3.1 Maximum Minimum Degree Algorithm

---

for $v \in V$ **do**
  $M(v) := N(v)$
end for
$S := V$
$MMD := 0$
while $S \neq \emptyset$ **do**
  $v^* := argmin_{v \in S} |M(v)|$
  if $|M(v^*)| > MMD$ **then**
    $MMD := |M(v^*)|$
  end if
  for $v \in M(v^*)$ **do**
    $M(v) := M(v) - \{v^*\}$
  end for
end while

The Improved Graph with Common Neighbors. For a graph $G = (V, E)$, let the *neighbor improved graph* $G' = (V, E')$ of G be the graph obtained by adding an edge $[v, u]$ to E for all pairs $v, u \in V$ such that v and u have at least $k + 1$ common neighbors in G.

Proposition 3.2. *(See [4]) If the treewidth of G is at most k, then the treewidth of the* neighbor improved graph G' *of G is at most k. Moreover, any tree-decomposition of G with treewidth at most k is also a tree-decomposition of the neighbor improved graph with treewidth at most k, and vice-versa.*

We can use proposition 3.2 to modify the initial graph and to compute a better lower bound by using the following algorithm (Cf. algorithm 3.2).

1. Assume that LB is a lower bound for the treewidth of G, i.e. $tw(G) \geq LB$.
2. Let us suppose that $tw(G) = LB$. We add an edge between each pair of vertices which have $LB+1$ common neighbors without modifying the treewidth of the graph.
3. Compute a lower bound LB' for the treewidth of the improved graph G'.
4. If $LB' > LB$, we have a contradiction. So, the initial assumption $tw(G) = LB$ was false.
5. We deduce $LB < tw(G)$, so $LB + 1$ is also a lower bound.
6. Repeat the process while there is a contradiction.

When there is no more contradiction, we cannot deduce any more information. The bound used is MMD, as it returns good results within a very small computing time. We denote this algorithm LB_N.

The Improved Graph and Vertex Disjoint Paths. The same technique can be applied to another improved graph. We now consider vertex disjoint paths instead of common neighbors. For a graph $G = (V, E)$, let the *paths improved* *graph* $G'' = (V, E'')$ of G be the graph obtained by adding an edge $[v, u]$ to E for all pairs $v, u \in V$ such that there are at least $k + 1$ vertex disjoint paths between v and u.

Proposition 3.3. *(See [4]) If the treewidth of G is at most k, then the treewidth of the paths improved graph G'' of G is at most k. Moreover, any tree-decomposition of G with treewidth at most k is also a tree-decomposition of the path improved graph with treewidth at most k, and vice-versa.*

To compute this bound, we have to create a new oriented graph D from G as follows.

1. Each edge $[v, u] \in E$ is replaced by two arcs (v, u) and (u, v).
2. Each vertex v is replaced by two vertices v_1 and v_2
3. Each arc (x, v) is replaced by an arc (x, v_1)
4. Each arc (v, x) is replaced by an arc (v_2, x)
5. An arc (v_1, v_2) with weight 1 is added.
6. A weight $+\infty$ is associated to each arc added at steps 3 and 4.

We add the edge $[v, u]$ in G if the maximum flow $f(v_2, u_1)$ between the two corresponding vertices v_2 and u_1 in D is strictly greater than k.

The computing time of this method is far larger, as we have to solve several network flow problems, but the results are improved dramatically because the edge set added is far larger than the previous one. We denote this algorithm *LB_P*.

Algorithm 3.2 Improving the Lower Bound

$LB := lower\_bound(G)$
repeat
    $LB := LB + 1$
    $G' := improve\_graph(G, LB)$
    $LB' := lower\_bound(G')$
until $LB' = LB$
$LB := LB - 1$

---

4 A New Heuristic to Compute the Treewidth

4.1 Triangulated Graphs

Computing the treewidth of a triangulated graph is linear in time. So, many heuristics use the properties of triangulated graphs to find upper bounds for the treewidth. A graph G is *triangulated* if for every cycle of length $k > 3$, there is a chord joining two non-consecutive vertices [16].

Proposition 4.1. *(See [9]) If G is triangulated, $tw(G) = \omega(G) - 1$. Moreover, if G is triangulated, computing $\omega(G)$ and thus $tw(G)$ has for complexity $O(n+m)$.*

So, given an arbitrary graph G, it is interesting to find a triangulated graph which contains G to get an upper bound for $tw(G)$. The triangulated graph $H = (V, E + E_t)$, with $E \cap E_t = \emptyset$, is called a *triangulation* for G. From the following property, we know that the optimal value of $tw(G)$ can be found using this technique:

Proposition 4.2. *(See [17]) For every graph G, there exists a triangulation H^* such that $tw(H^*) = tw(G)$. Moreover, any tree-decomposition for G is a tree-decomposition for H^* and vice-versa.*

Computing the treewidth of G is equivalent to finding such a graph H^* (and also equivalent to finding the triangulation with the minimum maximum clique size). This problem is NP-Hard as computing the treewidth is NP-Hard, but each triangulation for G gives an upper bound for $tw(G)$.

A vertex is said *simplicial* if its neighborhood is a clique [16]. Let $G = (V, E)$ be a graph, an ordering (or elimination ordering) $\sigma(1, 2, \ldots, n)$ of V is a *perfect elimination ordering* if and only if $\forall i \in [1, \ldots, n]$, $\sigma(i)$ is a simplicial vertex in $G[\{\sigma(i), \ldots, \sigma(n)\}]$, the subgraph induced by the higher ordered vertices.

Proposition 4.3. *(See [8]) G is triangulated if and only if it has a perfect elimination ordering.*

Let $G = (V, E)$ be a non-triangulated graph and σ be an elimination ordering. We denote $H(\sigma) = (V, E + E_t(\sigma))$ the triangulation of G obtained by applying the following algorithm [17]: the vertices are eliminated in the elimination ordering σ. At each step i of the algorithm, the necessary edges to make $v = \sigma(i)$ to be a simplicial vertex are added to the current graph (i.e. an edge is added between each pair of neighbors of v). Then the vertex is deleted (Cf. algorithm 4.1). Let $m' = |E + E_t(\sigma)|$. Algorithm 4.1 can be implemented in $O(n + m')$ time [18]. We denote $N_t(v)$ the set of neighbors of v in H.

So, computing the treewidth can be solved this way: find an elimination ordering which minimizes the maximum clique size of the triangulated graph obtained by algorithm 4.1. The treewidth is equal to $max_{v \in V}|\{u \in N_t(v)/\sigma^{-1}(v) < \sigma^{-1}(u)\}|$, the maximum degree of a vertex when it is eliminated.

4.2   The New Heuristic

The algorithm *min degree* is a classical triangulation heuristic. The method starts from the initial graph G. At each step of the algorithm, the vertex v with minimum degree in the current graph is chosen. First, the edges necessary to make v to be a simplicial vertex are added. Then v is deleted from the graph. The process is repeated for the remaining graph until the vertex set is empty. Locally, the choice of the vertex with minimum degree is suitable. Indeed, the size of the clique induced by the elimination of v is minimized, but this strategy is not optimal.

Algorithm 4.1 Graph Triangulation Associated to an Elimination Ordering σ

---

for $v \in V$ **do**
$\quad N_t(v) := N(v)$
end for
$E_t := \emptyset$
for $i := 1$ to n **do**
$\quad v := \sigma(i)$
$\quad$**for** $u, z \in N_t(v)$ **do**
$\quad\quad E_t := E_t \cup [u, z]$ {*add edge to triangulation*}
$\quad\quad N_t(u) := N_t(u) \cup \{z\}$ {*update neighborhood*}
$\quad\quad N_t(z) := N_t(z) \cup \{u\}$
$\quad$**end for**
$\quad$**for** $u \in N_t(v)$ **do**
$\quad\quad N_t(u) := N_t(u) \setminus \{v\}$ {*v is eliminated*}
$\quad$**end for**
end for

As the degree of the vertex eliminated is not sufficient, we add a more global criterion to improve the quality of the results. The idea is to compute a lower bound for the treewidth of the graph obtained after elimination of v. We want our algorithm to be a fast one, so we use the MMD bound [12].

Let $G_v = (V_v, E_v)$ be the graph obtained after eliminating the vertex v. It is computed with the two following operations:

- connect all current neighbors of v in G
- remove v from G

As the treewidth of the remaining graph is larger than the bound computed, the global criterion has the larger weight. The vertex v^* chosen is the one which minimizes the function $2 * lower\_bound(G_v) + |N_t(v)|$. We denote this algorithm D_LB (algorithm 4.2).

5   Computational Analysis

We have tested our methods on the DIMACS benchmarks for vertex coloring. For many of these graphs, $\omega(G)$ is known, and the lower bounds are far from the best heuristic results. We compare our methods with the most recent ones ($\omega(G) - 1$ and MMD [12] for the lower bound, Lex and MSVS [12] and the *Tabu PEO* [7] for the upper bound). Note that some values of $\omega(G)$ are unknown. Our algorithms are implemented in C on a Pentium III 1GHz. There is no randomization in our algorithms, so each method is launched one time. In comparison, the computer used for the tabu search is the same and the computation time of Lex and MSVS are obtained by Koster et al. with C++ implementation on a Pentium III 800 Mhz.

Our lower bounds can improve the results of the previous best bound (MMD) by a wide range. The results for the 62 instances are reported in Table 1. For 22

Algorithm 4.2 Greedy Algorithm for the Treewidth

$k := 0$ {*maximum degree found*}
for $v \in V$ **do**
   $N_t(v) := N(v)$
end for
for $i := 1$ to n **do**
   for $v \in V$ **do**
      $compute(G_v)$
   end for
   $v^* := argmin_{v \in V}(|N_t(v)| + 2 * lower\_bound(G_v))$
   $k = max(k, N_t(v))$ {*update width*}
   for $u, z \in N_t(v^*)$ **do**
      $N_t(u) := N_t(u) \cup \{z\}$ {*update neighborhood*}
      $N_t(z) := N_t(z) \cup \{u\}$
   end for
   for $u \in N_t(v^*)$ **do**
      $N_t(u) := N_t(u) \setminus \{v^*\}$ {v^* *is eliminated*}
   end for
end for

---

of them, the LB_N method improves the value of the lower bound. This value increases to 53 with the LB_P method. The computing times of our methods are larger, so they should not be used all along a *branch & bound* search method. They can be used at top levels, in order to cut a very large number of branches. Moreover, as some lower bounds reach the upper bounds, there is no need to launch an enumerative method for several instances (thanks to our method, the value of treewidth is known for 8 new benchmarks: FPSOL2.I.1, INITHX.I.1, MILES500, MULSOL.I.1, MULSOL.I.2, MULSOL.I.3, MULSOL.I.4, and ZEROINI.I.1).

The results of the heuristic are in average as good as those returned by MSVS and Lex, but the computing times are far smaller. It is a method which finds good bounds using a very small computing time. Furthermore, it seems to be more stable than Lex, and never returns too bad results as Lex does for example with graphs INITHX (the values returned by Lex are more than six times greater than those returned by the three others methods). Our computational experiments for the upper bounds are reported in Table 2.

6 Conclusion

We have proposed new methods to compute lower and upper bounds for the treewidth of graphs. The improvement of the lower bound is very significant, and allows us to establish new exact values of treewidth for several graphs of the DIMACS benchmark for graph coloration. The heuristic is a really fast method to find good quality upper bounds.

The gap between the best upper bounds and lower bounds remains large. Indeed, the exact value of treewidth is not known for most of the benchmarks. So, to reduce the gap between the bounds, we have to improve the value of the lower bound again before working on an exact method.

References

[1] *The second dimacs implementation challenge: NP-hard problems: Maximum clique, graph coloring, and satisfiability*, 1992-1993. 71

[2] S. Arnborg and A. Proskurowki, *Characterisation and recognition of partial 3-trees*, SIAM J. Alg. Disc. Meth. **7** (1986), 305–314. 70

[3] H. Bodlaender, *A linear time algorithm for finding tree-decompositions of small treewidth*, SIAM J. Comput. **25** (1996), 1305–1317. 70

[4] _____, *Necessary edges in k-chordalisation of graphs*, Tech. Report UU-CS-2000-27, Institute for Information and Computing Sciences, Utrecht University, 2000. 71, 72, 73, 74

[5] H. Bodlaender and R. Möhring, *The pathwidth and treewidth of cographs*, SIAM J. Disc. Math. **6** (1993), 181–188. 72

[6] J. Carlier and C. Lucet, *A decomposition algorithm for network reliability evaluation*, Discrete Applied Mathematics **65** (1993), 141–156. 70

[7] F. Clautiaux, A. Moukrim, S. Nègre, and J. Carlier, *A tabu search minimising upper bounds for graph treewidth*, Report, 2003. 76

[8] D. Fulkerson and O. Gross, *Incidence matrices and interval graphs*, Pacific J. Math. **15** (1965), 835–855. 75

[9] F. Gavril, *Algorithms for minimum coloring, maximum clique, minimum coloring cliques and maximum independent set of a chordal graph*, SIAM J. Comput. **1** (1972), 180–187. 75

[10] F. Jensen, S. Lauritzen, and K. Olesen, *Bayesian updating in causal probabilistic networks by local computations*, Computational Statistics Quaterly **4** (1990), 269–282. 70

[11] A. Koster, *Frequency assignment, models and algorithms*, Ph.D. thesis, Universiteit Maastricht, 1999. 70

[12] A. Koster, H. Bodlaender, and S. van Hoesel, *Treewidth: Computational experiments*, Fundamenta Informaticae **49** (2001), 301–312. 71, 72, 76

[13] C. Lucet, J.F. Manouvrier, and J. Carlier, *Evaluating network reliability and 2-edge-connected reliability in linear time for bounded pathwidth graphs*, Algorithmica **27** (2000), 316–336. 70

[14] C. Lucet, F. Mendes, and A. Moukrim, *Méthode de décomposition appliquée à la coloration de graphes*, ROADEF, 2002. 70

[15] N. Robertson and P. Seymour, *Graph minors. ii algorithmic aspects of tree-width*, Journal of Algorithms **7** (1986), 309–322. 70, 71

[16] D. Rose, *Triangulated graphs and the elimination process*, J Math. Anal. Appl. **32** (1970), 597–609. 74, 75

[17] _____, *A graph-theoretic study of the numerical solution of sparse positive definite systems of linear equations*, Graph Theory and Computing (R.C. Reed, ed.), Academic Press (1972), 183–217. 75

[18] D. Rose, E. Tarjan, and G. Lueker, *Algorithmic aspects of vertex elimination on graphs*, SIAM J. Comput. **5** (1976), 146–160. 75

Table 1. Lower bound

instance	n	m	ub	lb				CPU time		
			Tabu	$\omega(G) - 1$	MMD	LB_N	LB_P	MMD	LB_N	LB_P
anna	138	986	12	10	10	10	11	0.01	0.01	8.25
david	87	812	13	10	10	10	11	0.00	0.01	7.08
huck	74	602	10	10	10	10	10	0.00	0	1.09
homer	561	3258	31	12	12	14	21	0.04	2.19	1202.78
jean	80	508	9	9	9	9	9	0.00	0	2.76
games120	120	638	33	8	8	8	12	0.00	0.77	11.07
QUEEN5_5	25	160	18	4	12	12	12	0.00	0	0.88
QUEEN6_6	36	290	25	6	15	15	15	0.00	0.05	2.22
QUEEN7_7	49	476	35	6	18	18	20	0.01	0.33	1.57
QUEEN8_8	64	728	46	8	21	21	23	0.00	0.22	6.50
QUEEN9_9	81	1056	58	9	24	24	26	0.01	0.06	21.97
QUEEN10_10	100	1470	72	-	27	27	31	0.01	0.16	16.78
QUEEN11_11	121	1980	88	10	30	30	34	0.01	0.33	52.07
QUEEN12_12	144	2596	104	-	33	33	37	0.02	1.1	144.16
QUEEN13_13	169	3328	122	12	36	36	42	0.04	1.05	111.91
QUEEN14_14	196	4186	141	-	39	39	45	0.07	1.76	283.24
QUEEN15_15	225	5180	163	-	42	42	48	0.11	3.19	653.83
QUEEN16_16	256	6320	186	-	45	45	53	0.21	4.50	519.60
FPSOL2.I.1	269	11654	66	64	64	66	66	1.03	0.44	1003.86
FPSOL2.I.2	363	8691	31	29	31	31	31	0.51	0.38	3295.79
FPSOL2.I.3	363	8688	31	29	31	31	31	0.67	0.39	2893.60
INITHX.I.1	519	18707	56	53	55	56	56	5.83	0.88	23995.88
INITHX.I.2	558	13979	35	30	31	31	31	2.37	3.62	67708.76
INITHX.I.3	559	13969	35	30	31	31	31	2.48	4.01	54247.00
MILES1000	128	3216	49	41	41	44	48	0.07	0.66	268.73
MILES1500	128	5198	77	72	72	76	76	0.10	0.06	117.10
MILES250	125	387	9	7	7	8	8	0.00	0.05	26.80
MILES500	128	1170	22	19	19	21	22	0.01	0	42.44
MILES750	128	2113	36	30	31	32	33	0.02	0.06	438.35
MULSOL.I.1	138	3925	50	48	48	50	50	0.07	0.22	302.88
MULSOL.I.2	173	3885	32	30	31	32	32	0.09	0.11	256.44
MULSOL.I.3	174	3916	32	30	31	32	32	0.09	0	250.10
MULSOL.I.4	175	3946	32	30	31	32	32	0.09	0.06	252.96
MULSOL.I.5	176	3973	31	30	31	31	31	0.09	0	363.64
MYCIEL3	11	20	5	3	3	3	4	0.00	0	0
MYCIEL4	23	71	10	4	5	5	6	0.00	0	0.48
MYCIEL5	47	236	19	5	8	8	12	0.00	0.33	0.52
MYCIEL6	95	755	35	6	12	15	20	0.01	0.33	6.98
MYCIEL7	191	2360	66	7	18	25	34	0.03	0.77	98.38
SCHOOL1	385	19095	188	-	73	88	116	4.79	62.4	5960.29
SCHOOL1_NSH	352	14612	162	-	61	72	100	3.08	32.12	4142.04
ZEROIN.I.1	126	4100	50	48	48	50	50	0.17	0.11	53.82
ZEROIN.I.2	157	3541	32	29	29	31	31	0.09	0.28	290.09
ZEROIN.I.3	157	3540	32	29	29	31	31	0.07	0.11	283.54
LE450_5A	450	5714	256	4	17	17	33	0.35	28.4	686.56
LE450_5B	450	5734	254	4	17	17	33	0.46	29.38	608.12
LE450_5C	450	9803	272	4	33	33	51	1.38	38.77	1588.40
LE450_5D	450	9757	278	4	32	32	51	2.30	40.37	34947.64
LE450_15A	450	8168	272	14	24	24	56	0.39	35.87	1465.50
LE450_15B	450	8169	270	14	24	24	55	0.80	36.75	1678.40
LE450_15C	450	16680	359	14	49	49	92	7.58	79.43	5558.45
LE450_15D	450	16750	360	14	51	51	91	8.76	77.94	5422.94
LE450_25A	450	8260	234	24	26	27	62	1.31	36.41	1723.47
LE450_25B	450	8263	233	24	25	25	59	1.20	33.39	1692.10
LE450_25C	450	17343	327	24	52	52	100	8.13	74.48	5893.32
LE450_25D	450	17425	336	24	51	51	98	5.31	79.59	5635.75
DSJC125.1	125	736	66	-	8	8	16	0.01	0.83	6.28
DSJC125.5	125	3891	109	-	53	53	62	0.07	0.76	105.36
DSJC125.9	125	6961	119	-	103	107	108	0.20	0.28	77.51
DSJC250.1	250	3218	173	-	18	18	32	0.09	5.16	131.33
DSJC250.5	250	15668	232	-	109	109	125	3.62	11.31	3021.03
DSJC250.9	250	27897	243	-	211	213	218	8.20	6.31	1905.30

Table 2. Upper bound

instance	n	m	lb	ub				CPU time			
			LB_P	LEX	MSVS	D_LB	Tabu	LEX	MSVS	D_LB	Tabu
anna	138	986	11	12	12	12	12	1.24	18.39	0.880	2776.93
david	87	812	11	13	13	13	13	0.56	7.77	0.220	796.81
huck	74	602	10	10	10	10	10	0.24	2.30	0.130	488.76
homer	561	3258	21	37	31	32	31	68.08	556.82	27.380	157716.56
jean	80	508	9	9	9	9	9	0.29	1.98	0.130	513.76
games120	120	638	12	37	51	41	33	5.20	65.97	1.620	2372.71
QUEEN5_5	25	160	12	18	18	18	18	0.04	0.22	0.340	100.36
QUEEN6_6	36	290	15	26	28	27	25	0.16	1.16	0.140	225.55
QUEEN7_7	49	476	20	35	38	38	35	0.51	4.66	0.090	322.40
QUEEN8_8	64	728	23	46	49	50	46	1.49	16.38	0.350	617.57
QUEEN9_9	81	1056	26	59	66	64	58	3.91	47.35	0.740	1527.13
QUEEN10_10	100	1470	31	73	79	80	72	9.97	128.30	1.670	3532.78
QUEEN11_11	121	1980	34	89	101	102	88	23.36	310.83	3.160	5395.74
QUEEN12_12	144	2596	37	106	120	117	104	49.93	702.29	6.720	10345.14
QUEEN13_13	169	3328	42	125	145	141	122	107.62	1589.77	10.940	16769.58
QUEEN14_14	196	4186	45	145	164	164	141	215.36	3275.75	20.300	29479.91
QUEEN15_15	225	5180	48	167	192	194	163	416.25	6002.33	31.070	47856.25
QUEEN16_16	256	6320	53	191	214	212	186	773.09	11783.30	63.890	73373.12
FPSOL2.I.1	269	11654	66	66	66	66	66	319.34	4220.91	176.110	63050.58
FPSOL2.I.2	363	8691	31	52	31	31	31	622.22	8068.88	174.930	78770.05
FPSOL2.I.3	363	8688	31	52	31	31	31	321.89	8131.78	144.650	79132.70
INITHX.I.1	519	18707	56	223	56	56	56	3144.95	37455.10	2966.020	101007.52
INITHX.I.2	558	13979	31	228	35	35	35	5567.96	37437.20	1004.340	121353.69
INITHX.I.3	559	13969	31	228	35	35	35	5190.39	36566.80	884.430	119080.85
MILES1000	128	3216	48	49	53	53	49	14.39	229.00	3.420	5696.73
MILES1500	128	5198	76	77	83	77	77	29.12	268.19	3.470	6290.44
MILES250	125	387	8	10	9	9	9	1.12	10.62	0.350	1898.29
MILES500	128	1170	22	22	28	28	22	4.37	87.18	0.960	4659.31
MILES750	128	2113	33	37	38	43	36	8.13	136.69	1.850	3585.68
MULSOL.I.1	138	3925	50	66	50	50	50	17.77	240.24	12.700	3226.77
MULSOL.I.2	173	3885	32	69	32	32	32	34.06	508.71	15.290	12310.37
MULSOL.I.3	174	3916	32	69	32	32	32	34.58	527.89	14.010	9201.45
MULSOL.I.4	175	3946	32	69	32	32	32	35.53	535.72	14.100	8040.28
MULSOL.I.5	176	3973	31	69	31	31	31	36.25	549.55	12.920	13014.81
MYCIEL3	11	20	4	5	5	5	5	0.00	0.01	0.000	72.50
MYCIEL4	23	71	6	11	11	11	10	0.02	0.13	0.310	84.31
MYCIEL5	47	236	12	23	20	20	19	0.28	2.00	0.300	211.73
MYCIEL6	95	755	20	47	35	35	35	4.56	29.83	2.410	1992.42
MYCIEL7	191	2360	31	94	74	70	66	109.86	634.32	28.640	19924.58
SCHOOL1	385	19095	116	252	244	242	188	3987.64	41141.10	273.620	137966.73
SCHOOL1_NSH	352	14612	100	192	214	200	162	2059.52	28954.90	161.700	180300.10
ZEROIN.I.1	126	4100	50	50	50	50	50	17.78	338.26	10.680	2595.92
ZEROIN.I.2	157	3541	31	40	33	33	32	24.82	448.74	26.760	4825.51
ZEROIN.I.3	157	3540	31	40	33	33	32	24.69	437.06	24.780	8898.80
LE450_5A	450	5714	33	310	317	323	256	7836.99	73239.66	274.490	130096.77
LE450_5B	450	5734	33	313	320	321	254	7909.11	73644.28	260.290	187405.33
LE450_5C	450	9803	51	348	340	329	272	10745.70	103637.17	525.350	182102.37
LE450_5D	450	9757	51	349	326	318	278	10681.29	96227.40	566.610	182275.69
LE450_15A	450	8168	56	296	297	300	272	6887.15	59277.90	273.700	117042.59
LE450_15B	450	8169	55	296	307	305	270	6886.84	65173.20	230.900	197527.14
LE450_15C	450	16680	92	379	376	379	359	12471.09	122069.00	356.610	143451.73
LE450_15D	450	16750	91	379	375	380	360	12481.22	127602.00	410.350	117990.30
LE450_25A	450	8260	62	255	270	267	234	4478.30	53076.40	243.290	143963.41
LE450_25B	450	8263	59	251	264	266	233	4869.97	52890.00	248.610	184165.21
LE450_25C	450	17343	100	355	365	361	327	10998.68	109141.00	344.360	151719.58
LE450_25D	450	17425	98	356	359	362	336	11376.02	111432.25	434.120	189175.40
DSJC125.1	125	736	16	70	67	67	66	12.90	171.54	2.500	1532.93
DSJC125.5	125	3891	62	110	110	110	109	38.07	254.90	3.870	2509.97
DSJC125.9	125	6961	108	119	120	120	119	55.60	70.79	56.630	1623.44
DSJC250.1	250	3218	32	183	179	176	173	528.10	5507.86	32.730	28606.12
DSJC250.5	250	15668	125	233	233	233	232	1111.66	7756.38	48.510	14743.35
DSJC250.9	250	27897	218	243	244	244	243	1414.58	1684.83	15.600	30167.70

Search Data Structures for Skewed Strings

Pilu Crescenzi[1], Roberto Grossi[2] *, and Giuseppe F. Italiano[3] **

[1] Dipartimento di Sistemi e Informatica, Università degli Studi di Firenze, Italy
piluc@dsi.unifi.it
[2] Dipartimento di Informatica, Università di Pisa, Italy
grossi@di.unipi.it
[3] Dipartimento di Informatica, Sistemi e Produzione,
Università di Roma "Tor Vergata", Italy
italiano@disp.uniroma2.it

Abstract. We consider skewed distributions of strings, in which any two such strings share a common prefix much longer than that expected in uniformly distributed (random) strings. For instance, this is the case of URL addresses, IP addresses, or XML path strings, all representing paths in some hierarchical order. As strings sharing a portion of the path have a quite long common prefix, we need to avoid the time-consuming repeated examination of these common prefixes while handling the linked data structures storing them. For this purpose, we show how to implement search data structures that can operate on strings with long prefixes in common. Despite the simplicity and the generality of the method, our experimental study shows that it is quite competitive with several optimized and tuned implementations currently available in the literature.

1 Introduction

In many applications keys are arbitrarily long, such as strings, multidimensional points, multiple-precision numbers, or multi-key data, and are modelled as k-dimensional keys for a given positive integer $k > 1$, or as variable-length keys. The latter can be virtually padded with a sufficient number of string terminators so that they can be considered having all the same length k. When dealing with skewed distributions (shortly, *skewed strings*), such as URL addresses, IP addresses, or XML path strings representing paths in some hierarchical order, we can observe that they share typically long prefixes. It is more realistic to assume that the average length of the common prefix of any two such strings is much longer than that expected in the case of uniformly distributed (random) strings.

A reasonable measure of "skewness," denoted $\mathscr{L}(S)$, for a given set S of n strings $x_1, \ldots, x_n$ over an alphabet Σ can be formalized as follows. Let m_i be the longest matched prefix of x_i against the previous strings $x_1, \ldots, x_{i-1}$. Define $m = \sum_{i=2}^{n} m_i/(n-1)$ as the average length of matched prefixes in S. Note that

* Partially supported by the Italian MIUR Project ALINWEB.
** Partially supported by the EU Programme ALCOM-FT (n. IST-1999-14186) and by the Italian MIUR Project ALINWEB.

the definition of m does not depend on the order of the keys in S, as $(m + 1)$ is equivalent to the internal path length of the $|\Sigma|$-ary trie built on S, which is independent of the order of insertions. We use the fact that $(m + 1) \approx \log_{|\Sigma|} n$ when the strings in S are independent and identically distributed [31, Chap.4]. Hence, we measure the skewness (or non-randomness) of the strings in S by the ratio $\mathscr{L}(S) = (m + 1)/\log_{|\Sigma|} n$. In a certain sense, a large value of m is a good indicator of the skewness of the strings, which is normalized by that same indicator for random strings. Alternatively, since $\log_{|\Sigma|} n$ is proportional to the height of a random $|\Sigma|$-ary tree with n nodes [31, Chap.4], we can see $\mathscr{L}(S)$ as the average number of characters matched per level in a random trie. The intuition is that $\mathscr{L}(S)$ is small in random strings, and the dominant cost is given by traversing the access path from the root to a node. For skewed strings, $\mathscr{L}(S)$ is large and the dominant cost is due to scanning the long prefixes of the strings.

Fast searching on a dynamic set S of skewed strings can be supported by choosing from a vast repertoire of basic data structures, such as AVL-trees [2], red-black trees [4, 14, 32], (a,b)-trees [16], weight-balanced BB[α]-trees [22], finger search trees, self-adjusting trees [30], and random search trees [27], just to name a few. Many of these data structures exhibit interesting combinatorial properties that make them attractive both from the theoretical and from the practical point of view. They are defined on an ordered set of (one-dimensional or one-character) keys, so that searching is driven by comparisons against the keys stored in their nodes: it is usually assumed that any two keys can be compared in $O(1)$ time. Their operations can be extended to long keys by performing string comparisons while routing the keys for searching and updating. This approach may work well in the case of uniformly distributed strings, since a mismatching character is found in expected constant time. However, it is not satisfactory for skewed strings, since a string comparison may take as much as $O(k)$ time per operation, thus slowing down the performance of carefully designed data structures by a multiplicative $O(k)$ factor. To improve on this simple-minded approach, we would like to avoid repeated examinations of long common prefixes while handling the linked data structures storing them efficiently.

Several approaches for solving this problem are possible, such as using tries (see e.g. [17]) or *ad hoc* data structures for k-dimensional keys, for which we have many examples [5, 6, 8, 10, 11, 13, 19, 30, 33, 34, 35]. The height of all these data structures is $O(k + \log n)$, where k is the maximum key length and n the total number of keys. A more general paradigm, described in [12], makes use of algorithmic techniques capable of augmenting many kinds of (heterogeneous) linked data structures so that they can operate on k-dimensional keys. Consequently, the richness of results on data structures originally designed for one-dimensional keys is available for k-dimensional keys in a smooth, simple and efficient way, without incurring in the previously mentioned $O(k)$-slowdown.

A potential drawback of general techniques, such as the ones in [12], is that they may produce data structures that are difficult to use in real-life applications, which could take some advantages from their efficient implementation for skewed strings. Just to name one, Internet "interceptor" software blocks access to undesirable Internet sites, storing huge lists of URLs to be blocked, and checking by prefix searches in these URLs. Other applications may be in networking for filtering purposes and in databases for storing and searching XML records. Note that hashing is of little help in these situations as it cannot perform the search of prefixes of the keys, being scattered throughout the memory. Instead, the above data structures are more powerful than hashing, since they allow for prefix and one-dimensional range searching and for sorting keys.

In this paper, we pursue an algorithm engineering avenue to show that the techniques in [12] may be of practical value. We therefore discuss the implementing issues from the programmers' perspective. In particular, we show that the general transformation from one-character to k-character keys can be carried out by following a rather general and simple scheme, so that a given implementation of a data structure can be easily modified in order to handle k-character keys without incurring in the $O(k)$-slowdown. In principle, we can start from several basic data structures for one-character keys, and show how to make them work for k-character keys. This can be easily done as the technique itself is not invasive: if we start from a given data structure (e.g., a binary search tree, an AVL tree, an (a, b)-tree, or a skip list) or a related piece of code for one-character keys, we can produce a data structure for k-character keys which retains exactly the same topology (and the same structural properties) as the original structure.

We start out from the techniques in [12] for re-engineering the core of the search and update algorithms, and we compare the resulting codes to more optimized and tuned algorithms. In our experiments we consider the Patricia tries [17, 21], the ternary search trees [5] and a cache-aware variant of tries [1], which are known to be among the fastest algorithms for long keys. Timings were normalized with the average time of linear probing hashing [17]. Our experiments were able to locate two different thresholds on our data sets, based on the parameter $\mathscr{L}(S)$ for skewness. When the parameter is small, the adaptive tries of [1] give the fastest implementation. When it is large, our techniques deliver the fastest codes, which thus become relevant in the case of skewed strings. For intermediate values of $\mathscr{L}(S)$, the experimented algorithms have very close behavior: this is the range where the ternary search trees of [5] become very competitive. In any case, the fastest algorithms obtain a speedup factor that is bounded by 2 or 3. As a matter of fact this is not a limitation, since the data structures are highly tuned for strings and our *general* technique compares favorably in several cases. This may allow for treating 2–3 times more queries per time unit on a high-performance server, which is effective in practice. Our data structures are therefore a valid alternative to ubiquitous Patricia tries and compact tries in all their applications (indexing, sorting, compressing, etc. etc.).

2   Engineering the Algorithms for Long Keys and Strings

Our technique can be illustrated by running an example on the classical lookup procedure adopted in binary search trees. The reader can follow our example and apply the same methodology to other data structures fulfilling the requirements described in [12]. Namely, they must be linked data structures whose keys undergo a total order and whose search and update operations are driven by comparison of the keys (i.e., without hashing or bit manipulation of the keys, for which our technique does not work). With this approach, we have implemented the ANSI C code of several multi-character key data structures, starting from available code for binary search trees, treaps, AVL trees [2], (a, b)-trees [16], and skiplists [24]. In particular, for binary search trees, treaps, and AVL trees we started from the implementation described in [36], for skiplists we started from the implementation described in [24], which is available via anonymous ftp at [25], while for (a, b)-trees we started from our own implementation of [16, 20].

2.1 Preliminaries

In our description we identify the nodes with their stored keys, and assume that the keys fulfill a total order with two special keys $-\infty$ and $+\infty$ that are always the smallest and the largest ones. Given a node t, let π be the access path from the root to t. We define the *successor* π_t^+ of t along path π to be the smallest ancestor that is greater than t (i.e., we find π_t^+ by going upward in π until we cross a left link). The *predecessor* π_t^- is defined analogously: namely, π_t^- is the greatest ancestor that is smaller than t. Note that both π_t^+ and π_t^- are well defined because of the special keys. We store in each node t the maximum number of initial characters that t shares with π_t^+ and π_t^-, that is, their *longest common prefix length* or, shortly, *lcp*. As it will be clear, the *lcp* of two strings permits to compare them in $O(1)$ time by simply comparing their leftmost mismatching symbols, which occur in position *lcp* (we follow the convention of numbering string positions starting from 0). We assume that the keys are chosen from a universe `keytype`, such as a sequence of integers, characters, or reals, terminated by the special null value 0.

   An example of binary search tree and its augmented version with long keys is shown in Figure 1. In the example, the first number in each node is the value of the *lcp* with its predecessor while the second number is the value of the *lcp* with its successor. For example, `seafood` has `seacoast` as predecessor and `surf` as successor: hence, the two *lcp* values are 3 and 1, respectively.

2.2 The Case Study of Binary Search Trees

We begin by taking an available implementation of binary search trees, such as the one described in [36], where each node has the regular fields `key`, `left` and `right`. We may extend the functionalities to strings by using the C library function `strcmp` to perform comparisons of keys. Our first step is instead that of obtaining a preliminary version of the search procedure, called `FindA`, by augmenting each node with two fields `pred_lcp` and `succ_lcp` storing the longest common prefix of the strings pointed by the fields `key` of node t and its predecessor π_t^-, and that of node t and its successor π_t^+, respectively. We do not need to store pointers to π_t^- and π_t^+. Then, we replace `strcmp` with function `fast_scmp`, which is aware of previous comparisons of single characters. This function is at the heart of our computation as it compares efficiently a search key x against the key in a node t. Both keys have the same predecessor and successor with respect to the access path from the root to t (excluded), whereas their *lcps* may be different. Function fast_scmp makes use of three global static int variables:

- m denotes the number of characters matched in the search key x so far;
- left_best is the *lcp* between x and its current predecessor;
- right_best is the *lcp* between x and its current successor.

The initial value of the variables is 0. At the generic comparison step, m characters have been matched, and at least one of left_best and right_best equals m. To compare x and t's key, we first set a local variable lcp to the proper *lcp* between the key in node t and either its predecessor or its successor. Namely, we select one of the fields pred_lcp and succ_lcp in the node, driven by the invariant that at least one of left_best and right_best equals m. If lcp is at least m in value, we may have to extend the m matched characters by comparing one character at a time, starting from the character in position m of both keys. In any case, we end up storing in lcp the *lcp* between x and t's key. At this point, the mismatch between the characters in position lcp of both keys yields the outcome of fast_scmp. We do not comment further the source code of function fast_scmp here as it is the mere implementation of the ideas in [12]:

```
int fast_scmp ( keytype x, node t ) {
    int lcp = ( left_best == m ) ? t->pred_lcp : t->succ_lcp;
    if ( m <= lcp ) {
        for ( ; (x[m] != 0) && (x[m] == t->key[m]); m++ ) ;
        lcp = m;
    }
    if ( x[lcp] <= t->key[lcp] )
        right_best = lcp;
    else
        left_best = lcp;
    return ( x[lcp] - t->key[lcp] ); }
```

As previously mentioned, once *lcp* is known, the comparison among x and t->key is trivially done in the last if statement. Although satisfactory from a theoretical point of view, fast_scmp is not always better than using strcmp; however, it can be reused for other kinds of data structures by introducing minimal changes in their original source codes.

2.3   Single-Shot Version

The resulting code of FindA and fast_scmp in Section 2.2 can be tailored to achieve greater efficiency, writing it in a single and compact form, called FindB and reported below. There, static variables become local variables. At each iteration in the main while loop, we perform the computation of lcp, originally in fast_scmp. Next, we suitably combine the comparison of the mismatching symbols at position lcp and the subsequent node branching to prepare for the next iteration.

```
node FindB( keytype x, node t ) {
    int m = 0,  left_best = 0,  right_best = 0, lcp = 0;
    while ( t ) {
        lcp = (left_best == m) ? t->pred_lcp : t->succ_lcp;
        if ( m <= lcp ) {
            for ( ; (x[m] != 0) && (x[m] == t->key[m]); m++ ) ;
            lcp = m;  }
        if ( x[lcp] == 0 ) {
            return t;
        } else if ( x[lcp] < t->key[lcp] ) {
            t = t->left; right_best = lcp;
        } else {
            t = t->right; left_best = lcp; } }
    return t; }
```

2.4   Code Tuning and Faster *lcp* Computation

By profiling the code in FindB, we discovered that one potential bottleneck of our implementation was the *lcp* computation. In particular, line profiling revealed us that we had to infer the value of lcp computed by line

```
        lcp = ( left_best == m ) ? t->pred_lcp : t->succ_lcp.
```

We remark that the purpose of the above line is to store temporarily in lcp the *lcp* between t's key and its predecessor (resp., successor) when x matches the first m characters of that predecessor (resp., successor). We wish to avoid a direct computation. Hence, let us assume that we have that value of lcp as a consequence of some inductive computation, and that lcp is 0 initially. We unroll the while loop of function FindB and run its iterations as previously done. Then we restate each iteration according to a new scheme driven by two cases formalized in function FindC below, in which x and t are nonempty:

```
    node FindC( keytype x, node t ) {
        int m = 0,  lcp = 0,  nextleft = 0;
        while ( 1 ) {
            if (m <= lcp) {          // CASE 1
                for ( ; *x == t->key[m]; m++ )
                    if ( !*x++ ) return t;
                if ( *x < t->key[m] ) {
                    if ( !(t = t->left) ) return NULL;
                    lcp = t->succ_lcp; nextleft = 0;
                } else {
                    if ( !(t = t->right) ) return NULL;
                    lcp = t->pred_lcp; nextleft = 1; }
            } else {                 // CASE 2
                if ( nextleft ) {
                    if ( !(t = t->left) ) return NULL;
                    lcp = t->pred_lcp;
                } else {
                    if ( !(t = t->right) ) return NULL;
                    lcp = t->succ_lcp; }}}}
```

1. Case m ≤ lcp. [1] We compare the characters in x and t->key as before (the inner for loop). After that, m (and lcp) stores their *lcp*. We branch accordingly with the second if statement met in the iteration. Going left, the key

---

[1] Profiling and experiments show that splitting case 1 into two further cases < and = does not pay off.

in the current node will become the successor of x matching its first m characters. So, lcp will equal t->succ_lcp. Analogously, going right, the current
key will become the predecessor of x matching its first m characters and
lcp will equal t->pred_lcp. This keeps the induction on lcp for the next
while iteration.

2. Case m > lcp. We do not enter the first if condition as lcp already stores the
   lcp between x and t->key. Then, we branch with the second if statement
   as in case 1. An important difference with case 1 is that, here, we have
   the opposite situation to keep the induction on lcp. Indeed, the key in the
   current node is not matching the first m characters of x. Hence, going left, we
   surely know that the predecessor of x matches its first m characters rather
   than the current key (the successor). So, lcp will equal t->pred_lcp. Going
   right, lcp will equal t->succ_lcp.

In order to better optimize the algorithm behind FindC, we avoid a three-way
character comparison in case 2 by introducing a variable nextleft that stores
the negation of the outcome of the comparison in the least recent execution of
case 1. In other words, if we branched left the last time we executed case 1 (so,
nextleft is 0), we have to go right in the following executions of case 2, until
another execution of case 1 comes into play. We have an analogous situation for
the right branch (nextleft is 1). Note that we do not need anymore variables
left_best and right_best.

2.5   Simplification and Re-engineering of the Final Version

Although being optimized, the code of function FindC looks rather cryptic. For
this reason, we completely restructure it by eliminating the need for the local
variables, except m, and by running the main while loop in classical three-way
branching of binary search trees. Namely, we divide the top level search in the
three standard cases for tree searching $[<, =, >]$, rather than cases 1–2 of Section 2.4, and design a new while loop. Even if variable lcp disappears, we refer
to cases 1–2 in equivalent terms. We say that case 1 holds if the key in current
node matches the first m characters of x or more; we say that case 2 holds if the
key matches strictly less than m characters of x.

We now describe the new search function shown in Figure 1. Let us assume that initially case 1 holds with the m-th character of the key. As the new
while loop goes on, we keep the invariant that case 1 holds at the beginning of
each iteration. That is, the key x shares *at least* m initial characters with t itself (and no ancestor can share more). Then, we restate cases 1–2 in terms of
the classical three-way branching, according to the outcome of the comparison
between the characters in position m of x and t->key:

[<] We branch to the left child. We start an inner while loop that branches
    rightward as long as case 2 holds. That is exactly what the more complicated
    procedure FindC does. We exit from the inner while loop when case 1 holds
    again, so that we can start a new iteration of the main while loop.

[>] We handle this case analogously to the previous one, except that we branch to the right child and start an inner `while` loop branching leftward.

[=] We extend the match for equality as done in function `FindC`. At that point, either we find a whole match and return, or a mismatch and case 1 holds.

For example, in the tree of Figure 1, suppose that we want to search for x = seashore. We start from the root seacoast with m = 0 and compare the characters in position 0, which are equal. We therefore are in the third case, where we match m = 3 characters (i.e., the prefix sea). We then run another iteration of the main `while` loop, which leads to the second case. We apply `goright` reaching surf and, since its `pred_lcp` is smaller than m, we apply `goleft` reaching seafood. At this point, we compare the characters in position m and then apply goright reaching seaside. We run another iteration of the main `while` loop matching one further character, so that we have m = 4 (i.e., the prefix seas). Finally, we `goleft` and find a NULL pointer, completing the search with a failure. Searching for x = seaside follows the same path, except for matching all characters, thus completing the search with a success.

2.6   Properties and Extensions

The code can be easily modified to handle *prefix searching*, namely, to check whether x is a prefix of one of the stored keys or to compute the longest matching prefix of x. That operation is useful, for example, in text indexing for performing full text searching and in hierarchical path strings to find a common path of two URLs. Despite its simplicity, the code is very efficient. Since the length of the access path in the tree is upper bounded by the height h of the tree in the worst case, the code reduces the cost of searching key x from $O(k \cdot h)$ to $O(k + h)$ in the worst case. To see why, we observe that each match increases the counter m and each mismatch causes traversing one or more nodes in the access path.

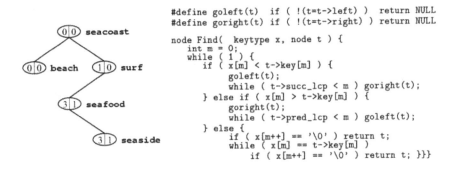

```
#define goleft(t)  if ( !(t=t->left) )  return NULL
#define goright(t) if ( !(t=t->right) ) return NULL

node Find( keytype x, node t ) {
    int m = 0;
    while ( 1 ) {
        if ( x[m] < t->key[m] ) {
            goleft(t);
            while ( t->succ_lcp < m ) goright(t);
        } else if ( x[m] > t->key[m] ) {
            goright(t);
            while ( t->pred_lcp < m ) goleft(t);
        } else {
            if ( x[m++] == '\0' ) return t;
            while ( x[m] == t->key[m] )
                if ( x[m++] == '\0' ) return t; }}}
```

Fig. 1.   An augmented binary search tree for strings and its search function Find

Fact 1 *Applying procedure* Find *for a search key x of length k in a root-to-node access path of length h of a linked data structure requires m successful comparisons of single characters and at most h unsuccessful comparisons of single characters, where $m \leq k$ is the length of the longest matching prefix of x.*

As previously remarked, the worst situation is for skewed strings, in which the value of m can be frequently much larger than the expected value of $O(\log_{|\Sigma|} n)$, typical of independent and identically distributed strings [31, Chap.4]. Since the average height of a tree is $O(\log n)$, we expect that procedure Find has a cost of $O(k + \log n)$ time for n keys. What if we use the standard strcmp routine of the C library instead? While this may work reasonably well for random strings, the theory suggests that our techniques can be competitive for skewed strings, since strcmp does not exploit previously made comparisons of single characters.

We note that performing insertions is not difficult in this contest. When creating a new node s storing the search key x, we must have followed a certain root-to-node access path π. A better look at the code of Find reveals that we can also compute the values of pred_lcp and succ_lcp for s. We need two extra variables for this purpose, initialized to 0. In the last else-branch in Figure 1, after extending a match, we have that m is the length of the longest common prefix between x and the key in the current node. The character in position m can tell if we will reach either the left child or the right child of the current node in π (the next iteration of the main while loop). In the former case, the current node is the best candidate for being the successor of s, and so we record m to be possibly stored in field succ_lcp of s; in the latter case, the current node is the best candidate for being the predecessor of s, and so we record m to be possibly stored in field pred_lcp of s. When s is actually created at the end of π, the best candidates are employed as the predecessor and the successor of s. Having computed the values of fields pred_lcp and succ_lcp, we proceed by inserting s according to what is required by the data structure at hand (e.g., restructuring it). If needed, the *lcp* values can be suitably updated in $O(1)$ time per node as discussed in [12].

3   The Experimental Setup

In this section we describe the experiments performed. We first briefly sketch the techniques on which we experimented. Next we describe the data sets used, and finally describe the experimental results. We ran our codes on several computing platforms, but we will present only the results relative to a processor AMD Athlon (1Ghz clock and 512Mb RAM) running Linux 2.4.0, since similar results have been obtained on the other platforms.

3.1   The Data Structures

We developed the ANSI C code for binary search trees [17], treaps [27], AVL trees [2], (a, b)-trees [16], and skiplists [24]. In particular, for binary search trees,

Table 1. The number of nodes and the height of the data structures considered in the experiments for a set S of n strings. The bounds are also in terms of m, the average length of matched prefixes in the formula for the skewness, $(m+1) = \mathscr{L}(S) \log_{|\Sigma|} n$ (see Section 1). The starred bounds hold on the average

	cat	tst	pat	bst	avl
#nodes	$O\big(n(m+1)\big)^*$	$O\big(n(m+1)\big)^*$	$O(n)$	$O(n)$	$O(n)$
height	$O(m+1)^*$	$O\big(\log n + (m+1)\big)^*$	$O(m+1)^*$	$O(\log n)^*$	$O(\log n)$

treaps, and AVL trees we started from the implementation described in [36], for skiplists we started from the implementation described in [24], which is available via anonymous ftp at [25], while for (a,b)-trees we started from our own implementation of [16, 20]. We used function fast_scmp in the same way as described in Section 2.2, leaving the tuning to a later stage. We extensively tested these five implementations on several data sets and computing platforms. We omit here the results of these preliminary tests and we only mention that the computational overhead is very small, validating the prediction of the theory as the key length k goes to infinity. At this stage, however, the resulting data structures were not always competitive with the original data structures using the standard strcmp, especially for small/intermediate key lengths. The purpose of these experiments was identifying the most qualified implementations to be tuned as described in Sections 2.3–2.5 and to be tested against already existing data structures. We singled out the binary search trees and the AVL trees, obtaining their tuning code, denoted bst and avl, respectively.

The string data structures on which we compared are the the adaptive tries of Acharya, Zhu, and Shen [1], the ternary search trees of Bentley and Sedgewick [5], the Patricia tries of Morrison [21], and the linear probing hashing described by Knuth [17]. To establish a connection between the measure of skewness $\mathscr{L}(S)$ given in Section 1 and the performance of the above data structures for a set S of n strings, Table 1 reports their number of nodes and their height in terms of m, the average length of matched prefixes in the formula for the skewness, $(m+1) = \mathscr{L}(S) \log_{|\Sigma|} n$. Note that $m \leq k$, the maximum string length.

The cache-aware version of tries (cat) uses multiple alternative data structures (partitioned arrays, B-trees and hashing arrays) for representing different trie nodes, according to their fanout and to the underlying cache. The height is $O(m+1)$ on the average and $O(k)$ in the worst case, with $O(n(m+1))$ nodes on the average and $O(nk)$ nodes in the worst case. We slightly modified their highly tuned and optimized C++ code at [1] to use our timing routines.

The ternary search trees (tst) are a blend of tries and binary search trees. A ternary search tree has $O(n(m+1))$ nodes on the average and $O(nk)$ nodes in the worst case. Its height is $O(m + \log n)$ on the average [9] and contributes to the cost of searching, $O(k + \log n)$ time. We downloaded their C code at [5] and slightly modified it with our timing routines.

The Patricia tries (**pat**) are compacted tries storing the strings in their leaves in a total of $O(n)$ nodes, and their height is $O(k)$ in the worst case and $O(m + 1)$ on the average. Searching in Patricia tries takes $O(k + \log n)$ time on the average [9, 17] by comparing just one character per node, at the price of a full comparison of the key in the node where the search ends. We employed the implementation described in [26] and removed the recursion.

Finally, we implemented linear probing hashing (**hsh**) by using the hash function in [18] as suggested in [29], with load factor 0.5 yielding $O(n)$ occupied space. Although hashing does not permit to perform prefix searching, we employed its average search time to *normalize* the timings of the other data structures, believing that it may reveal their qualities better than absolute timings.

3.2   The Data Sets

1. A word data set. It consists of two dictionaries **dictwords1** and **dictwords2**. The former is available at [5]: it contains 25,481 words of average length of 8.23 characters. The latter is the dictionary available on Unix machines in file **/usr/dict/words**, containing 45,407 words of average length 9.01.
2. A book data set. It consists of **mobydick** (i.e., Melville's "Moby Dick" available from [23]) with 212,891 words of average length 6.59 and of **warandpeace** (i.e., Tolstoj's "War and Peace" available from the same URL) with 565,551 words of average length 5.64.
3. A library call number data set. Used in one of the DIMACS Implementation Challenges, it is available from [28]. Each entry is a library card number (such as WGER_____2455___55___20). From this data set, we picked **circ2.10000**, which consists of 9,983 strings of average length 22.52, and **circ2.100000**, which consists of 100,003 strings of average length 22.54. We also generated **libcall** from other files, consisting of 50,000 strings of average length 22.71.
4. A source code data set. Here, keys are the code lines. We choose **gprof**, the source code of a profiler consisting of 7,832 code lines of average length 25.87, **kernel**, the source code of a Linux Kernel consisting of 48,731 code lines of average length 29.11, and **ld**, the source code of a linker containing 19,808 lines of average length 27.77.
5. A URL data set. Each key is a uniform resource locator in the Web, where we dropped the initial **http://** prefix. Files range from 25,000 to 1,500,000 URLS. The URLs were collected in the Web domain ***.*.it** by a spider for the search engine in [3] and by the proxy server at the University of Pisa [7].

The dictionary and book data sets in points 1 and 2 tend to have the smaller values of skewness $\mathcal{L}(S)$: they consist of many words having small average lengths. On the other extreme of the spectrum we have the source code and the URL data sets in points 4–5, which are characterized by large values of $\mathcal{L}(S)$. In between, we find the library call number data sets in point 3.

Table 2. Running times of the insert operation for the data sets S in order of skewness $\mathscr{L}(S)$. The running times are normalized with the corresponding ones of `hsh`, whose value is hence 1 in each entry (not reported)

$\mathscr{L}(S)$	cat	pat	tst	bst	avl
1.98	1.00	3.33	3.01	2.33	3.00
2.53	0.88	2.00	1.34	1.55	2.05
4.91	1.93	3.00	2.57	2.53	3.61
5.60	2.58	1.7	2.66	1.25	1.67
10.95	2.56	4.62	4.10	3.49	4.42
11.75	2.69	3.66	3.60	2.72	3.56
12.94	3.26	3.06	3.41	2.25	3.03
14.70	3.86	2.66	3.26	1.93	2.60
16.64	5.48	2.57	3.28	1.71	2.14
19.14	4.00	3.15	3.73	2.31	3.03
22.26	4.80	2.66	3.56	1.93	2.43
26.07	5.14	2.42	3.47	1.67	2.04

3.3   The Results

We ran several experiments on our data sets. One experiment was using two different data sets. The first data set (e.g., a dictionary, such as `dictcall1` or `dictcall2`) was inserted into the data structure at hand, one word at the time. Against this data set, we ran searches using all the words of the second data set (e.g., a book such as `mobydick` and `warandpeace`). The experiment thus consisted of carrying out a batch of insertions followed by a batch of lookups. We performed several experiments of this type, according to different data sets, and to whether the data sets were scrambled randomly or not. As a special case of this experiment, we also used the same data set twice: i.e., we first inserted the words of a data set (randomly scrambled or not) one at the time in the data structure at hand; next, we scrambled the same data set, and searched the data structure for each word in this random order. The goal of the data scrambling was to avoid sorted data, and to force the batches of insertions and lookups to follow completely different patterns in the data structure. We considered the searches both in the successful case (the key is stored in the trees) and in the unsuccessful case (the key is not stored). We ran each experiment several times and retained the average of their running times. Times were measured in milliseconds with the Unix command `getrusage()`.

The running times of the insert operations, the successful searches and the unsuccessful searches are reported in Tables 2–4, respectively. All these experiments yielded very similar results. Indeed for different experiments, there seemed to be very little difference in the relative behavior of our algorithms (even though, clearly, the absolute CPU times were substantially different). What we learned from our experiments is that `cat` appeared to be fast on large data sets with small keys (i.e., sets S with small skewness $\mathscr{L}(S)$). On the opposite side, `bst`

Table 3. Running times of the successful search for the data sets S in order of of skewness $\mathcal{L}(S)$. The running times are normalized with the corresponding ones of hsh

$\mathcal{L}(S)$	cat	pat	tst	bst	avl
1.98	0.30	3.33	3.04	2.33	2.67
2.53	0.55	2.11	1.27	1.47	1.94
4.91	0.96	2.85	2.44	2.62	3.14
5.60	0.92	1.33	1.93	1.00	1.40
10.95	1.15	4.16	3.63	3.19	3.58
11.75	1.21	3.29	3.18	2.52	2.87
12.94	1.49	2.62	2.91	2.05	2.43
14.70	1.77	2.11	2.61	1.66	2.11
16.64	2.10	1.55	2.34	1.33	1.55
19.14	1.65	2.46	2.92	1.92	2.36
22.26	1.77	1.89	2.64	1.53	1.97
26.07	2.18	1.70	2.55	1.33	1.74

and avl produced data structures which were effective on data sets with skewed keys (i.e., with large $\mathcal{L}(S)$). In between, the performance of these data structures were not so much different, depending on the underlying architecture: however, in all our experiments tst started being competitive with the other two approaches exactly in this range. Note that pat was sometimes less competitive because of the double traversal needed by the search and update operations.

An explanation of the experimental behavior of these data structures and their relation to the skewness $\mathcal{L}(S)$ can be found in Table 1, in which their number of nodes models the space occupancy and their height models the time performance. First, notice that we can classify the data structures in three classes: tries (cat,tst), compacted tries (pat), and lexicographic trees (tst,bst,avl). The members of the former two classes have depth proportional to the average length of matched prefixes, m, while the members of the latter class have a depth proportional to the logarithm of the number of strings, n, with tst being a hybrid member of two classes. Second, their performance is influenced by the skewness $\mathcal{L}(S)$, as larger values of $\mathcal{L}(S)$ cause larger values of m for fixed n.

With the two observations above, we now discuss the experimental behavior of data structures. When $\mathcal{L}(S)$ is small, the (compacted) tries have depth smaller than $O(\log n)$ and so their performance is superior, with the exception of pat, which requires a double traversal. Also the number of nodes is nearly $O(n)$. When $\mathcal{L}(S)$ is large, the lexicographic trees have depth smaller than $O(m)$, and they compare better. They always require $O(n)$ nodes, while the others do not guarantee this upper bound for skewed strings or have worse time performance. For intermediate values of $\mathcal{L}(S)$, tries and lexicographic trees are of comparable heights, and the data structures behave reasonably well. It turns out that cat is very fast for small skewness; tst is more competitive for middle skewness; avl and bst are the choice for large skewness; pat is sometimes less efficient for

Table 4. Running times of the unsuccessful search (not reported for cat) for the data sets S in order of of skewness $\mathcal{L}(S)$. The running times are normalized with the corresponding ones of hsh

$\mathcal{L}(S)$	pat	tst	bst	avl
1.98	3.00	3.02	3.00	4.02
2.53	3.02	2.50	3.01	3.01
4.91	2.25	1.75	2.50	3.00
5.60	1.33	1.66	1.00	2.01
10.95	3.88	3.17	3.30	3.80
11.75	2.79	2.45	2.39	2.87
12.94	2.43	2.43	2.12	2.68
14.70	1.77	1.88	1.33	1.88
16.64	2.00	2.33	2.00	2.66
19.14	2.43	2.43	2.10	2.66
22.26	2.06	2.25	1.62	2.18
26.07	1.63	1.91	1.36	1.90

searching and inserting; hsh is the fastest in all experimented cases, at the price of supporting less powerful operations. In general, cat takes significant more space than the other data structures when the strings are skewed.

In summary, for $m < \alpha_1 \log_{|\Sigma|} n$, for some constant α_1, we expect cat to be the fastest, and for $m > \alpha_2 \log_{|\Sigma|} n$, for some other constant $\alpha_2 > \alpha_1$, we expect avl and bst to take over, while tst are good in the middle range. These constants are highly machine-dependent, as expected. In our case, the search running times were from 1 to 4 times slower than hashing while the insert running times were slower from 1 to 5 times. The maximum speedup, computed as the maximum ratio between the running times of the slowest and the fastest algorithms for each data set (except for the first two in Table 3), is around 2 or 3. Although it may not appear as impressive as in other problems, we point out that the involved data structures are highly tuned. When employed in a high-performance server for indexing, sorting and compressing textual data, doubling the actual amount of served requests per time unit can make a difference in practice.

Some final remarks are in order. Search time in tst is rarely worst-case compared to that of bst, and tst has typically smaller height than that of bst. The extra work for keeping avl balanced apparently does not pay. As for the required space, cat and tst take more nodes than the others, which require to store explicitly the strings somewhere else. In some cases, tst may require globally less space (including that required by the strings) because it can store the strings in its structure. Being so simple, tst are the method of choice for tries [9] while our experiments indicate that bst are effective choices when compacted tries are needed, especially with skewed strings.

4   Conclusions

We described a general data structuring technique that, given a linked data structure for storing atomic keys undergoing comparisons, transforms it into a data structure for strings. Despite the simplicity and generality of our method, our experimental study shows that it is quite competitive with some implementations currently available in the literature. We can extend our technique to obtain text indexing data structures that are also able to perform suffix sorting. Hence, our augmented data structures provide a good alternative to popular text indexes such as suffix arrays and suffix trees [15]. They can also perform the suffix sorting of a text, which is the basis of text indexing and data compression.

Acknowledgements

We thank Salvatore Ruggieri for providing some of the data sets in [7] and Craig Silverstein for sending us a copy of [29].

References

[1] A. Acharya, H. Zhu, K. Shen. Adaptive algorithms for cache-efficient trie search. *Proc. ALENEX 99*. Source code in http://www.cs.rochester.edu/~kshen, 1999. 83, 90

[2] G. M. Adel'son-Vel'skii and E. M. Landis. An algorithm for the organization of information. *Soviet Mathematics Doklady*, 3 (1962), 1259–1263. 82, 84, 89

[3] Arianna Search Engine, http://arianna.libero.it. 91

[4] R. Bayer. Symmetric binary B-trees: Data structure and maintenance algorithms. *Acta Informatica* 1 (1972), 290–306. 82

[5] J. L. Bentley and R. Sedgewick. Fast algorithms for sorting and searching strings. *SODA 1997*, 360–369 (http://www.cs.princeton.edu/~rs/strings). 82, 83, 90, 91

[6] S. W. Bent, D. D. Sleator and R. E. Tarjan. Biased search trees. *SIAM Journal on Computing* 14 (1985), 545–568. 82

[7] F. Bonchi, F. Giannotti, C. Gozzi, G. Manco, M. Nanni, D. Pedreschi, C. Renso and S. Ruggieri, Web log data warehousing and mining for intelligent web caching. *Data and Knowledge Engineering*, to appear. 91, 95

[8] H. A. Clampett. Randomized binary searching with the tree structures. *Communications of the ACM* 7 (1964), 163–165. 82

[9] J. Clément, Ph. Flajolet and B. Vallée. The analysis of hybrid trie structures. *SODA 1999*. 90, 91, 94

[10] J. Feigenbaum and R. E. Tarjan. Two new kinds of biased search trees. *Bell Systems Technical Journal* 62 (1983), 3139–3158. 82

[11] T. F.Gonzalez. The on-line *d*-dimensional dictionary problem. *SODA1992*, 376–385. 82

[12] R. Grossi, G. F. Italiano. Efficient techniques for maintaining multidimensional keys in linked data structures. *ICALP 1999*, 372–381. 82, 83, 84, 85, 89

[13] R. H. Gueting and H.-P. Kriegel. Multidimensional B-tree: An efficient dynamic file structure for exact match queries. *10th GI Annual Conference*, 375–388. 82

[14] L. J. Guibas and R. Sedgewick. A dichromatic framework for balanced trees. *FOCS 1978*, 8–21. 82

[15] D. Gusfield, *Algorithms on Strings, Trees and Sequences: Computer Science and Computational Biology*, Cambridge University Press, 1997. 95

[16] S. Huddleston and K. Mehlhorn. A new data structure for representing sorted lists. *Acta Informatica* 17 (1982), 157–184. 82, 84, 89, 90

[17] D. E. Knuth. *The Art of computer programming, Vol. 3 : Sorting and Searching.* Second edition, Addison-Wesley, 1973, 1998. 82, 83, 89, 90, 91

[18] R. Jenkins, `http://burtleburtle.net/bob/hash/`. 91

[19] K. Mehlhorn. Dynamic binary search. *SIAM J. on Computing* 8 (1979), 175–198. 82

[20] K. Mehlhorn. *Data structures and algorithms:Searching and sorting*, Springer 1984. 84, 90

[21] D. R. Morrison. PATRICIA — Practical Algorithm To Retrieve Information Coded In Alphanumeric. *J. ACM*, 15, 514–534,1968. 83, 90

[22] J. Nievergelt and E. M. Reingold. Binary search trees of bounded balance. *SIAM Journal on Computing* 2 (1973), 33–43. 82

[23] J. M. Ockerbloom. The on-line books page. `http://digital.library.upenn.edu/books/` 91

[24] W. Pugh. Skip Lists: A probabilistic alternative to balanced trees. *Communications of the ACM* 33 (1990), 668–676. 84, 89, 90

[25] W. Pugh. Skip Lists ftp directory. `ftp://ftp.cs.umd.edu/pub/skipLists/`. 84, 90

[26] R. Sedgewick. *Algorithms in C*, Addison-Wesley, 1998. 91

[27] R. Seidel and C. R. Aragon. Randomized search trees. *Algorithmica* 16 (1996), 464–497. 82, 89

[28] C. Silverstein. Library call data set. `http://theory.stanford.edu/~csilvers/libdata`. 91

[29] C. Silverstein. A practical perfect hash algorithm. Manuscript, 1998. 91, 95

[30] D. D. Sleator and R. E. Tarjan. Self-adjusting binary search trees. *Journal of the ACM* 32 (1985), 652–686. 82

[31] W. Szpankowski, *Average Case Analysis of Algorithms on Sequences*, Wiley, 2001. 82, 89

[32] R. E. Tarjan. *Data structures and network algorithms*, SIAM (1983). 82

[33] V. K. Vaishnavi. Multidimensional height-balanced trees. *IEEE Transactions on Computers* C-33 (1984), 334–343. 82

[34] V. K. Vaishnavi. Multidimensional balanced binary trees. *IEEE Transactions on Computers* C-38 (1989), 968–985. 82

[35] V. K. Vaishnavi. On *k*-dimensional balanced binary trees. *Journal of Computer and System Sciences* 52 (1996), 328–348. 82

[36] Mark A. Weiss. *Data structures and algorithm analysis in C*, Addison Wesley, (1997). Source code in `http://www.cs.fiu.edu/~weiss/dsaa_c2e/files.html`. 84, 90

Evaluation of Basic Protocols
for Optical Smart Dust Networks[*]

Josep Díaz, Jordi Petit, and Maria Serna

Departament de Llenguatges i Sistemes Informàtics
Universitat Politècnica de Catalunya
Jordi Girona Salgado 1–3, 08034 Barcelona
{diaz,jpetit,mjserna}@lsi.upc.es

Abstract. In this paper we analyze empirically the performance of several protocols for a random model of sensor networks that communicate through optical links. We provide experimental evidence of the basic parameters and of the performance of the distributed protocols described and analyzed under assymptotic assumptions in [3] for relatively small random networks (1000 to 15000 sensors).

1 Introduction

The fast development of microelectronics and communications has opened a fast field in the development of low-cost, little power consumption and small size, multi-functional sensors. Large number of these sensors can be spread to form *networks of sensors* [1]. Sensor networks are supposed to be deployed in hostile environments, in order to permit monitoring and tracking of remote objects, detecting anomalous situations as fires in the woods, seismic activity, and others. Communication among the sensors is usually done by radio frequency or, in more recent projects, by optical transmission via small lasers (see i.e. [4, 5, 8]). Free-space optical links have the limitation of an uninterrupted line-of-sight path for communication, but avoid radio interference. In particular, the Smart Dust project at Berkeley [6, 8] has proposed the use of optical communication between sensors as an alternative to radio frequency. Both in the case of radio frequency and of optical communication, the energy consumption is a key factor in the evaluation of these networks.

Recent efforts have been made to formalize and to give more efficient algorithms for networks of sensors, but these efforts have mainly concentrated on models using radio frequency communication [2, 7]. In [3], the authors proposed a probabilistic model to analyze smart dust networks communicating through optical devices. Although simple, our model seems to incorporate the basic technological specifications of smart dust systems [5]. We also considered and analyzed there some basic communication protocols in the proposed model, taking into account power management as a main issue in their design. Such protocols

---

[*] Work partially supported by the IST Programme of the EU under contract number IST-2001-33116 (FLAGS) and by the Spanish CICYT project TIC-2001-4917-E.

K. Jansen et al. (Eds.): WEA 2003, LNCS 2647, pp. 97–106, 2003.

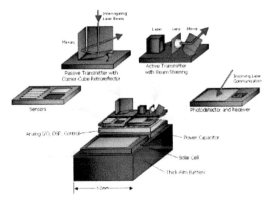

Fig. 1. Elements in the motes of the Smart Dust project (reproduced from [5])

represent a research challenge, due to the optical nature of these networks and because of the presence of faulty connections among sensors. However, the analytical results obtained in [3] were mostly of asymptotic nature. In the present work, we experimentally study whether these asymptotic results hold for smaller networks.

Recall that the setting of smart dust systems is to have a *base-station transceiver* (BTS) at a relative elevation (or in a small plane) monitoring periodically the information of a large amount of sensors (*motes*) that have been scattered on a terrain. Motes include sensor devices, a small laser cannon and a set of optical devices able to modulate and reflect the light they receive. Fig. 1 reproduces the design of such a mote.

The scenario considered in [3] is as follows. The motes are scattered massively at random from a vehicle. As a consequence some of them may break or fall in such a way that can not communicate with the BTS. Then, the communication will be initiated from the BTS: the BTS scans an area with its laser, and each mote passively modulates and reflects the beam. This is the preferred way to communicate, as the mote uses little power. If a mote is shadowed from the BTS, this mote must rely its information to the BTS through other motes (any mote can detect a failure of communication with the BTS by noticing that sufficient time has passed without communication). Communication between motes is done in an active way using their laser beams. This kind of transmission uses more power. To send information, they use an orientable low power laser beam, which current technology allows to move sidewards and upward about forty degrees. To receive information, motes have an optical device able to detect and interpret laser signals, as well as to evaluate the direction of the incoming beam.

The network is established in two phases. The *localization* phase consists in finding out the position of each mote; that is, at the end of this phase it is required that all operative motes know their approximate coordinates (GPS systems cannot be currently used because of their volume, price and power consumption). The second phase consists in establishing a route from the motes that cannot communicate directly with the BTS through the motes than can. Such phase is called *route establishment* and is attained by repeated operations of broadcasting and gathering protocols. Once these two main steps are performed, the network is able to enter in its exploitation phase, which will last until the power supply of the motes decays.

In this paper, we give empirical evidence that the system and protocols described and analyzed in [3] perform correctly, with only a small increase in the predicted constants, when dealing with networks of a few thousand sensors.

2   The Model

In this section, we recall the random model for smart dust systems that was proposed in [3].

According to [3], we deal with an ideal two dimensional squared region in which a mote falls with coordinates following the uniform distribution. Any mote can orient its laser cannon in any position of its scanning area, which covers a sector of a fixed angle of α radians. The sector is also oriented randomly with an angle b relative to some horizon, but the mote can receive light from any point, within distance r, which is "looking" to it. We assume that the terrain to be covered is an square area of size $D \times D$. An appropriate scaling can map it to the unit square $[0,1]^2$.

These considerations give raise to a directed graph model. We define a random digraph, where the vertices are the motes, and given vertices i and j, there is an arc (i,j) if j lies in the sector with center i defined by r, starting at angle α and ending at angle $\alpha + b_i$ (modulo 2π). More specifically, let $X = (X_i)_{i\geq 1}$ be a sequence of *independently and uniformly distributed* (i.u.d.) random points in $[0,1]^2$, let $B = (b_i)_{i\geq 1}$ be a sequence of i.u.d angles and let $(r_i)_{i\geq 1}$ be a sequence of numbers in $[0,1]$. For any natural n, we write $\chi_n = \{X_1, \ldots, X_n\}$ and $\beta_n = \{b_1, \ldots, b_n\}$. We call $\mathcal{G}_\alpha(\chi_n, \beta_n, r_n)$ the *random scaled sector graph* with n nodes.

The objective of the network is monitoring the terrain. To do so, we assume that the terrain is dissected in a grid of s cells, each of size $s^{-1/2} \times s^{-1/2}$. This grid serves as reference position for the motes, and the data reported from a mote refers to the cell in the grid that contains it. The parameter s is selected to represent the *sensing precision* of the network and controls its scalability.

In the process of scattering the motes, some of them may break down or fall upside down and become inoperative, some may be hidden from the BTS and for some pair of motes the line of sight may be blocked. We assume that motes fall uniformly at random in the terrain and that each mote can fail, independently at random, with probability $1 - p_o$. Operative motes communicate with the BTS with probability p_b. Also, we assume that p_c is the probability that the line of sight allowing communication from one operative mote to another operative mote is not interrupted by any obstacle.

We use the following parameters to describe a *smart dust network*: n is the number of motes; r is the laser range of the motes; α is the laser scanning angle; s is the number of cells in the grid; p_o is the probability that a mote is operative; p_b is the probability that an operative mote can communicate with the BTS (these are called *communicating motes*); p_c is the probability that the line of sight between two motes is not interrupted. We consider here that p_o, p_c and p_b are fixed probabilities.

As defined in [3], the *random smart dust network* $\mathcal{SD}_n(X, B, \alpha, r, s, p_o, p_b, p_c)$ is a network of n motes $X_1, \ldots, X_n$ where X_i is operative with probability p_o, X_i can communicate with the BTS with probability p_b, and X_i and X_j can communicate with probability p_c if X_j is contained in the sector of radius r and angle α starting at β_i. In the remaining of the paper, such system will be simply denoted as $\mathcal{SD}_n$.

We will use the term ϵ-*normalized random smart dust system* to refer to a random smart dust system $\mathcal{SD}_n$, with $n = (1 + \epsilon)(s \ln s)/p_o$ motes and such that $g = r/\sqrt{s}$ is a constant and $\epsilon \geq \epsilon_0$. Here, ϵ_0 is a constant that depends on g. This selection of n is made according to [3], where it is stated that, with high probability, a random smart dust system with n motes can effectively monitor s cells. Given the laser range r and the number of cells s, the value $g = r/\sqrt{s}$ is another relevant parameter that will appear in the analysis of random smart dust systems. In our experiments we take $g = 10$.

Most of the results relate to *interior motes*, defined as those motes whose distance to the boundaries of $[0, 1]^2$ is greater than r.

3   Connectivity

It is known that, with high probability, the interior motes of normalized smart dust systems are strongly connected:

Theorem 1 ([3]). *Let $\mathcal{SD}_n$ be a ϵ-normalized smart dust system. Then, w.h.p., for all pairs (x, y) of interior operative motes, there is a directed path from x to y.*

Recall that a sequence of events $(E_n)_{n \geq 0}$ occurs *with high probability (w.h.p.)*, if $\lim_{n \to \infty} \mathbf{Pr}[E_n] = 1$. As a consequence, the result in the above theorem is just asymptotic and does not provide any clue on how fast the probability converges to 1. Our first set of experiments provides an analysis of the network connectivity. The goal is to simulate smart dust networks in order to know whether the predicted asymptotic behavior holds in small networks for different values of its parameters.

To do so, we have generated different random networks with $\alpha = 40$ degrees and have measured the number of interior motes that belong to the largest strongly connected component for different values of n, p_o and p_c. The number of motes (n) has been selected in the range 1000 to 15000 with increments of 1000; we believe that this is a reasonable number of motes. The probability of communication between motes (p_c) has been selected in the range 0 to 1, with increments of 0.1, in order to cover a wide spectrum of possibilities. In this and all the following experiments, we have taken $r_n = \log n/\sqrt{n}$ for the radius; the rationale behind this selection is that the results in [3] hold for $r_n > c_0 \sqrt{(\log n)/n}$

for c_0 large enough, and our selection is just slightly greater than any c_0. In all the experiments, we have also restricted the values for the probability of being operative (p_o): We have selected three different values: 0.15, 0.5 and 0.85; these represent low, medium and high probabilities.

The obtained results are presented in Table 1, which shows the percentage of interior motes that belong to the largest strongly connected component related to the total number of interior motes. The experiments have been repeated 200 times, without noticing large deviations from the averages, which are shown in the tables.

As it could be expected, the results show that as n, p_c or p_o grow, the ratio of number of motes in the largest strongly connected component grows. More importantly, the results show that for certain values of the parameters, almost 100% of the pairs of interior motes are connected. This shows that the asymptotic predicted behavior in Theorem 1 can be observed even for relatively small values of the n, p_c and p_o parameters.

4 The Localization Algorithm

Once deployed, the first task for a smart dust network is to find the position of each operative interior mote. Motes that can communicate with the BTS will receive its coordinates from it (it suffices to include the emission angle in the BTS message [5]). The remaining motes will have to compute their coordinates based on the coordinates of other motes and the angle of incidence of the incoming laser beams. In this section, we simulate the localization algorithm localize proposed in [3] to solve the localization stage for optical smart dust systems. We refer to [3] for a complete description of this protocol. The expected performance and runtime of the localize protocol is given by the following result:

Theorem 2 ([3]). *Let $\mathcal{SD}_n$ be an ϵ-normalized random smart dust system. Then, as s grows, w.h.p., after a constant number of phases of the localize protocol, all operative interior motes know their position.*

Again, the result is asymptotic. Moreover, the required number of steps is unknown. Therefore, our goal in this section is two-fold: we want to know whether the localization phase can be performed with the localize algorithm when the number of motes is in the order of a few thousands, and we want to know how many phases are needed.

In this case, the probability p_b of reaching the BTS matters. Therefore, the parameters for these simulations involve n, p_b, p_o and p_c. Again, we take $p_b \in \{0.15, 0.50, 0.85\}$. Tables 2, 3 and 4 show the empirical results obtained using the localize protocol. From these values, it can be observed that the localization phase will succeed when p_c, p_b or p_o are high, but that for a low number of motes and a low values of p_c, p_b or p_o the algorithms fails to assign coordinates to each interior operative mote. In theses cases, though, more than 90% of the interior motes could compute their position. With regards to the number of phases needed to complete the algorithm, our results indicate that, in the average, 6 phases are enough and that never more than 20 phases have been required.

5   Broadcasting from the BTS to the Motes

Another of the issues to be considered is the broadcasting of messages originated in the BTS. We consider the simple-bro protocol given in [3], which is a classical flooding algorithm with multiple source points. The theoretical analysis of the broadcasting protocol in the random smart dust network is the following:

Theorem 3 ([3]). *Let $\mathcal{SD}_n$ be an ϵ-normalized random smart dust system. Then, as s grows, w.h.p., after performing $4\lceil \sqrt{2p_o/(p_b(1+\epsilon)g^2)} \rceil$ phases of the simple-bro protocol, a message originated in the BTS will be broadcasted to all interior motes.*

Using the same parameters as in the experiments done in the previous section, Tables 5, 6 and 7 show the percentage of interior motes which receive the message transmitted by the BTS using the simple-bro protocol, within the number of phases indicated in Theorem 3. Again, for reasonable values of p_o and p_b, we can observe that as n and p_c grow, the percentage of interior motes receiving the message originated in the BTS in the prescribed number of phases soon goes over 95%.

6   Route Establishment from the Motes to the BTS

Before exploiting the network of sensors, a smart dust system must establish a routing so that any operative interior mote can send its information to the BTS. Motes that can communicate directly with the BTS will simply transmit passively their information when the BTS queries them. The remaining motes will have to send their information actively by the way of multiple hops. Since the communication between motes is not bidirectional, we can not simply reverse the paths found by the previous algorithm.

To establish a route from the mote to the BTS, we use the protocol simple-link described in [3]. That protocol computes a set of the routes from the motes to the BTS form an oriented forest with roots in the communicating motes.

Theorem 4 ([3]). *Let $\mathcal{SD}_n$ be an ϵ-normalized random smart dust system. Then, as s grows, w.h.p., after performing $4/\lceil \sqrt{(p_o p_b g)} \rceil$ phases of the simple-link protocol, all interior motes have selected a neighbour to whom send the information to be forwarded to thr BTS.*

The important part in the analysis of the previous algorithm done in [3] was the fact that, asymptotically, almost all pairs of motes at Euclidean distance less than r are at distance at most 4 in the network. Therefore, in this case, our experiments focus on this property rather than in the whole simulation of the simple-link protocol. Using the same parameters as in the previous experiments, Tables 8 and 9 present the measured results. (Unfortunately, we could not compute shortest paths in networks with more than 5000 motes when $p_o = 0.85$.) In any case, these show that for the smaller number of nodes that we are dealing with the number of phases of broadcasting used in each simple-link phase must be increased to 6.

7   Conclusions

In this paper, we have investigated whether the asymptotic results we draw in [3] hold for normalized smart dust systems with thousands of motes.

Our experiments have focussed on the simulation of the basic protocols presented in [3] and span several basic tasks to exploit networks of sensors. The parameters we have used to study our protocols are the size of the network (n) and the different failure probabilities (p_o, p_c and p_b). We have reported results for $\alpha = 40^o$, but simulations with $\alpha = 20^o$ present a similar behavior.

The results we have reported show that the predicted asymptotic behavior can be detected on small networks by setting the probabilities to reasonably high values, at the expense of a slight increase of the constants. Observe that this increase of the constants only delays the length of the broadcasting, but does not result in any additional increment of energy consumption that remains as one full scan per mote and per phase of the simple-link protocol.

References

[1] I. Akyildiz, W. Su, Y. Sankarasubramaniam, and E. Cayirci. Wireless sensor networks: a survey. *Computer Networks*, 38:393–422, 2002. 97
[2] I. Chatzigiannakis, S. Nikoletseas, and P. Spirakis. Smart dust local detection and propagation protocols. In *ACM Workshop on Principles of Mobile Computing*, Toulouse, 2002. 97
[3] J. Díaz, J. Petit, and M. Serna. A random graph model for optical networks of sensors. Technical report, LSI-02-72-R, Universitat Politecnica Catalunya, 2002. 97, 98, 99, 100, 101, 102, 103
[4] D. Estrin, R. Govindan, J. Heidemann, and S. Kumar. Next century challenges: Scalable coordination in sensor networks. In *ACM/IEEE Int. Conf. on Mobile Computing and Networking*, pages 263–270, Seattle, 1999. 97
[5] J. Kahn, Katz, R. H., and K. Pister. Mobile networking for smart dust. In *ACM/IEEE Int. Conf. on Mobile Computing and Networking*, pages 176–189, Seattle, 1999. 97, 98, 101
[6] Smart dust project. http://www.robotics.eecs.berkeley.edu/~pister/SmartDust. 97
[7] S. N. Simic and S. Sastry. Distributed localization in wireless ad hoc networks. Technical report, Dept. Electrical Engineering and Computer Science, University of California at Berkeley, 2001. 97
[8] B. Warneke, B. Liebowitz, K. S. J. Pister, and E. Cayirci. Smart dust: communicating with a cubic-millimiter computer. *IEEE Computer*, January:2–9, 2001. 97

Table 1. Percentage of interior motes that belong to the largest strongly connected component

$p_o = 0.15, \alpha = 40°$

n					p_c					
	0.1	0.2	0.3	0.4	0.5	0.6	0.7	0.8	0.9	1.0
1000	2	2	3	4	9	17	23	29	43	52
3000	0	1	2	7	20	40	61	72	81	88
5000	0	1	3	9	31	59	70	87	89	92
7000	0	0	2	12	45	71	83	90	94	95
9000	0	0	3	27	64	77	85	91	94	97
11000	0	0	4	32	68	82	88	92	95	97
13000	0	0	5	32	70	83	90	94	96	98
15000	0	0	7	43	72	84	91	94	97	98

$p_o = 0.5, \alpha = 40°$

n					p_c					
	0.1	0.2	0.3	0.4	0.5	0.6	0.7	0.8	0.9	1.0
1000	1	27	77	90	96	98	99	99	99	99
3000	2	69	91	97	99	99	99	100	100	100
5000	5	78	95	98	99	99	99	100	100	100
7000	12	83	96	99	99	99	100	100	100	100
9000	20	86	97	99	99	100	100	100	100	100
11000	29	87	97	99	99	100	100	100	100	100
13000	35	89	98	99	99	100	100	100	100	100
15000	37	90	98	99	99	100	100	100	100	100

$p_o = 0.15, \alpha = 40°$

n					p_c					
	0.1	0.2	0.3	0.4	0.5	0.6	0.7	0.8	0.9	1.0
1000	1	27	77	90	96	98	99	99	99	99
3000	2	69	91	97	99	99	99	100	100	100
5000	5	78	95	98	99	99	99	100	100	100
7000	12	83	96	99	99	99	100	100	100	100
9000	20	86	97	99	99	100	100	100	100	100
11000	29	87	97	99	99	100	100	100	100	100
13000	35	89	98	99	99	100	100	100	100	100
15000	37	90	98	99	99	100	100	100	100	100

Table 2. Percentage of interior motes that get their coordinates after 20 steps

$p_o = 0.15, p_b = 0.15, \alpha = 40°$

n					p_c					
	0.1	0.2	0.3	0.4	0.5	0.6	0.7	0.8	0.9	1.0
1000	16	14	15	20	14	28	20	23	59	52
3000	14	16	16	24	56	63	73	82	77	94
5000	15	15	30	40	72	80	92	93	95	97
7000	15	20	43	43	82	88	85	96	97	98
9000	14	19	40	76	86	90	95	96	98	98
11000	15	21	52	71	87	92	95	97	98	98
13000	15	28	48	79	88	92	96	97	98	99
15000	16	22	59	81	89	94	96	97	98	99

$p_o = 0.5, p_b = 0.15, \alpha = 40°$

n					p_c					
	0.1	0.2	0.3	0.4	0.5	0.6	0.7	0.8	0.9	1.0
1000	14	37	75	88	98	99	99	100	100	100
3000	35	87	96	99	99	99	100	100	100	100
5000	54	91	97	99	99	100	100	100	100	100
7000	64	93	98	99	99	100	100	100	100	100
9000	66	94	98	99	100	100	100	100	100	100
11000	69	95	99	99	100	100	100	100	100	100
13000	71	95	99	99	100	100	100	100	100	100
15000	74	96	99	99	100	100	100	100	100	100

$p_o = 0815, p_b = 0.15, \alpha = 40°$

n					p_c					
	0.1	0.2	0.3	0.4	0.5	0.6	0.7	0.8	0.9	1.0
1000	39	94	98	99	100	100	100	100	100	100
3000	81	97	99	100	100	100	100	100	100	100
5000	86	98	99	100	100	100	100	100	100	100
7000	89	99	99	100	100	100	100	100	100	100
9000	90	99	99	100	100	100	100	100	100	100
11000	91	99	100	100	100	100	100	100	100	100
13000	92	99	100	100	100	100	100	100	100	100
15000	93	99	100	100	100	100	100	100	100	100

Table 3. Percentage of interior motes that get their coordinates after 20 steps

$p_o = 0.15, p_b = 0.5, \alpha = 40°$

n					p_c					
	0.1	0.2	0.3	0.4	0.5	0.6	0.7	0.8	0.9	1.0
1000	51	52	55	64	80	80	85	87	83	93
3000	52	55	74	79	88	92	94	95	97	97
5000	51	70	78	87	91	94	95	96	97	98
7000	52	73	80	86	93	94	96	97	97	98
9000	58	73	81	88	93	95	97	97	99	99
11000	59	74	84	90	93	96	97	98	99	99
13000	58	75	84	90	94	96	98	98	99	99
15000	61	76	85	91	94	96	98	98	99	99

$p_o = 0.5, p_b = 0.5, \alpha = 40°$

n					p_c					
	0.1	0.2	0.3	0.4	0.5	0.6	0.7	0.8	0.9	1.0
1000	64	87	95	97	99	99	99	100	99	100
3000	79	93	97	99	99	99	100	100	100	100
5000	82	95	98	99	99	100	100	100	100	100
7000	84	96	99	99	99	100	100	100	100	100
9000	84	97	99	99	99	100	100	100	100	100
11000	85	97	99	99	100	100	100	100	100	100
13000	87	97	99	99	100	100	100	100	100	100
15000	87	97	99	99	100	100	100	100	100	100

$p_o = 0.85, p_b = 0.5, \alpha = 40°$

n					p_c					
	0.1	0.2	0.3	0.4	0.5	0.6	0.7	0.8	0.9	1.0
1000	85	96	99	99	100	100	100	100	100	100
3000	91	98	99	99	100	100	100	100	100	100
5000	93	99	99	100	100	100	100	100	100	100
7000	94	99	99	100	100	100	100	100	100	100
9000	95	99	99	100	100	100	100	100	100	100
11000	95	99	99	100	100	100	100	100	100	100
13000	96	99	99	100	100	100	100	100	100	100
15000	96	99	99	100	100	100	100	100	100	100

Table 4. Percentage of interior motes that get their coordinates after 20 steps

$p_o = 0.15, p_b = 0.85, \alpha = 40°$

n	0.1	0.2	0.3	0.4	0.5	0.6	0.7	0.8	0.9	1.0
1000	86	84	87	86	92	94	97	98	97	97
3000	85	86	93	95	96	97	98	98	99	99
5000	86	91	94	97	97	98	98	99	99	99
7000	87	91	95	96	97	98	99	99	99	99
9000	88	92	95	97	98	98	99	99	99	99
11000	89	93	95	97	98	98	99	99	99	99
13000	89	93	96	97	98	99	99	99	99	99
15000	90	94	96	97	98	99	99	99	99	99

$p_o = 0.5, p_b = 0.85, \alpha = 40°$

n	0.1	0.2	0.3	0.4	0.5	0.6	0.7	0.8	0.9	1.0
1000	90	96	98	99	99	99	99	99	100	100
3000	94	98	98	99	100	100	100	100	100	100
5000	95	98	98	99	100	100	100	100	100	100
7000	95	98	99	99	100	100	100	100	100	100
9000	96	99	99	99	100	100	100	100	100	100
11000	96	99	99	100	100	100	100	100	100	100
13000	96	99	99	100	100	100	100	100	100	100
15000	96	99	99	100	100	100	100	100	100	100

$p_o = 0.85, p_b = 0.85, \alpha = 40°$

n	0.1	0.2	0.3	0.4	0.5	0.6	0.7	0.8	0.9	1.0
1000	95	99	99	99	100	100	100	100	100	100
3000	97	99	99	100	100	100	100	100	100	100
5000	98	99	99	100	100	100	100	100	100	100
7000	98	99	99	100	100	100	100	100	100	100
9000	98	99	100	100	100	100	100	100	100	100
11000	98	99	100	100	100	100	100	100	100	100
13000	98	99	100	100	100	100	100	100	100	100
15000	99	99	100	100	100	100	100	100	100	100

Table 5. Percentage of interior that receive a message from the BTS

$p_o = 0.15, p_b = 0.15, \alpha = 40°$

n	0.1	0.2	0.3	0.4	0.5	0.6	0.7	0.8	0.9	1.0
1000	18	27	30	33	36	48	44	59	67	73
3000	20	28	33	41	46	59	63	71	74	82
5000	20	27	39	45	54	63	70	76	81	84
7000	22	30	40	49	58	66	74	78	85	87
9000	22	31	43	53	61	72	78	82	85	90
11000	22	32	43	53	64	71	80	84	88	90
13000	23	33	45	56	65	74	80	83	89	91
15000	23	35	45	56	66	74	82	86	90	93

$p_o = 0.5, p_b = 0.15, \alpha = 40°$

n	0.1	0.2	0.3	0.4	0.5	0.6	0.7	0.8	0.9	1.0
1000	28	49	61	85	90	92	97	98	99	99
3000	37	62	83	92	96	98	99	99	99	100
5000	41	71	88	95	98	99	99	100	100	100
7000	43	72	89	96	99	99	99	100	100	100
9000	44	75	90	97	99	99	100	100	100	100
11000	48	78	92	98	99	99	100	100	100	100
13000	49	79	93	98	99	99	100	100	100	100
15000	50	81	94	98	99	99	100	100	100	100

$p_o = 0.85, p_b = 0.15, \alpha = 40°$

n	0.1	0.2	0.3	0.4	0.5	0.6	0.7	0.8	0.9	1.0
1000	45	74	89	98	99	99	100	100	100	100
3000	55	87	97	99	100	100	100	100	100	100
5000	62	91	98	99	100	100	100	100	100	100
7000	67	93	99	99	100	100	100	100	100	100
9000	69	95	99	100	100	100	100	100	100	100
11000	70	95	99	100	100	100	100	100	100	100
13000	72	96	99	100	100	100	100	100	100	100
15000	73	97	99	100	100	100	100	100	100	100

Table 6. Percentage of interior that receive a message from the BTS

$p_o = 0.15, p_b = 0.5, \alpha = 40°$

n	0.1	0.2	0.3	0.4	0.5	0.6	0.7	0.8	0.9	1.0
1000	57	65	71	72	82	81	87	90	92	93
3000	59	68	75	81	87	89	92	94	95	97
5000	60	69	76	84	87	92	94	96	97	97
7000	60	71	79	85	89	93	95	96	98	99
9000	62	72	79	87	91	94	96	97	98	99
11000	62	72	81	88	91	94	96	97	98	99
13000	62	74	82	89	92	95	97	98	98	99
15000	62	73	83	89	93	96	97	98	99	99

$p_o = 0.5, p_b = 0.5, \alpha = 40°$

n	0.1	0.2	0.3	0.4	0.5	0.6	0.7	0.8	0.9	1.0
1000	66	77	84	91	92	95	97	97	98	99
3000	70	84	90	94	97	98	98	99	99	99
5000	73	85	92	96	98	98	99	99	99	99
7000	74	87	93	96	98	99	99	99	99	100
9000	75	88	94	97	98	99	99	99	99	100
11000	76	89	94	97	98	99	99	99	100	100
13000	76	89	94	97	99	99	99	99	100	100
15000	78	90	95	98	99	99	99	99	100	100

$p_o = 0.85, p_b = 0.5, \alpha = 40°$

n	0.1	0.2	0.3	0.4	0.5	0.6	0.7	0.8	0.9	1.0
1000	73	87	94	97	98	99	99	99	100	99
3000	80	93	97	98	99	99	99	100	100	100
5000	82	94	96	99	99	99	100	100	100	100
7000	84	95	98	99	99	99	100	100	100	100
9000	85	95	98	99	99	99	100	100	100	100
11000	85	96	98	99	99	99	100	100	100	100
13000	86	96	98	99	99	99	100	100	100	100
15000	87	96	99	99	99	100	100	100	100	100

Table 7. Percentage of interior that receive a message from the BTS

$p_o = 0.85, p_b = 0.85, \alpha = 40°$

n						p_c				
	0.1	0.2	0.3	0.4	0.5	0.6	0.7	0.8	0.9	1.0
1000	98	99	99	100	100	100	100	100	100	100
3000	96	99	99	100	100	100	100	100	100	100
5000	97	99	99	100	100	100	100	100	100	100
7000	98	99	99	100	100	100	100	100	100	100
9000	98	99	99	100	100	100	100	100	100	100
11000	98	99	100	100	100	100	100	100	100	100
13000	98	99	100	100	100	100	100	100	100	100
15000	98	99	100	100	100	100	100	100	100	100

$p_o = 0.5, p_b = 0.85, \alpha = 40°$

n						p_c				
	0.1	0.2	0.3	0.4	0.5	0.6	0.7	0.8	0.9	1.0
1000	93	95	98	99	99	99	99	100	99	100
3000	94	97	99	99	99	100	100	100	100	100
5000	94	98	99	99	100	100	100	100	100	100
7000	95	98	99	99	100	100	100	100	100	100
9000	95	98	99	99	100	100	100	100	100	100
11000	96	98	99	99	100	100	100	100	100	100
13000	95	99	99	99	100	100	100	100	100	100
15000	96	99	99	99	100	100	100	100	100	100

$p_o = 0.15, p_b = 0.85, \alpha = 40°$

n						p_c				
	0.1	0.2	0.3	0.4	0.5	0.6	0.7	0.8	0.9	1.0
1000	85	89	93	96	92	95	97	98	98	98
3000	91	94	95	96	97	98	98	99	99	99
5000	89	92	95	96	97	98	98	99	99	99
7000	89	93	96	96	98	98	99	99	99	99
9000	89	93	95	97	98	98	99	99	99	99
11000	90	93	96	97	98	99	99	99	99	99
13000	90	93	96	97	98	99	99	99	99	99
15000	90	93	96	97	98	99	99	99	99	99

Table 8. Percentage of pairs of interior motes that are at Euclidean distance ≤ r and their graph distance is ≤ 4

$p_o = 0.85, \alpha = 40°$

n						p_c				
	0.1	0.2	0.3	0.4	0.5	0.6	0.7	0.8	0.9	1.0
1000	5	51	89	98	99	100	100	100	100	100
3000	13	80	98	99	100	100	100	100	100	100
5000	19	88	99	100	100	100	100	100	100	100
7000	24	93	99	100	100	100	100	100	100	100
9000	27	95	99	100	100	100	100	100	100	100

$p_o = 0.5, \alpha = 40°$

n						p_c				
	0.1	0.2	0.3	0.4	0.5	0.6	0.7	0.8	0.9	1.0
1000	2	12	44	73	88	95	98	99	99	100
3000	3	30	74	93	98	99	99	100	100	100
5000	4	41	82	96	99	99	100	100	100	100
7000	5	46	87	98	99	100	100	100	100	100
9000	6	53	91	98	99	99	100	100	100	100
11000	7	57	93	99	99	100	100	100	100	100
13000	7	62	94	99	99	100	100	100	100	100
15000	8	65	95	99	99	100	100	100	100	100

$p_o = 0.15, \alpha = 40°$

n						p_c				
	0.1	0.2	0.3	0.4	0.5	0.6	0.7	0.8	0.9	1.0
1000	1	3	5	9	17	22	33	36	40	50
3000	1	8	16	25	37	52	60	70	78	82
5000	1	4	10	19	33	47	56	70	78	82
7000	1	4	11	20	37	57	65	75	83	88
9000	1	4	12	24	39	57	67	79	84	91
11000	1	5	12	26	43	60	71	82	87	92
13000	1	5	13	28	47	63	75	84	90	94
15000	1	5	14	30	49	65	78	86	91	94

Table 9. Percentage of pairs of interior motes that are at Euclidean distance ≤ r and their graph distance is ≤ 6

$p_o = 0.85, \alpha = 40°$

n						p_c				
	0.1	0.2	0.3	0.4	0.5	0.6	0.7	0.8	0.9	1.0
1000	3	22	51	79	91	96	98	99	99	100
3000	6	37	76	92	98	99	99	100	100	100
5000	7	46	83	96	99	99	100	100	100	100
7000	9	52	87	97	99	99	100	100	100	100
9000	10	55	89	98	99	99	100	100	100	100

$p_o = 0.5, \alpha = 40°$

n						p_c				
	0.1	0.2	0.3	0.4	0.5	0.6	0.7	0.8	0.9	1.0
1000	2	9	22	45	63	78	85	93	96	97
3000	3	15	38	63	82	91	96	98	99	99
5000	3	18	46	72	88	94	98	99	99	99
7000	3	21	51	77	91	96	98	99	99	100
9000	4	23	55	80	92	97	99	99	99	100
11000	4	25	58	83	93	98	99	99	99	100
13000	4	26	61	84	94	98	99	99	100	100
15000	4	28	63	86	95	98	99	99	100	100

$p_o = 0.15, \alpha = 40°$

n						p_c				
	0.1	0.2	0.3	0.4	0.5	0.6	0.7	0.8	0.9	1.0
1000	1	3	5	8	13	21	26	31	32	43
3000	1	3	7	13	20	28	36	44	52	60
5000	1	4	8	16	22	30	42	50	57	66
7000	1	4	8	15	24	36	45	55	63	72
9000	1	4	9	17	26	37	49	59	67	74
11000	1	4	9	19	28	40	52	61	68	75
13000	1	4	10	18	29	41	52	65	72	78
15000	1	5	10	20	31	43	54	65	73	79

Linear Time Local Improvements for Weighted Matchings in Graphs[*]

Doratha E. Drake and Stefan Hougardy

Institut für Informatik
Humboldt-Universität zu Berlin, 10099 Berlin, Germany
{drake,hougardy}@informatik.hu-berlin.de

Abstract. Recently two different linear time approximation algorithms for the weighted matching problem in graphs have been suggested [5][17]. Both these algorithms have a performance ratio of 1/2. In this paper we present a set of local improvement operations and prove that it guarantees a performance ratio of 2/3. We show that a maximal set of these local improvements can be found in linear time.

To see how these local improvements behave in practice we conduct an experimental comparison of four different approximation algorithms for calculating maximum weight matchings in weighted graphs. One of these algorithms is the commonly used Greedy algorithm which achieves a performance ratio of 1/2 but has $O(m \log n)$ runtime. The other three algorithms all have linear runtime. Two of them are the above mentioned 1/2 approximation algorithms. The third algorithm may have an arbitrarily bad performance ratio but in practice produces reasonably good results. We compare the quality of the algorithms on a test set of weighted graphs and study the improvement achieved by our local improvement operations. We also do a comparison of the runtimes of all algorithms.

1   Introduction

A *matching* M in a graph $G = (V, E)$ is defined to be any subset of the edges of G such that no two edges in M are adjacent. If $G = (V, E)$ is a weighted graph with edge weights given by a function $w : E \rightarrow \mathbb{R}_+$ the *weight of a matching* is defined as to be $w(M) := \sum_{e \in M} w(e)$. The weighted matching problem is to find a matching M in G that has maximum weight. Calculating a matching of maximum weight is an important problem with many applications. The fastest known algorithm to date for solving the weighted matching problem in general graphs is due to Gabow [7] and has a runtime of $O(|V||E| + |V|^2 \log |V|)$.

Many real world problems require graphs of such large size that the runtime of Gabow's algorithm is too costly. Examples of such problems are the refinement of FEM nets [15], the partitioning problem in VLSI-Design [16], and the gossiping problem in telecommunications [2]. There also exist applications were the weighted matching problem has to be solved extremely often on only moderately large graphs. An example of such an application is the virtual screening of

[*] Supported by DFG research grant 296/6-3.

K. Jansen et al. (Eds.): WEA 2003, LNCS 2647, pp. 107–119, 2003.

protein databases containing the three dimensional structure of the proteins [6]. The graphs appearing in such applications only have about 10,000 edges. But the weighted matching problem has to be solved more than 100,000,000 times for a complete database scan.

Therefore, there is considerable interest in approximation algorithms for the weighted matching problem that are very fast, having ideally linear runtime, and that nevertheless produce very good results even if these results are not optimal.

The quality of an approximation algorithm for solving the weighted matching problem is measured by its so-called *performance ratio*. An approximation algorithm has a performance ratio of c, if for all graphs it finds a matching with a weight of at least c times the weight of an optimal solution. Recently two different linear time approximation algorithms for the weighted matching problem in graphs have been suggested [5][17]. Both these algorithms have a performance ratio of 1/2. In this paper we present a set of local improvement operations and prove that it guarantees a performance ratio of 2/3. We show that a maximal set of these local improvements can be found in linear time.

The performance ratio only gives information about the worst case behaviour of an algorithm. In this paper we will also study how several algorithms for the weighted matching problem behave in practice. We make an experimental comparison of the three known approximation algorithms for the weighted matching problem that have a performance ratio of $\frac{1}{2}$. We also include a simple, extremely fast heuristic that cannot guarantee any performance ratio but behaves reasonably well in practice. In addition we apply our local improvement operations to all these algorithms and test how much improvement they yield in practice.

2   Local Improvements

The idea of local improvements has been used in several cases to improve the performance ratio of approximation algorithms. See [10, 3] for such examples.

In the case of the unweighted matching problem which is usually called the maximum matching problem it is well known that by local improvements a given matching can be enlarged. In this case the local improvements are augmenting paths, i.e. paths that alternately consist of edges contained in a matching M and not contained in M such that the first and the last vertex of the path are not contained in an edge of M. From a result of Hopcroft and Karp [11] it follows that if M is a matching such that a shortest augmenting path has length at least l then M is an $\frac{l-1}{l+1}$ approximation of a maximum matching.

We extend the notion of augmenting paths to weighted matchings in a natural way. Let $G = (V, E)$ be a weighted graph with weight function $w : E \to \mathbb{R}_+$ and $M \subseteq E$ be an arbitrary matching in G. A path or cycle is called M-*alternating* if it uses alternately edges from M and $E \setminus M$. Note that alternating cycles must contain an even number of edges. Let P be an alternating path such that if it ends in an edge not belonging to M then the endpoint of P is not covered by an edge of M. The path P is called M-*weight-augmenting* if

$$w\left(E(P) \cap M\right) \;<\; w\left(E(P) \setminus M\right) \;.$$

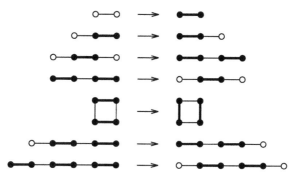

Fig. 1. The seven local improvements. Edges belonging to the matching are shown in bold. Hollow vertices are vertices not contained in any matching edge

If P is an M-weight-augmenting path then $M \triangle P$ (the symmetric difference between M and P) is again a matching with strictly larger weight than M. The notion of M-weight-augmenting cycles is defined similarly.

We will consider in the following M-weight-augmenting cycles of length 4, M-weight-augmenting paths of length at most 4 and M-weight-augmenting paths of length 5 that start and end with an edge of M. See Fig. 1 for all possibilities of such weight augmenting paths and cycles.

The following result shows that the non-existence of short M-weight-augmenting paths or cycles guarantees that the weight of M is at least $2/3$ of the maximum possible weight.

Theorem 1. *Let M_{opt} be a maximum weight matching in a weighted graph $G = (V, E)$ with weight function $w : E \to \mathbb{R}_+$. If M is a matching in G such that none of the seven operations shown in Fig. 1 increases the weight of M then*

$$w(M) \geq \frac{2}{3} \cdot w(M_{opt}) .$$

Proof. Consider the graph induced by the symmetric difference $M \triangle M_{opt}$. It consists of even alternating cycles C_i and alternating paths P_i. We will show that

$$w(C_i \cap M) \geq \frac{2}{3} \cdot w(C_i \cap M_{opt}) \qquad \forall i \tag{1}$$

and

$$w(P_i \cap M) \geq \frac{2}{3} \cdot w(P_i \cap M_{opt}) \qquad \forall i . \tag{2}$$

Equation (1) and (2) imply

$$w(M) = w(M \cap M_{opt}) + \sum w(C_i \cap M) + \sum w(P_i \cap M)$$
$$\geq \frac{2}{3} \cdot w(M \cap M_{opt}) + \frac{2}{3} \cdot \sum w(C_i \cap M_{opt}) + \frac{2}{3} \cdot \sum w(P_i \cap M_{opt})$$
$$= \frac{2}{3} \cdot w(M_{opt}) .$$

We start proving (1). If C_i is a cycle of length 4 then by the assumptions of the theorem

$$w(C_i \cap M) \geq w(C_i \cap M_{opt}) \geq \frac{2}{3} \cdot w(C_i \cap M_{opt}) \ .$$

Now assume C_i is a cycle of length at least 6. Let $e_1, e_2, e_3, \ldots$ be the edges of C_i in a consecutive order such that $e_j \in M$ for j odd. Consider a subpath of type $e_{2k+1}, e_{2k+2}, e_{2k+3}, e_{2k+4}, e_{2k+5}$ in C_i. By the assumptions of the theorem we have

$$w(e_{2k+1}) + w(e_{2k+3}) + w(e_{2k+5}) \geq w(e_{2k+2}) + w(e_{2k+4}) \ .$$

By summing this inequality over all possible values for k we see that each edge of $C_i \cap M$ appears 3 times on the left side and each edge of $C_i \cap M_{opt}$ appears 2 times on the right side of the inequality. Therefore

$$3 \cdot w(C_i \cap M) \geq 2 \cdot w(C_i \cap M_{opt}) \ .$$

This proves (1).

We use a similar idea to prove (2). Let e_j be the edges of a path P_i such that $e_j \in M$ if and only if j is odd. The path may start with e_1 or e_0 as we do not know whether it starts with an edge of M or an edge of $E \setminus M$. Consider subpaths of P_i of type $e_{2k+1}, e_{2k+2}, e_{2k+3}, e_{2k+4}, e_{2k+5}$. We have

$$w(e_{2k+1}) + w(e_{2k+3}) + w(e_{2k+5}) \geq w(e_{2k+2}) + w(e_{2k+4}) \ .$$

Now extend the path P_i artificially to both sides by four edges of weight 0. So we have edges $e_{-1}, e_{-2}, \ldots$ on the left side of P_i. Now add the above inequality for all k where k starts at -2. Then each edge of $P_i \cap M$ appears in three inequalities on the left and each edge of $P_i \cap M_{opt}$ appears in 2 inequalities on the right. The artificial edges may appear in arbitrary number. As they have weight 0, it does not matter. Therefore

$$3 \cdot w(P_i \cap M) \geq 2 \cdot w(P_i \cap M_{opt}) \ .$$

This proves (2). $\qquad\square$

Theorem 2. . Let $G = (V, E)$ be a weighted graph with weight function $w : E \to \mathbb{R}_+$ and let $M \subseteq E$ be a matching. A maximal set of any of the seven operations shown in Fig. 1 that are pairwise node disjoint and such that each operation increases the weight of M can be found in linear time.

Proof. To achieve linear runtime in a preprocessing step each vertex that is covered by M gets a pointer to the edge of M it belongs to. Consider an arbitrary edge $e \in M$. To decide whether e belongs to an M-weight-augmenting C_4 run over all edges incident to one endpoint of e and mark all edges of M that are adjacent to such edges. Now run over the edges incident to the other endpoint of e and see whether they are incident to the other endpoint of a marked edge. If yes, an alternating C_4 has been found. Check whether it is M-weight-augmenting. If yes remove it from the graph.

Similarly paths of length at most 4 and paths of length 5 which have an edge of M at both ends can be found in linear time. $\qquad\square$

3   The Algorithms

In this section we briefly describe four different approximation algorithms for the weighted matching problem in graphs. For all these algorithms we have tested their performance and runtime with and without our local improvements. The results of these tests are given in Section 5.

The first of the approximation algorithms is Greedy Matching [1] shown in Fig. 2. Greedy Matching repeatedly removes the currently heaviest edge e and all of its adjacent edges from the input graph $G = (V, E)$ until E is empty. In each iteration e is added to the matching M which is returned as the soltution. It is easy to see that Greedy Matching has a performance ratio of $\frac{1}{2}$ [1]. If the edges of G are sorted in a preprocessing step then the runtime is $O(|E| \log |V|)$.

The second algorithm is the LAM algorithm by Preis [17]. A sketch of this algorithm is shown in Fig. 3. Preis improved upon the simple greedy approach by making use of the concept of a so called *locally heaviest edge*. This is defined as an edge for which no other edge currently adjacent to it has larger weight. Preis proved that the performance ratio of the algorithm based on this idea is $\frac{1}{2}$. He also showed how a locally heaviest edge in a graph can be found in amortized constant time. This results in a total runtime of $O(|E|)$ for the LAM algorithm. See [17] for more details.

The third algorithm is the Path Growing Algorithm (PGA) by Drake and Hougardy [5] shown in Fig. 4. PGA constructs node disjoint paths in the input graph G along heaviest edges by lengthening the path one node at a time. Each time a node is added to a path it and all of its incident edges are removed from the graph. This is repeated until the graph is empty. By alternately labelling the edges of the paths 1 and 2 one obtains two matchings M_1 and M_2, the larger of which is returned as the solution. The PGA algorithm has a performance ratio of $\frac{1}{2}$ and a runtime of $O(|E|)$ [5]. There are two simple improvements that are proposed in [5] which can be applied to the PGA algorithm without changing its runtime. The first is to compute a maximum weight matching along the paths constructed by the algorithm and return this as the solution. The second is to add any remaining edges in the graph to the solution until the solution becomes a maximal matching. Neither of these improvements can guarantee a better worst case behaviour for the algorithm, but in practice these improvements can make

---

Greedy Matching $(G = (V, E), w : E \to \mathbb{R}_+)$

1 $M := \emptyset$
2 **while** $E \neq \emptyset$ **do begin**
3   let e be the heaviest edge in E
4   add e to M
5   remove e and all edges adjacent to e from E
6 **end**
7 **return** M

Fig. 2. The greedy algorithm for finding maximum weight matchings

```
LAM (G = (V, E), w : E → ℝ₊)

1 M := ∅
2 while E ≠ ∅ do begin
3     find a locally heaviest edge e ∈ G
4     remove e and all edges adjacent to e from G
5     add e to M
6 end
7 return M
```

Fig. 3. The LAM approximation algorithm for finding maximum weight matchings

```
PathGrowingAlgorithm (G = (V, E), w : E → ℝ₊)

1  M₁ := ∅, M₂ := ∅
2 while E ≠ ∅ do begin
3     choose x ∈ V of degree at least 1 arbitrarily
4     grow a path from x along heaviest edges added alternately to M₁ and M₂
5     remove the path from G
6 end
7 return max(w(M₁), w(M₂))
```

Fig. 4. The Path Growing Algorithm for finding maximum weight matchings

```
MM (G = (V, E), w : E → ℝ₊)

1 M := ∅
2 while E ≠ ∅ do begin
3     choose x ∈ V arbitrarily
4     add to M the heaviest edge e incident to x
5     delete all edges adjacent to e from E
6 end
7 return M
```

Fig. 5. The MM heuristic for finding maximum weight matchings

a considerable difference. Therefore we test both versions of this algorithm. We call the second version where both improvements are applied PGA'.

Finally we include a trivial heuristic called MM (for Maximal Matching) shown in Fig. 5. This heuristic computes a maximal matching in a graph in a greedy manner. For each vertex x a heaviest edge e which is incident to x is added to the solution. All edges adjacent to e are removed. The runtime of algorithm MM is $O(|E|)$. It is easy to construct examples where the weighted matching returned by MM can be arbitrarily bad. Yet it is interesting to see how this heuristic behaves in practice. Therefore we have included it in this study.

4   Test Instances

We test our implementations of the above algorithms and local improvements against four classes of data sets: random graphs, two dimensional grids, complete graphs, and randomly twisted three dimensional grids. We do not include any geometric instances in this study as none of the algorithms considered here was designed to take advantage of any geometric properties.

Within each class instances with different parameters such as number of vertices, edge densities etc. have been generated. For each specific parameter set ten instances were randomly generated. Each of the algorithms was run on all ten instances and then the average value of the runtime and percentage of the deviation from an optimal solution was computed. We calculated the optimal solutions to these instances using LEDA [13]. The size of the test instances was restricted to graphs with at most 100,000 vertices and at most 500,000 edges because for these instances the runtime of the exact algorithm was already several hours.

The random graphs are based on the $G_{n,p}$ model. This means that the graphs have n vertices and the possibe $\binom{n}{2}$ edges are chosen independently with probability p. We have chosen $n = 10,000$ and p in the range from 5/10000 to 100/10000 resulting in graphs from about 25,000 to 500,000 edges. The edge weights are integer values chosen randomly from the range of 1 to 1000. The graphs are labelled as "R10000.5" through "R10000.100".

The two dimensional grids are grids of dimension $h \times 1000$ with h chosen in the range 10 to 100. The edge weights for these grids have been assigned integer values chosen randomly from the range between 0 and 999. These graphs are labelled "G10" through "G100".

The complete graphs are graphs on n vertices containing all possible $\binom{n}{2}$ edges. We have generated these graphs for $n = 200$ to 2000. The integer edge weights have been randomly assigned from the range of 0 to 999. These graphs are labelled "K200" through "K2000".

For the randomly twisted three dimensional grid we have used the RMFGEN graph generator introduced in [9] which was used to generate flow problems. The graphs created consist of a square grid of dimension a called a frame. There are b such frames $F_1, \ldots, F_b$ which are all symmetric. There are a^2 edges connecting the nodes of F_i to a random permutation of the nodes of F_{i+1} for $1 \le i < b$. The edges within a frame all have weight 500. Those between frames have weights randomly chosen between 1 and 1,000. The only changes we have made to the RMFGEN generator besides making the graph undirected is concerning the weights of the in-frame edges. We have assigned 500 to these edges instead of $a^2 * 1000$ assigned by RMFGEN as the latter value did not seem to produce instances that were as interesting for the weighted matching problem. We have created three such tests on graphs of dimension $a = 4$, $b = 1250$; $a = 27$, $b = 27$; and $a = 70$, $b = 4$. These three instances are labelled "a", "b" and "c".

5   Experimental Results

The following two subsections contain the experimental results we obtained on the test instances described in Section 4. We have compared the five algorithms described in Section 3 with and without additionally performing our local improvement operations which where applied as follows: For each of the seven local improvement operations shown in Fig. 1 we have computed a maximal set of disjoint improvements in time $O(m)$ as indicated in Theorem 2 and then augmented along this set. We used the same ordering of the operations as shown in Fig. 1.

For each row in a table ten different test instances have been generated and the average value has been taken. The variance was in all cases below 1%, in many cases even below 0.5%. Due to space restrictions we list the results for some part of the instances only.

5.1   Performance

Tables 1, 2, and 3 show the performances of the five algorithms on the different classes of test sets. The first column of the tables contains the name of the test instance as described in Section 4. The next two columns "n" and "m" denote the number of vertices and edges of the graph. In case of the graphs "R10000.x" the number of edges is the average value of the ten test instances that were computed for each row of the table. In all other cases the ten test instances have the same number of edges, only the weight of the edges differs. The next five columns show the difference in % of the solution found by the algorithms to the optimum solution. The names of the algorithms are abbreviated as in Section 3. Each row contains one value in bold which is the best value. The worst value is given in gray.

As can be seen from Table 1 there is a great difference in the quality of the PGA and PGA′ algorithm. The simple heuristics added to the PGA algorithm drastically improve its performance in practice. This observation also holds for all other test instances. On all random graph instances the PGA′ algorithm performs best. The LAM and Greedy algorithms have almost the same quality which

Table 1. Performances of the five algorithms on weighted random graphs with different densities. The values denote the difference from the optimum in %

graph	n	m	Greedy	MM	LAM	PGA	PGA′
R10000.5	10000	25009	8.36	14.66	8.38	14.70	**7.38**
R10000.10	10000	49960	9.18	12.59	9.18	12.56	**8.31**
R10000.20	10000	100046	7.73	9.40	7.74	9.52	**7.20**
R10000.30	10000	150075	6.28	7.55	6.29	7.55	**5.95**
R10000.40	10000	200011	5.46	6.33	5.50	6.30	**5.08**
R10000.60	10000	299933	4.19	4.83	4.23	4.82	**3.99**
R10000.80	10000	399994	3.52	3.99	3.54	4.02	**3.39**
R10000.100	10000	499882	3.05	3.39	3.08	3.45	**2.95**

Table 2. Performances of the five algorithms on weighted grid like graphs. The values denote the difference from the optimum in %

graph	n	m	Greedy	MM	LAM	PGA	PGA'
G10	10000	18990	5.87	12.38	5.87	12.76	**4.51**
G20	20000	38980	5.97	13.04	5.98	12.50	**4.53**
G40	40000	78960	5.99	13.45	5.99	12.52	**4.60**
G60	60000	118940	5.96	13.68	5.97	12.49	**4.66**
G80	80000	158920	5.99	13.79	5.99	12.61	**4.67**
G100	100000	198900	6.05	13.82	6.05	12.60	**4.66**
a	20000	49984	5.71	10.22	**5.50**	12.91	5.59
b	19683	56862	5.79	8.79	**5.25**	13.14	6.14
c	19600	53340	6.13	8.00	**4.93**	12.71	6.22

Table 3. Performances of the five algorithms on weighted complete graphs. The values denote the difference from the optimum in %

graph	n	m	Greedy	MM	LAM	PGA	PGA'
K200	200	19900	1.75	1.63	1.77	1.65	**1.55**
K600	600	179700	0.72	0.82	**0.69**	0.75	0.72
K1000	1000	499500	**0.46**	0.55	0.49	0.53	0.51
K1400	1400	979300	0.39	0.38	0.39	0.37	**0.35**
K2000	2000	1999000	**0.27**	0.29	0.29	0.29	0.29

is slightly worse than that of PGA'. For all algorithms the performance improves as the random graphs get denser. The only exception are the extremely sparse graphs "R10000.5". Such an effect also has been observed in the unweighted case [12].

Table 2 shows that the performances of all algorithms are independent of the size of the test instances. For the two dimensional grids the PGA' algorithm achieves the best solutions. Again the Gredy algorithm and the LAM algorithm have almost the same quality which is significantly worse than that of PGA'. This situation changes in the case of the randomly twisted three dimensional grids. Here the LAM algorithm achieves the best result and the Greedy algorithm is slightly better than PGA'.

On weighted complete graphs all five algorithms have almost the same quality. For large complete graphs the performances of all algorithms tends to one. This is of course not surprising as in complete graphs an algorithm can barely choose a 'wrong' edge.

Tables 4, 5, and 6 show the performances of the five algorithms on the different classes of test sets with local improvements applied. This means that we have taken the solution returned by the algorithms and then computed a maximal set of pairwise disjoint local improvements for each of the seven local improvements shown in Fig. 1. The quality of all algorithms is drastically improved by the local improvements. The deviation from the optimum solution is reduced by a factor

Table 4. Performances of the five algorithms on weighted random graphs with different densities with local improvements applied. The values denote the difference from the optimum in %

graph	n	m	Greedy	MM	LAM	PGA	PGA'
R10000.5	10000	25009	3.15	4.57	3.15	5.21	**3.07**
R10000.10	10000	49960	4.62	5.52	4.62	6.14	**4.53**
R10000.20	10000	100046	4.39	4.90	**4.39**	5.42	4.44
R10000.30	10000	150075	**3.75**	4.18	3.76	4.56	3.87
R10000.40	10000	200011	**3.22**	3.60	3.26	3.84	3.37
R10000.60	10000	299933	**2.53**	2.84	2.55	3.00	2.75
R10000.80	10000	399994	**2.11**	2.38	2.15	2.52	2.28
R10000.100	10000	499882	**1.85**	2.08	1.87	2.15	2.00

Table 5. Performances of the five algorithms on weighted grid like graphs with local improvements applied. The values denote the difference from the optimum in %

graph	n	m	Greedy	MM	LAM	PGA	PGA'
G10	10000	18990	2.02	5.58	2.02	4.86	**1.79**
G20	20000	38980	2.11	6.06	2.11	4.94	**1.87**
G40	40000	78960	2.11	6.31	2.11	5.01	**1.91**
G60	60000	118940	2.12	6.47	2.12	5.01	**1.94**
G80	80000	158920	2.11	6.57	2.11	5.05	**1.94**
G100	100000	198900	2.16	6.56	2.16	5.05	**1.93**
a	20000	49984	2.52	6.00	**2.49**	6.42	3.00
b	19683	56862	2.81	6.43	**2.61**	7.31	3.48
c	19600	53340	2.99	6.01	**2.62**	7.54	3.77

Table 6. Performances of the five algorithms on weighted complete graphs with local improvements applied. The values denote the difference from the optimum in %

graph	n	m	Greedy	MM	LAM	PGA	PGA'
K200	200	19900	0.81	0.98	**0.78**	0.92	0.94
K600	600	179700	0.38	0.48	**0.35**	0.45	0.42
K1000	1000	499500	**0.23**	0.32	0.26	0.31	0.31
K1400	1400	979300	**0.19**	0.22	0.19	0.21	0.20
K2000	2000	1999000	**0.14**	0.17	0.15	0.18	0.18

between 1.5 and 3. The performances of the Greedy algorithm and the LAM and PGA' algorithms are very similar after performing the local improvements.

We also tested how much improvement can be achieved by applying the local improvement operations as long as they are possible. Usually computing three rounds of maximal sets of these local improvements led to a matching satisfying the conditions of Theorem 1. In some cases up to 5 iterations were necessary.

Using this approach the deviation from an optimum solution can be reduced by additional 30% on average. The runtime increases by a factor of 2 to 3.

5.2   Runtimes

We compared the runtimes of all five algorithms on all instances against each other and also compared it to LEDA's exact algorithm. Table 7 shows the relative runtimes of the algorithms LAM, PGA and PGA′ compared to the MM algorithm which is clearly the fastest. As can be seen there is no big difference between the runtimes of algorith PGA and PGA′. The PGA′ algorithm is about a factor 2, the LAM algorithm about a factor 3 slower than the MM algorithm. The factor by which the local improvement operations are slower than the MM algorithm is about 4. This means that the LAM and PGA′ algorithm with local improvements applied are about a factor of 7 and 6 slower than MM.

Table 8 shows the relative runtimes of the Greedy algorithm and LEDA's exact algorithm compared to the MM algorithm. The Greedy algorithm has a worst case runtime of $O(m \log m)$. The algorithm implemented in LEDA has a runtime of $O(mn \log n)$. Therefore one should expect that the Greedy algorithm is within a factor of $\log m$ and LEDA's algorithm is within a factor of $n \log n$ of the runtime of the MM algorithm. This behaviour can be roughly confirmed by the data.

To give an impression of the absolute running times we mention these for algorithm MM on some large graphs. It took 0.16 seconds on R10000.100, 0.06 seconds on G100 and 0.62 seconds on K2000. All times were measured on a 1.3GHz PC.

Table 7. Relative runtimes expressed as a range of the factor within which the runtime of the algorithm is slower than the MM algorithm

MM	LAM	PGA	PGA′	local improvements
1.00	2.68 - 3.55	1.58 - 2.06	1.88 - 2.10	3.45 - 5.12

6   Conclusion

We have suggested a set of seven local improvement operations for weighted matchings in graphs which guarantee a performance ratio of $\frac{2}{3}$. A maximal set of such operations can be found in linear time. We compared five different approximation algorithms for the weighted matching problem. The algorithms MM, LAM, PGA, and PGA′ have linear runtime while the Greedy algorithm has runtime $O(m \log n)$. The PGA′ algorithm is significantly better than PGA. The computation of a maximum weight matching on the paths generated by PGA can easily be incorporated in the generation of these paths. Therefore the PGA′

Table 8. Relative runtimes expressed as a factor by which the runtime of the Greedy algorithm and LEDA's exact algorithm is slower than the MM algorithm

graph	n	m	Greedy	MM	LEDA	graph	n	m	Greedy	MM	LEDA
a	20000	49984	7.28	1.0	26119	G10	10000	18990	6.64	1.0	14873
b	19683	56862	6.98	1.0	27181	G20	20000	38980	9.19	1.0	27667
c	19600	53340	5.21	1.0	28944	G40	40000	78960	11.64	1.0	53737
R10000.5	10000	25009	5.71	1.0	11156	G60	60000	118940	13.11	1.0	79321
R10000.10	10000	49960	8.15	1.0	8255	G80	80000	158920	14.10	1.0	104473
R10000.20	10000	100046	10.55	1.0	6004	G100	100000	198900	15.13	1.0	129857
R10000.30	10000	150075	12.11	1.0	5096	K200	200	19900	11.59	1.0	266
R10000.40	10000	200011	13.44	1.0	4660	K600	600	179700	20.67	1.0	514
R10000.60	10000	299933	15.05	1.0	4286	K1000	1000	499500	23.54	1.0	775
R10000.80	10000	399994	15.68	1.0	3949	K1400	1400	979300	22.05	1.0	1006
R10000.100	10000	499882	15.99	1.0	3832	K2000	2000	1999000	18.65	1.0	1402

algorithm requires almost no additional expense in coding or runtime and is definitely the better choice.

Only in the case of complete graphs are there a few instances where the Greedy algorithm achieved the highest quality. But in these cases the LAM and PGA' algorithm are very close to these results. Therefore the higher runtime required by the Greedy algorithm does not justify its application. The LAM or PGA' algorithm is a better choice. They usually should guarantee a solution within 5% of the optimum.

The local improvement operations introduced in this paper yield better performances for all five algorithms. On average the deviation from the optimum solution is reduced by a factor of 2. This improvement is achieved by more than doubling the runtimes of the linear time algorithms. Still these runtimes are dramatically smaller than those required by exact algorithms. Therefore the linear time LAM and PGA' algorithms are definitely the best choice for applications where runtimes are of crucial importance. If better quality is needed our local improvement operations should be applied which increase the runtime of these two algorithms by roughly a factor of 2-3 only. By applying our local improvement operations as long as they are possible the distance from an optimal solution can be decreased by additional 30% while the runtime grows by a factor of 4.

References

[1] D. Avis, A Survey of Heuristics for the Weighted Matching Problem, Networks, Vol. 13 (1983), 475–493   111

[2] R. Beier, J. F. Sibeyn, A Powerful Heuristic for Telephone Gossiping, Proc. 7th Colloquium on Structural Information and Communication Complexity, Carleton Scientific (2000), 17–35   107

[3] B. Chandra, M. M. Halldórsson, Greedy local improvement and weighted set packing approximation, Journal of Algorithms, 39 (2000), 223–240   108

[4] W. Cook, A. Rohe, Computing minimum-weight perfect matchings, INFORMS Journal on Computing 11 (1999), 138–148

[5] D. E. Drake, S. Hougardy, A Simple Approximation Algorithm for the Weighted Matching Problem, Information Processing Letters 85 (2003), 211–213 107, 108, 111

[6] C. Frömmel, A. Goede, C. Gröpl, S. Hougardy, T. Nierhoff, R. Preissner, M. Thimm, Accelerating screening of 3D protein data with a graph theoretical approach, Humboldt-Universität zu Berlin, September 2002 108

[7] H. N. Gabow, Data Structures for Weighted Matching and Nearest Common Ancestors with Linking, SODA 1990, 434–443 107

[8] H. N. Gabow, R. E. Tarjan, Faster Scaling Algorithms for General Graph-Matching Problems, JACM 38 (1991), 815–853

[9] G. Goldfarb, M. D. Grigoriadis, A Computational Comparison of the Dinic and Network Simplex Methods for Maximum Flow, Annals of Operations Research 13 (1988), 83-123 113

[10] M. M. Halldórsson, Approximating Discrete Collections via Local Improvements, In Proc. of Sixth SIAM/ACM Symposium on Discrete Algorithms, San Francisco (1995), 160–169 108

[11] J. E. Hopcroft, R. M. Karp, An $n^{5/2}$ algorithm for maximum matchings in bipartite graphs, SIAM Journal on Computing 2 (1973), 225–231 108

[12] J. Magun, Greedy Matching Algorithms, an Experimental Study, ACM Journal of Experimental Algorithms, Volume 3, Article 6, 1998 115

[13] K. Mehlhorn, S. Näher, LEDA: A Platform for Combinatorial and Geometric Computing, ACM Press (1995), New York, NY 113

[14] S. Micali and V. V. Vazirani, An $O(\sqrt{V}E)$ Algorithm for Finding Maximum Matching in General Graphs, Proc. 21st Annual IEEE Symposium on Foundations of Computer Science (1980), 17–27

[15] R. H. Möhring, M. Müller-Hannemann, Complexity and Modeling Aspects of Mesh Refinement into Quadrilaterals, Algorithmica 26 (2000), 148–171 107

[16] B. Monien , R. Preis , R. Diekmann, Quality Matching and Local Improvement for Multilevel Graph-Partitioning, Parallel Computing, 26(12), 2000, 1609–1634 107

[17] R. Preis, Linear Time 1/2-Approximation Algorithm for Maximum Weighted Matching in General Graphs, Symposium on Theoretical Aspects of Computer Science, STACS 99, C. Meinel, S. Tison (eds.), Springer, LNCS 1563, 1999, 259–269 107, 108, 111

[18] V. V. Vazirani, A Theory of Alternating Paths and Blossoms for Proving Correctness of the $O(\sqrt{V}E)$ Maximum Matching Algorithm, Combinatorica 14:1 (1994), 71–109

Experimental Studies of
Graph Traversal Algorithms

Rudolf Fleischer and Gerhard Trippen[*]

Department of Computer Science
The Hong Kong University of Science and Technology, Hong Kong
{rudolf,trippen}@cs.ust.hk

Abstract. We conduct an experimental evaluation of all major online graph traversal algorithms. This includes many simple natural algorithms as well as more sophisticated strategies. The observations we made watching the animated algorithms explore the graphs in the interactive experiments motivated us to introduce some variants of the original algorithms. Since the theoretical bounds for deterministic online algorithms are rather bad and no better bounds for randomized algorithms are known, our work helps to provide a better insight into the practical performance of these algorithms on various graph families. It is to observe that all the tested algorithm have a performance very close to the optimum offline algorithm in a huge family of random graphs. Only few very specific lower bound examples cause bad results.

1 Introduction

The exploration problem is a fundamental problem in online robotics. The robot starts in an unknown terrain and has to explore it completely (i.e., draw a map of the environment). Since all computations can only be based on local (or partial) information, the exploration problem falls naturally into the class of *online algorithms* [4, 7]. The quality of an online algorithm is measured by the *competitive ratio* which means we compare the length of the path the robot travels to explore the environment to the length of the shortest path that can perform the same task. Of course, the former should be as short as possible to achieve a good competitive ratio. More formally, for any given scene S let $L_A(S)$ be the length of the path traversed by an (online) algorithm A to explore S. Let $L_{OPT}(S)$ be the length of the path of an optimum algorithm that knows the scene in advance. Then we call an online exploration algorithm A c-competitive if for all scenes S, $L_A(S) \leq c \cdot L_{OPT}(S)$ (see [10, 7, 4]).

The graph exploration problem is the problem of constructing a complete map of an unknown strongly connected digraph using a path that is as short as possible. Describing the environment to be explored by the robot by a graph

[*] The work described in this paper was partially supported by a grant from the Research Grants Council of the Hong Kong Special Administrative Region, China (Project No. HKUST6010/01E).

K. Jansen et al. (Eds.): WEA 2003, LNCS 2647, pp. 120–133, 2003.
© Springer-Verlag Berlin Heidelberg 2003

leaves aside all the geometric features of a real environment. So we can focus on the combinatorial aspects of the exploration problem. We usually denote a graph by $G = (V, E)$, where V is the set of n nodes and E is the set of m edges. We denote by R the number of edge traversals of an algorithm to determine a map of G, i.e., the adjacency matrix of G.

We assume the robot can memorize all visited nodes and edges and it can recognize them if it reaches them again. Standing on a node, it can sense all the outgoing edges but it does not know where the unexplored edges are leading to. It cannot recognize how many edges are coming into the node. The directed edges can only be traversed from tail to head, not vice versa. We assume w.l.o.g. that traversing an edge has unit cost. Thus, the total cost of an exploration tour is equal to the number of traversed edges.

The robot always has the choice of leaving the currently visited node by traversing an already known edge or by exploring an unvisited outgoing edge. If the current node does not have any unvisited outgoing edges, we say the robot is *stuck* at the current node.

The well-known Chinese Postman Problem describes the offline version of the graph exploration problem [6]. Here, the postman has to find the shortest tour visiting all edges of a graph. For either undirected or directed graphs the problem can be solved in polynomial time. If the postman needs to serve a mixed graph that has both undirected and directed edges, the problem becomes NP-complete [9].

The online problem for undirected graphs can easily be solved by depth first search which visits every edge exactly twice. Since the optimal algorithm also needs to visit every edge at least once, the depth first search strategy is 2-competitive.

The problem becomes more difficult for strongly connected directed graphs. It was firstly investigated by Deng and Papadimitriou [5] who gave an online exploration algorithm with R exponential in the deficiency d of the graph (the deficiency is the minimum number of edges that must be added to make the graph Eulerian). Although it was conjectured that algorithms with R polynomial in d might exist no such algorithm has been found. Even worse, no randomized algorithm with better performance than the deterministic ones is known. Therefore, we study in this paper the performance of deterministic and randomized algorithms experimentally on several families of random graphs to get a better understanding of their strengths and weaknesses. Besides from simply running experiments we used algorithm animation as a further tool to gain deeper insight into the behavior of existing and new algorithms, hoping this would lead to preliminary ideas for a theoretical analysis. It turns out that all algorithms show a very good performance (better that 2-competitive) on all of our graph families, except on the family of graphs specially designed to show worst case behavior of the common deterministic algorithms. Experimental comparison of online algorithms has been done before for other problems, for example for the list update problem [3] and for the 2-processor scheduling problem [2].

This paper is organized as follows. In Section 2, we give an overview of existing deterministic online algorithms for the graph exploration problem. In Section 3 and 4, we propose new deterministic and randomized online algorithms. In Section 5, we describe the setup of our experiments. All source code listings are omitted due to space limitations but can be found together with more detailed tables and graphics at http://www.cs.ust.hk/~trippen/graphtrav/source/. In Section 6, we present the results of our experiments. We discuss these results in Section 7, and we close with some open problems in Section 8.

2   Background

A graph is *Eulerian* if there exists a path that visits each edge precisely once. An Eulerian graph can be explored online with only $2m$ edge traversals [5]. The robot simply takes an unexplored edge whenever possible. If it gets stuck, it considers the closed walk ("cycle") of unexplored edges just visited and retraces it, stopping at nodes that have unexplored edges, applying this algorithm recursively from each such node. It is easy to see that this algorithm achieves a competitive ratio of two in Eulerian graphs because no edge will be traversed more than twice. The reason for this is that the recursions will always take place in completely new parts of the graph. In fact, this algorithm is optimal.

For non-Eulerian graphs, Deng and Papadimitriou [5] showed that the deficiency d of the graph is a crucial parameter of the difficulty of the problem. The *deficiency* d of G is the minimum number of edges that have to be added to make G Eulerian. More formally, let $in(v)$ and $out(v)$ denote the in-degree and the out-degree of the node v. The *deficiency* of a node v, is $d(v) = out(v) - in(v)$. The *deficiency* d of the graph G is $d = \sum_{v \in V, d(v) > 0} d(v)$. A node with negative deficiency is called a *sink*, a node with positive deficiency is called a *source*.

If an algorithm knew the d missing edges $(s_1, t_1), (s_2, t_2), \ldots, (s_d, t_d)$ to make the graph Eulerian, then a modified version of the Eulerian online exploration algorithm could be executed [1]. Whenever the original Eulerian algorithm traverses an edge (s_i, t_i), the modified Eulerian algorithm traverses the corresponding path from s_i to t_i. This gives an additional factor of at most d for each edge, so that the robot traverses each edge at most $2d + 2$ times.

Deng and Papadimitriou suggested to study the dependence of R on m and d and showed the first upper and lower bounds. They gave a graph for which any algorithm, deterministic or randomized, needs at least $\Omega(d^2 m / \log d)$ edge traversals. Koutsoupias later improved the lower bound for deterministic algorithms to $\Omega(d^2 m)$.

Apart from the optimal algorithm for $d = 0$ which has a competitive ratio of two, Deng and Papadimitriou also gave an optimal algorithm for $d = 1$ with a competitive ratio of four. The generalization of this algorithm for greater d seems highly complicated, so instead they gave a different algorithm that explores a graph with deficiency d using $d^{O(d)}$ edge traversals. They posed the question whether the exponential gap between the upper and lower bounds can be bridged. They conjectured the existence of an online exploration algorithm whose competitive ratio is polynomial in d for all graphs with deficiency d. No such algorithm has been found so far.

Albers and Henziger [1] did a first step in giving a positive answer to that question. They presented the Balance Algorithm (described below) which can explore a graph of deficiency d with $d^{O(\log d)}m$ edges traversals. They also showed that this bound is tight for their algorithm.

They also gave lower bounds of $2^{\Omega(d)}m$ edge traversals for several natural exploration algorithms as Greedy, Depth-First, and Breadth-First. For Generalized Greedy they gave a lower bound of $d^{\Omega(d)}m$.

We now give a short description of these algorithms.

- Greedy (Gr):
  If stuck at a node y, the robot moves to the nearest known node z that has unexplored outgoing edges. As every edge has unit cost we are looking for a shortest path to an unfinished node.
- Generalized-Greedy (G-G):
  At any time, for each path in the subgraph explored so far, define a lexicographical vector as follows. For each edge on the path, determine its current cost, which is the number of times the edge was traversed so far. Sort these costs in non-increasing order and assign this vector to the path. If stuck at a node y, out of all paths to nodes with unexplored outgoing edges the robot traverses the path whose vector is lexicographical minimum.
- Depth-First Search Strategy (DFS):
  If stuck at a node y, the robot moves to the most recently discovered node z that can be reached and that has unexplored outgoing edges.
- Breadth-First Search Strategy (BFS):
  Let v be the node where the exploration started. If stuck at a node y, the robot moves to the node z that has the smallest distance from v among all nodes with unexplored outgoing edges that can be reached from y.
- Balance Algorithm (Bal):
  A graph with deficiency d can be made Eulerian by adding d edges (s_1, t_1), ..., (s_d, t_d). The Balance algorithm is too involved to give a full description here, see [1] for details. Basically, it tries to find the above mentioned missing edges by maintaining d edge-disjoint chains such that the end-node of chain i is s_i and the start-node of chain i is the current guess of t_i. The current guess of t_i will be marked with a token τ_i. As the algorithm progresses, paths can be appended at the start of each chain or inserted into chains. At termination, the start-node of chain i is indeed t_i. To mark chain i all edges on chain i are colored with color i.
  The algorithm consists of two phases.
  Phase 1: The robot traverses unexplored edges until getting stuck at a sink s. The algorithm switches to Phase 2. The cycle starting and ending at s is the initial chain C_0.
  Phase 2: Phase 2 consists of sub-phases. From a high-level point of view, at any time, the subgraph explored so far is partitioned into chains, namely C_0 and the chains generated in Phase 2. During the actual exploration in the

sub-phases, the robot travels between chains. While doing so, it generates or extends fresh chains, which will be taken into progress later, and finishes the chains currently in progress. During each sub-phase the robot visits a current node x of a current chain C and makes progress towards finishing the nodes of C. The current node of the first sub-phase is s, its current chain is C_0. The current node and current chain of sub-phase j depend on the outcome of sub-phase $j - 1$. All the chains introduced in Phase 2 are marked with a token.

A tree T is maintained by the algorithm such that each chain C corresponds to a node $v(C)$ of T, and $v(C')$ is a child of $v(C)$ if the last sub-path appended to C' was explored while C was the current chain.

While the robot always explores unvisited edges until it gets stuck, it relocates using the tree T. While the current chain is not finished it relocates to the next unfinished node on it. Simply spoken, the robot relocates to an unfinished chain C' whose subtree of the vertex $v(C')$ has the smallest number of tokens. In this sense the algorithm always tries to balance out the tree.

3   New Deterministic Algorithms

We also studied other deterministic algorithms. We considered different variations of Greedy and a new algorithm called List.

- Cheapest-Edge-Greedy (E-G):
  If stuck at a node y, the robot follows a path — leading to a node that still has unexplored outgoing edges — that chooses the edge at each node that has been used least often. In case of a tie the edge with the earliest exploration time will be chosen.

  This algorithm is similar to Generalized Greedy. However, for this algorithm we can give a simple worst case example with deficiency 0 that makes the robot oscillate between two ends of the graph, so it has a rather bad competitive ratio. Figure 1 shows the behavior of the robot on the graph. We string together many cycles of length two. The robot starts in the middle of this long chain (node 1), and it will alternately explore one new cycle at the right end (node 2, 4, 6, ...) and then at the left end (node 3, 5, 7, ...) always traversing all the cycles in between again and again.

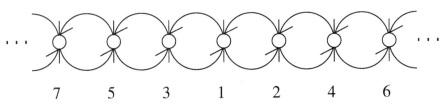

Fig. 1. The numbers represent the time the node has been explored

- Path-Cost-Greedy (C-G):
  This algorithm determines the cost of a path as the sum of all edge costs which is the number of times the edge was traversed so far. If stuck at a node y, the robot will follow the cheapest path to a node that still has unexplored outgoing edges.
- List Algorithm (List):
  Like Balance, List builds chains. However, the chains are stored in a list and every chain has an energy value associated with it. Chains will be considered in the order they appear in the list. If the robot gets stuck in a chain, this chain loses energy. If the energy of a chain drops down to zero, it will be moved from its current position in the list to the end with full energy again. A full and comprehensive description seems not possible here. No theoretical upper bound for this algorithm has been proven yet.

4   Randomized Algorithms

- Random Balance (R-B):
  Whereas the Balance algorithm relocates to a subtree which has minimum number of tokens, Random Balance chooses this subtree randomly.
- Harmonic (Har):
  When getting stuck, the robot will relocate to a node that still has unexplored outgoing edges with a probability that is inversely proportional to its distance from the current robot position.
- Random (Rand):
  After determining the paths to nodes with unexplored edges which do not pass any other unfinished node, the robot chooses one of these paths uniformly at random.

5   Experimental Setup

As we do not compare running times of the different algorithms but the number of edge traversals to explore a graph we do not give the machine details but we describe the testbeds we used. Naturally, more sophisticated algorithms need more computing time and the robot movements do not really take any time. So even when simple algorithms need more edge traversals they might finish their computations faster than more complicated algorithms.

We generated a huge testbed of random graphs. All programs can be found at http://www.cs.ust.hk/~trippen/graphtrav/source/. After creating one single cycle with m nodes and m edges we randomly merged nodes until only n nodes were left. Merging two nodes v_1 and v_2 means that the newly emerging node v has an incoming edge from a node u if at least one of the nodes v_1 or v_2 had an incoming edge from node u, and v has an outgoing edge to a node w if at least one of the nodes v_1 or v_2 had an outgoing edge to node w. This created a "random" graph of deficiency zero. To obtain a graph of deficiency $d > 0$, we added d edges ensuring that every new edge really increased the current deficiency by one (so the graphs have actually $m + d$ edges). We used four different methods (named $V1$ to $V4$ in the following) to add these d edges to obtain families of graphs with different structural properties where the various algorithms might show different behavior (as it turned out, their behavior did not differ). We either introduced the d additional edges randomly ($V1$), or we restricted them to starting at one common source ($V2$) or ending at one common sink ($V3$), or we chose one common source and one common sink ($V4$) for all these edges.

Although our algorithm to create the test graphs does not really create graphs according to a uniform distribution on all graphs with a certain deficiency (we do not know how to do that), our graphs do not seem to have hidden structural properties that could be unfairly exploited by any of the online algorithms.

We considered graphs with a deficiency from 0 up to 30. In our first series of experiments we considered graphs with $n = 30$ nodes. We generated graphs of different sizes with $m = 60$, $m = 100$, $m = 200$, and $m = 400$ edges. In the second series of experiments we considered graphs with $n = 100$ nodes and with $m = 500$, $m = 2,000$, and $m = 5,000$ edges.

For each choice of parameters we generated 100 random graphs which consist of the graphs with $n = 30$ nodes, and since the exploration of the big graphs with $n = 500$ and $m = 5,000$ is too time consuming we only generated ten random graphs for this family. While graphs of our first series were explored very quickly our second series took much more time. A single algorithm usually took a few minutes for graphs with $m = 500$ edges. On a graph with $m = 2,000$ edges it needed about 15 minutes. For the large graphs with $m = 5,000$ edges, the algorithms usually traversed only one edge per minute, on the average.

For the smaller graphs in the first series, we started each algorithm on every node of the graph. This way, each graph gave rise to 30 different experiments for each algorithm. Since we used 100 random graphs, every point in the diagram (for a particular parameter setting) for each algorithm is the average value of 3,000 runs.

Due to the long running times of the exploration algorithms on graphs of our second series we started every algorithm only once on each graphs. Thus, for these graphs every point in the diagram (for a particular parameter setting) for each algorithm is only the average value of 100 runs (or ten runs for the large graphs).

We also generated the families of lower bound graphs for the old deterministic algorithms to show their bad performance on these examples and to test how our new algorithms would perform on these graphs.

Additionally to our own random graphs, we used random graphs obtained from the Internet published by the University "Roma Tre", Italy
http://www.dia.uniroma3.it/~gdt/editablePages/test.htm.

These graphs have size ranging from 10 up to 100 nodes. Those sparse graphs were biconnected and directed. We only added edges to make them strongly connected.

The results of the experiments on these graphs will not be listed in Section 6 but they were quite similar compared to our random graphs with similar numbers of nodes, edges, and deficiencies.

6   Results

Table 1 and Fig. 2 show the performance of some old exploration algorithms – different variations of them will be discussed later – and our new List algorithm on the lower bound example for the Greedy algorithm [1]. Indeed, the same graphs can also be used as lower bound graphs for DFS and BFS. For the size of these graphs see Table 1. All algorithms clearly show a competitive ratio growing superpolynomially in the deficiency d (the y-axis is in logscale!). In consequence of the exponential growth of the size of these graphs, Fig. 2 only shows a smaller range of the x-axis compared to the following figures.

Table 2 and Fig. 3 compare the performance of all algorithms on three graphs of the family for the lower bound example for the Balance algorithm [1]. Because of the special recursive structure of this lower bound example there are only graphs with deficiency 2, 7, and 22. These three graphs have sizes $m = 16$ and $n = 14$, $m = 76$ and $n = 64$, and $m = 886$ and $n = 739$, respectively.

Table 3 and Fig. 4 show the results of experiments over 100 different random graphs with 100 nodes and $500 + d$ edges, where the d edges were added randomly ($V1$). Reflecting the fact that all natural exploration algorithms and their variations behave very similarly to each other, the curves in the diagram would be mostly overlapping. This is why we only show a few curves in the diagrams. Larger graphics with finer granularity can be downloaded at

Table 1. Number of edge traversals on the lower bound graph for Greedy, DFS, and BFS [1]

d	m	n	OPT	Greedy	DFS	BFS	Balance	List
1	13	10	15	26	26	26	26	26
2	31	24	43	94	94	94	86	86
3	71	56	124	329	329	329	330	258
4	159	128	316	1182	1182	1182	908	688
5	351	288	785	4002	4082	4002	2909	2102
6	767	649	1858	12808	13354	12808	8211	4635
7	1663	1408	4276	37997	41362	37997	20156	13479

Table 2. Number of edge traversals on the lower bound graph for Balance [1]

d	OPT	Gr	G-G	C-G	E-G	DFS	BFS	List	Bal	R-B	Har	Rand
2	33	39	39	39	39	39	39	39	63	63	39	39
7	370	474	475	475	475	518	474	480	752	968	430	474
22	11597	38828	45344	38360	45327	42175	48333	29235	28608	97779	39758	40284

http://www.cs.ust.hk/~trippen/graphtrav/source/.

Due to space limitations we cannot include the results of experiments with all other graph classes (28 different sizes, and different types of adding the d edges, $V1$-$V4$) in this extended abstract, but the algorithms behaved very similar on all the other graph classes. In Table 4 and Fig. 5, we show in comparison the results for random graphs with 30 nodes and $400 + d$ edges, $V4$.

7   Evaluation of the Results

The large testbed of random graphs shows that actually in most of the cases the performance of the algorithm is quite close (better that 2-competitive) to the optimum. However, for all the natural deterministic algorithms presented here it is possible to give a lower bound example where only $\exp(d)$-competitiveness can be achieved. But we cannot give a general lower bound that is greater than $\Omega(d^2)$ for *any* online algorithm.

Albers and Henzinger [1] mentioned that with respect to the current knowledge of the *s-t* connectivity problem it seems unlikely that one can prove superpolynomial lower bounds for a *general* class of graph exploration algorithms.

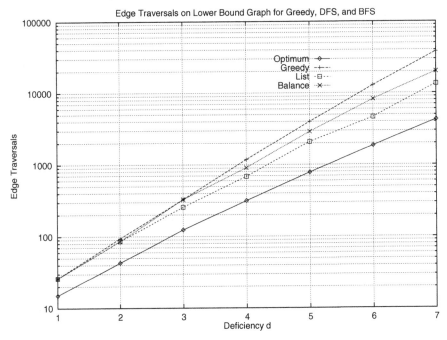

Fig. 2. Number of edge traversals on the lower bound graph for Greedy, DFS, and BFS [1]. See Table 1 for sizes of the graphs. DFS and BFS behave very similar to Greedy and therefore they are omitted here

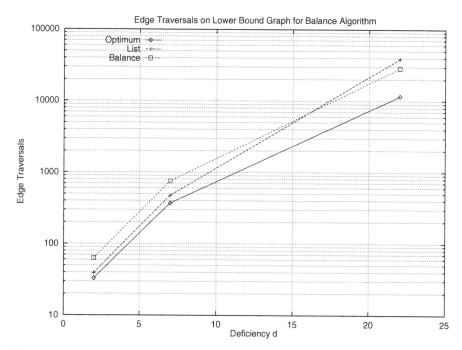

Fig. 3. Number of edge traversals on the lower bound graph for `Balance` [1]. The three graphs have sizes $m = 16$ and $n = 14$, $m = 76$ and $n = 64$, and $m = 886$ and $n = 739$, respectively. Here we only show three curves because the curves of all other algorithms are overlapped by the curve of the `List` algorithm

Indeed, every deterministic algorithm usually focuses on a specific part of the graph. This behavior is well known to the adversary, and he can exploit it to let the algorithm run large detours.

`Greedy` is always sticking to its nearby environment and it never considers changing to a different part of the graph unless the current local area is completely explored.

`DFS` has high relocation costs due to the fact that it always relocates to the "deepest" node where it could continue.

`BFS` changes to different parts of the graph when getting stuck but these parts always have about the same distance from the initial start node of the exploration. In doing so it might swap too often between different areas of the graph which implies high relocation costs.

The `Balance` algorithm might be forced to move tokens inside "small" areas. It then neglects other parts of the graph for too long time.

All the lower bound graphs for individual algorithms have one feature in common. They exploit the weakness of the specific algorithm and lead it again and again through the same bottleneck node. There are usually subproblems con-

Table 3. Average number of edge traversals on random graphs with 100 vertices and 500 + d edges (V1)

d	OPT	Gr	G-G	C-G	E-G	DFS	BFS	List	Bal	R-B	Har	Rand
3	510	521	522	521	578	539	533	600	616	616	530	533
6	519	532	532	532	610	561	552	629	663	663	546	552
9	528	542	542	542	638	584	572	652	705	706	560	568
12	535	553	553	552	664	600	592	686	762	763	578	584
15	541	561	564	562	689	619	608	727	823	825	590	598
18	549	573	571	569	715	631	622	746	882	886	602	611
21	557	577	584	579	743	647	641	775	897	899	610	620
24	563	587	592	589	746	677	661	787	904	908	629	642
27	569	593	602	595	757	691	674	816	976	980	638	651
30	575	600	614	605	789	708	690	837	1031	1033	652	669

Table 4. Average number of edge traversals on random graphs with 30 vertices and 400 + d edges (V4)

d	OPT	Gr	G-G	C-G	E-G	DFS	BFS	List	Bal	R-B	Har	Rand
3	408	414	415	415	415	418	416	424	433	433	424	418
6	414	425	427	426	428	430	426	448	466	466	452	432
9	424	438	457	439	441	445	441	464	492	492	468	443
12	431	441	466	448	478	452	449	467	499	499	448	452
15	444	460	511	461	503	466	466	480	524	524	463	466
18	452	467	627	472	524	475	473	492	546	546	472	473
21	459	478	686	483	553	489	488	490	549	549	484	486
24	465	481	779	494	576	497	487	504	569	569	490	490
27	468	490	810	501	586	503	490	519	586	586	497	497
30	478	494	917	517	629	521	505	510	588	588	506	511

structed in a recursive manner forcing the algorithm to traverse the bottleneck a super-polynomial(d) number of times.

Because of the simple structure of the "natural" exploration algorithms, Greedy and its variations, DFS, and BFS, it is easy to give exponential lower bound examples. Only the more sophisticated Balance algorithm by Albers and Henzinger [1] achieves a sub-exponential number of edge traversals, namely $d^{O(\log d)}m$.

The List algorithm uses the similar basic concept as Balance. But it considers the chains in a different order. List will not move the tokens of the same group of chains for a long time within the same part of the graph if this always implies relocations via the sink. Instead, it uses an energy value associated with each chain to preempt the further exploration on some chains to change to some other chains that have not caused so many relocations through their sinks.

For randomized algorithms, it seems harder to construct lower bound examples. The adversary cannot "hide" some part of the graph as easily as he

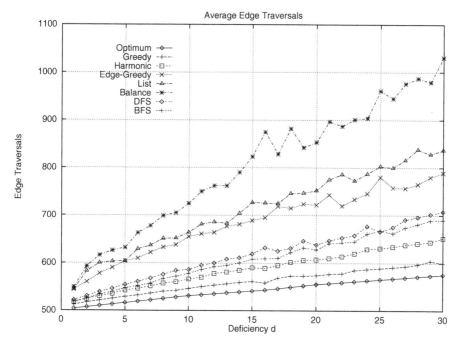

Fig. 4. Average number of edge traversals on random graphs with 100 vertices and $500 + d$ edges ($V1$). Omitted curves are mostly overlapped by the curve of the Greedy algorithm. Random Balance behaves very similar to Balance

can do when he knows where the algorithm will continue its exploration after getting stuck. The adversary can only "determine" which of the unexplored outgoing edges the robot will choose after relocating to a vertex with unexplored outgoing edges.

Despite their bad worst-case performance bounds, all natural online exploration algorithms showed near-optimal performance on our set of random graphs. Apparently, the randomly distributed edges in the graphs make relocation paths for these algorithms very short. The simplest algorithms, Greedy (Gr) and Rand, found the shortest tours. Generalized Greedy (G-G) and Path-Cost Greedy (C-G) are very similar to Greedy. This is implied by the fact that edges leading to nodes with unexplored outgoing edges are traversed not very often — so the path-vectors of paths leading to such nodes are very small, thus Generalized Greedy would prefer such a path —, and that a path with edges that are not traversed very often has low costs, and in general short paths often have lower costs than longer paths — so Path-Cost Greedy would also choose the same path as Greedy. However, in a very dense graph especially of variant $V4$ Generalized Greedy rather follows a "detour" on rarely traversed edges than a short path of often used edges. This results in a large number of total edge traversals which can be seen in Fig. 5.

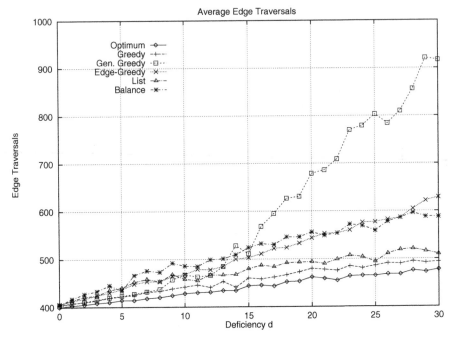

Fig. 5. Average number of edge traversals on random graphs with 30 vertices and $400 + d$ edges ($V4$). Omitted curves are mostly overlapped by the curve of the `Greedy` algorithm. `Random Balance` behaves very similar to `Balance`

The algorithms that work with chains (`Balance`, `Random Balance`, and `List`) have a little disadvantage in these cases as they are forced to follow their chains even if there are shorter ways to relocate. However, this only leads to a small overhead. We note that `List` performs better than `Balance`, although we do not have any theoretical analysis of this phenomenon. The standard exploration strategies `DFS` and `BFS` perform somewhere in the middle between `Greedy` and `Balance`.

On the other hand (and not surprisingly), the sophisticated algorithms do much better on the very special graphs that provide lower bounds for the simple algorithms.

8   Conclusions and Open Problems

In our experiments, we observe that all algorithms show a very good performance (better that 2-competitive) on all random graph families we tested, except on the family of graphs specially designed to show worst case behavior of the common deterministic algorithms where we observe an exponential number of edge traversals.

The main open problem remains unsolved. Is there an exploration algorithm for strongly connected directed graphs with deficiency d that achieves an upper bound on the number of edge traversals that is polynomial in d?

Randomization seems to be a promising way as our experimental results have shown that randomized algorithms do not show any disadvantage compared to deterministic ones. On the contrary, it seems to be hard to design good lower bound examples. Future work should include the analysis of randomized algorithms under different models of adversaries.

References

[1] S. Albers and M. R. Henzinger. Exploring unknown environments. *SIAM Journal on Computing*, 29(4):1164–1188, 2000. 122, 123, 127, 128, 129, 130

[2] S. Albers and B. Schröder. An experimental study of online scheduling algorithms. In *Proceedings of the 4th Workshop of Algorithms and Engineering (WAE'00)*. Springer Lecture Notes in Computer Science 1982, pages 11–22, 2000. 121

[3] R. Bachrach and R. El-Yaniv. Online list accessing algorithms and their applications: recent empirical evidence. In *Proceedings of the 8th ACM-SIAM Symposium on Discrete Algorithms (SODA'97)*, pages 53–62, 1997. 121

[4] A. Borodin and R. El-Yaniv. *Online Computation and Competitive Analysis*. Cambridge University Press, 1998. 120

[5] X. Deng and C. H. Papadimitriou. Exploring an unknown graph. In *Proceedings of the 31st Symposium on Foundations of Computer Science (FOCS'90)*, pages 355–361, 1990. 121, 122

[6] J. Edmonds and E. L. Johnson. Matching, Euler tours and the Chinese postman. *Mathematical Programming*, 5:88–124, 1973. 121

[7] A. Fiat and G. J. Woeginger, editors. *Online Algorithms — The State of the Art*. Springer Lecture Notes in Computer Science 1442, 1998. 120

[8] Jesus M. Salvo Jr. Openjgraph - java graph and graph drawing project. 2000. http://openjgraph.sourceforge.net/.

[9] C. H. Papadimitriou. On the complexity of edge traversing. *Journal of the ACM*, 23(3):544–554, 1976. 121

[10] D. Sleator and R. Tarjan. Amortized efficiency of list update and paging rules. *Communications of the ACM*, 28(2):202–208, 1985. 120

[11] Harold Thimbleby. The directed Chinese postman problem. Technical report. http://www.uclic.ucl.ac.uk/harold/cpp/index.html.

A Nondifferentiable Optimization Approach to Ratio-Cut Partitioning

Kārlis Freivalds

University of Latvia
Institute of Mathematics and Computer Science
Raina blvd. 29, LV-1459, Riga, Latvia
Karlis.Freivalds@mii.lu.lv

Abstract. We propose a new method for finding the minimum ratio-cut of a graph. Ratio-cut is NP-hard problem for which the best previously known algorithm gives an $O(\log n)$-factor approximation by solving its dually related maximum concurrent flow problem. We formulate the minimum ratio-cut as a certain nondifferentiable optimization problem, and show that the global minimum of the optimization problem is equal to the minimum ratio-cut. Moreover, we provide strong symbolic computation based evidence that any strict local minimum gives an approximation by a factor of 2. We also give an efficient heuristic algorithm for finding a local minimum of the proposed optimization problem based on standard nondifferentiable optimization methods and evaluate its performance on several families of graphs. We achieve $O(n^{1.6})$ experimentally obtained running time on these graphs.

1 Introduction

Balanced cuts of a graph are hard computation problems important both in theory and practice. Ratio-cut is the most fundamental one since most of the others including minimum quotient cut, minimum bisection, multi-way cuts can be easily approximated using it [13, 20]. Also several other important approximation algorithms like crossing number and minimum cut linear arrangement are based on the ratio-cut [13]. Ratio-cut has many practical applications, most important being VLSI design, clustering and partitioning [23, 14, 1].

Since ratio-cut is a NP-hard problem [15] we must seek for approximation algorithms to solve it in practically reasonable time. Many purely heuristic algorithms were developed [23, 25, 21, 8] most of them relying on simulated annealing, spectral methods or iterative movement of nodes from one side of the partition to the other. A common idea exploited by several authors [25, 2, 9, 21, 22] to improve their quality is using multi-scale graph representation usually obtained by edge contraction. At first a partition at the coarsest scale is obtained and then refined to a more detailed one by one of mentioned algorithms. Although these algorithms may perform well in practice no optimality bounds are known for them.

K. Jansen et al. (Eds.): WEA 2003, LNCS 2647, pp. 134–147, 2003.

The best previously known algorithm with proven optimality bounds finds an $O(\log n)$-factor approximation, where n is the number of nodes in the graph. It is based on reduction of the ratio-cut problem to the multi-commodity flow problem, which can be solved with polynomial time linear programming methods. Unfortunately this method is not practical since the resulting linear program is of quadratic size of the number of nodes in the graph and cannot be solved efficiently. Then, approximation algorithms [16, 17, 12, 24] were discovered for the multi-commodity flow problem itself making this approach usable in practice. Several heuristic implementations [16, 11, 24] are based on this idea, some of them quite effective and practical. The most elaborate one [11] can deal with up to 100000-node graphs in reasonable time.

In this paper we propose a new way of finding the minimum ratio-cut of a graph. We construct a nondifferentiable optimization problem whose minimum solution equals the minimum ratio-cut value and use nonlinear programming methods to search for it. Since the problem is non-convex, we may find only a local minimum. However, we show that any strict local minimum gives us a factor of 2 approximate cut. For that purpose we introduce a notion of locally minimal ratio cut for which no subset of nodes taken from one side of the cut and moved to the other side decrease the cut value. We establish one-to-one correspondence between strictly locally minimal cuts and strict local minima of the proposed optimization problem.

The reduction of a NP-hard discrete problem to a continuous one is not a novel idea. For example the maximum clique problem of a graph can also be stated as an optimization problem [6] and numerical optimization methods for finding the optimum may be used. However for the maximum clique problem no optimality bounds of a local minimum are known. To show that our method is practical we present an efficient heuristic algorithm for finding a local minimum of the proposed problem, which is based on the standard methods of nondifferentiable optimization and analyze its performance on several families of graphs. With the proposed method we can find a good partition of a 200000-node graphs in less than one hour.

2   Problem Formulation

We are dealing with an undirected graph $G = (V, E)$, where V is its node set and E is its edge set. The nodes of the graph are identified by natural numbers from 1 to n. Each node i has a weight $d_i = 0$, and each edge (i, j) has a capacity $c_{ij} = 0$, satisfying the properties $c_{ij} = c_{ji}$, $c_{ii} = 0$. We define $c_{ij} = 0$ when there is no edge (i, j) in G. We assume that there are at least two nodes with non-zero weights. (A, A') denotes a cut that separates a set of nodes A from its complement $A' = V \backslash A$. The capacity of the cut $C(A, A')$ is the sum of edge capacities between A and A'. The ratio of the cut $R(A, A')$ is defined as follows

$$R(A, A') = \frac{C(A, A')}{\sum_{i \in A} d_i \cdot \sum_{i \in A'} d_i} \tag{1}$$

We will focus on finding the minimum ratio-cut i.e. the cut (A, A') with the minimum ratio over all nonempty A, A'.

Definition 1. *A cut (A, A') is locally minimal if for all non-empty $U \subset A, U \neq A : R(A, A') \leq R(A \setminus U, A' \cup U)$ and for all non-empty $U' \subset A', U' \neq A' : R(A, A') \leq R(A \cup U', A' \setminus U')$. Similarly we call a ratio-cut strictly locally minimal if strong inequalities hold in this definition.*

3   Ratio-Cut as an Optimization Problem

We can assign a variable $x_i \in \mathbb{R}$ to each node i and consider the following optimization problem over x from $\mathbb{R}^n$.

$$\min F(x) = \sum_{i,j \in V} c_{ij} |x_i - x_j| \tag{2}$$

$$\text{subject to } H(x) = \sum_{i,j \in V} d_i d_j |x_i - x_j| = 1, \tag{3}$$

$$\sum_{i \in V} x_i = 0. \tag{4}$$

This optimization problem is equivalent to the ratio-cut problem in the sense described below.

Definition 2. *A characteristic vector x^A for a cut (A, A') is defined such that its components*

$$x_i^A = \begin{cases} a, i \in A \\ b, i \in A' \end{cases}, \text{ where} \tag{5}$$

$$a = -\frac{1}{2 \sum\limits_{i \in A} d_i \cdot \sum\limits_{i \in A'} d_i} \cdot \frac{|A'|}{|V|}, b = \frac{1}{2 \sum\limits_{i \in A} d_i \cdot \sum\limits_{i \in A'} d_i} \cdot \frac{|A|}{|V|}.$$

It is straightforward from this definition that for a cut (A, A') x^A satisfies the constraints (3, 4), and $F(x^A) = R(A, A')$.

Definition 3. *For some feasible x and some $p \in \mathbb{R}$ we call the cut (P, P') positional, if $P = \{i | x_i \leq p\}$, $P' = V \setminus P$, and both P and P' are non-empty. The ratio of this cut $R_p(x) = R(P, P')$. For a fixed x we can speak of minimum positional cut $R_{\min}(x)$ over all possible positional cuts obtained for different p:*

$$R_{\min}(x) = \min_{p \in \mathbb{R}} R_p(x).$$

Theorem 1. *For each feasible x^* of (2, 3, 4) $F(x^*) \geq R_{\min}(x^*)$. Also $F(x^*) > R_{\min}(x^*)$ if there are at least two positional cuts with different values.*

Proof. Here we only sketch the main steps of the proof, for details omitted due to space limitation refer to [7]. Let us partition all nodes into three sets U_1, U_2, U_3 as follows.

$$y_1^* = \min_i x_i^*, \quad y_2^* = \min_{i|x_i^*>y_1} x_i^*,$$

$$U_1 = \{i|x_i^* = y_1^*\}, \quad U_2 = \{i|x_i^* = y_2^*\}, \quad U_3 = \{i|x_i^* > y_2^*\}$$

If U_3 is empty there is exactly one positional cut, x^* is in the form of characteristic vector and $F(x^*) = R_{\min}(x^*)$ what concludes this proof, else define

$$y_3^* = \min_{i|x_i^*>y_2} x_i^*.$$

We create a sub-problem of (2, 3, 4) by reducing the original one to a new variable $y = (y_1, y_2, y_3)$. Next, we further restrict it with $y_1 \leq y_2 \leq y_3$ and can drop the absolute value signs getting:

$$\min F_2(y) = 2 \sum_{i\in U_1, j\in U_2} c_{ij}(y_2 - y_1) + \sum_{i\in U_1, j\in U_3} c_{ij}(y_3 + l_j - y_1) +$$

$$\sum_{i\in U_2, j\in U_3} c_{ij}(y_3 + l_j - y_2) + K, \quad (6)$$

$$\text{subject to } H_2(y) = 2 \sum_{i\in U_1, j\in U_2} d_i d_j(y_2 - y_1) + \sum_{i\in U_1, j\in U_3} d_i d_j(y_3 + l_j - y_1) +$$

$$\sum_{i\in U_2, j\in U_3} d_i d_j(y_3 + l_j - y_2) + P, \quad (7)$$

$$|U_1|y_1 + |U_2|y_2 + |U_2|y_2 = 0, \quad (8)$$

$$y_1 \leq y_2 \leq y_3, \quad (9)$$

where P and K are appropriately calculated constants.

We have obtained a locally equivalent linear program in the sense that for y^* the constraints (7, 8, 9) are satisfied and $F(x^*) = F_2(y^*)$. Also, if we can find a better solution for (6 − 9) we can substitute the result back to the original problem giving a better feasible solution for it. From the linear programming theory it is known that we can find the optimal solution by examining the vertices of the polytope defined by the constraints − in our case that means when one of the inequalities in (9) is satisfied as equality. Let us examine both cases.

In case $y_1 = y_2$ after some calculations we get

$$F_2(y_1, y_1, y_3) = (1 - L)R(U_1 \cup U_2, U_3) + B \quad (10)$$

for appropriate constants L and B. And similarly in case $y_2 = y_3$ we get

$$F_2(y_1, y_2, y_2) = (1 - L)R(U_1, U_2 \cup U_3) + B. \quad (11)$$

We chose the solution y with $y_1 = y_2$ if $R(U_1 \cup U_2, U_3) < R(U_1, U_2 \cup U_3)$, otherwise choose the solution with $y_2 = y_3$. It is evident that if both these ratio costs are non-equal we get a strictly smaller function value. We substitute the solution back into the original problem obtaining a new x. x is a feasible solution of (2, 3, 4) with a smaller or equal function value and the set U_2 merged to U_1 or U_3. We repeat the described process until U_3 is empty. Let us analyze the resulting x and sets U_1 and U_2. We have $F(x) = R(U_1, U_2)$, where (U_1, U_2) is some positional cut of x^* (in fact the minimal one), hence $F(x^*) = F(x) = R_{\min}(x^*)$. If we had some non-equal cuts compared during the process, we have a strict decrease in the function and hence the second statement of the theorem holds.   □

Definition 4. *A feasible x^* is a local minimum of (2, 3, 4) if there exists $\varepsilon > 0$ such that $F(x^*) \leq F(x)$ for each x satisfying (3, 4) and $||x - x^*|| < \varepsilon$.*

Definition 5. *A feasible x^* is a strict local minimum of (2, 3, 4) if there exists $\varepsilon > 0$ such that $F(x^*) < F(x)$ for each $x \neq x^*$ satisfying (3, 4) and $||x - x^*|| < \varepsilon$.*

Theorem 2. *Each strict local minimum x^* of (2, 3, 4) is in the form $x^* = x^A$ for some cut (A, A').*

Proof. We need to prove that in a strict local minimum the expression (5) holds for some a and b, the correct values for a and b are guaranteed by the constraints (3) and (4). Assume to the contrary that there are more than two distinct values for the components of x^*. We form the reduced problem like in Theorem 1 obtaining equations (6 − 9). We are able to do that since U_3 is non-empty due to our assumption. From the linear programming theory a strict local minimum can only be on the vertex of the polytope defined by the constraints (7 − 9), however in our case y^* cannot be on the vertex since the constraints where equality holds at y^* are less then the number of variables. Consequently, x^* cannot be a strict local minimum for the original problem. Hence our assumption is false and the theorem is proven.   □

There is one-to-one correspondence between strictly locally minimal cuts and strict local minimums of (2, 3, 4) as stated in the following theorem.

Theorem 3. *x is a strict local minimum of (2, 3, 4) if and only if x is a characteristic vector of some strictly locally minimal cut (A, A').*

The proof is rather technical and is omitted due to space constraints. See [7] for details.

We shall note that also for each non-strictly minimal ratio cut its characteristic vector x^A gives a local minimum of the function, however the converse is not true. There exist non-strict local minima of (2, 3, 4) with the function value not equal to any locally minimal cut value.

Theorem 4. *The minimum ratio-cut R is equal to the optimum solution of (2, 3, 4).*

Proof. From Theorem 1 $F(x) \geq R_{\min}(x) \geq R$. For the characteristic vector x^A of the minimum ratio cut $F(x^A) = R$. The claim follows immediately.   □

Theorem 5. *Each locally minimal cut* (A, A') *is not grater than two times the minimum ratio cut.*

Proof. Let us denote the optimal cut (B, B'). We form four sets $A \cap B$, $A \cap B'$, $A' \cap B$, $A' \cap B'$ shown in Fig. 1. From Definition 1 the ratio of each of the cuts $C_1 = (A \cap B, V - A \cap B)$, $C_2 = (A \cap B', V - A \cap B')$, $C_3 = (A' \cap B', V - A' \cap B')$, $C_4 = (A' \cap B, V - A' \cap B)$ is at least as large as ratio of $C_{12} = (A, A')$ and $C_{23} = (B, B')$. We form a full graph by taking each of the sets $A \cap B$, $A \cap B'$, $A' \cap B$, $A' \cap B'$ as the nodes of the graph and assign edge capacities as the sum of all edge capacities in the original graph between the corresponding sets. We obtain

$$R_1 = \frac{c_1 + c_4 + c_6}{d_1(d_2 + d_3 + d_4)}, R_2 = \frac{c_1 + c_2 + c_5}{d_2(d_1 + d_3 + d_4)},$$

$$R_3 = \frac{c_2 + c_3 + c_6}{d_3(d_1 + d_2 + d_4)}, R_4 = \frac{c_3 + c_4 + c_5}{d_4(d_1 + d_2 + d_3)},$$

$$R_{12} = \frac{c_2 + c_4 + c_5 + c_6}{(d_1 + d_2)(d_3 + d_4)}, R_{23} = \frac{c_1 + c_3 + c_5 + c_6}{(d_2 + d_3)(d_1 + d_4)}.$$

$$R_1 \geq R_{12}, R_2 \geq R_{12}, R_3 \geq R_{12}, R_4 \geq R_{12}.$$

And the statement of the theorem to be proven translates to $\frac{R_{12}}{R_{23}} \leq 2$. We verified this using symbolical computation in Mathematica 4.0. See [7] for details.   □

Corollary 1. *Each strict local minimum of (2, 3, 4) is not grater than two times the minimum ratio cut.*

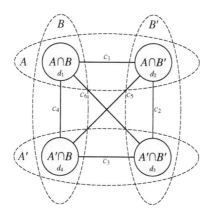

Fig. 1. Illustration to the proof of Theorem 5

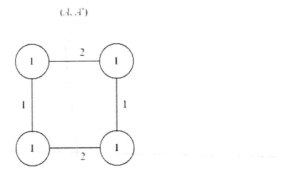

Fig. 2. A graph achieving the bound of Theorem

The proof is straightforward from Theorem 3 and Theorem 5.

The bound of Theorem 5 is tight; the graph with 4 nodes shown in Fig. 2 achieves this bound. One can make larger examples easily by substituting any connected graph with sufficiently high edge capacities in place of the nodes of the given graph.

4   The Algorithm

In this section we present a heuristic algorithm based on standard nondifferentiable optimization methods for finding a local minimum of (2, 3). Finding a minimum of a nondifferentiable function is one of well- explored nonlinear programming topics [5, 18]. One of the possibilities is to approximate the nondifferentiable function with a smooth one and apply one of the well-known algorithms to find its local minimum [5, 3]. Often a better approach is to handle it directly. Indeed, in our case we obtain a very simple and fast algorithm.

Most of the optimization theory deals with convex problems for which algorithms with proven convergence can be developed. Many of these methods also work for non-convex functions finding a local minimum. However, then the convergence cannot be shown or can be shown only in a local neighborhood of some local minimum what is not satisfactory in our case. The very basic algorithm of nondifferentiable optimization is a *subgradient* algorithm [5, 18]. We will adopt it for solving our problem. Since we apply it to a non-convex problem, it should be considered mostly as a heuristic, however practice shows that it actually converges to a local minimum of our problem.

The algorithm is iterative one. The iteration of the algorithm consists of going in the negative direction of a subgradient of the function by a fixed step and then performing a projection onto the constraint (3). The constraint (4) was introduced for technical reasons required in proofs and may be not considered in the algorithm. An appropriate subgradient q of F can be calculated with the following equation for its components

$$q_i = \sum_{j \in V} \text{sign}(x_i - x_j),$$

Choosing the right step size is crucial for the convergence speed. Our heuristic observation is that it should be proportional to the node spread. We choose the step equal to $\texttt{stepFactor} \cdot (x_{\max} - x_{\min})$, where $\texttt{stepFactor}$ is some parameter. Initially $\texttt{stepFactor} = 1/14$. Its update during the algorithm is be discussed later. The projection is performed by going in the direction of a subgradient of the constraint till the constraint is reached. For the constraint H we can write its subgradient r in a different form to allow its faster evaluation

$$r_i = d_i \sum_{j \in V} d_j \text{sign}(x_i - x_j) = d_i \left(\sum_{j:x_j < x_i} d_j - \sum_{j:x_j > x_i} d_j \right).$$

If we sort the x_i values and consider them in increasing order, then the needed sums can be updated incrementally leading to linear time evaluation (not counting the sorting). To perform the projection we need to calculate the step length towards the constraint. To simplify the calculations we will assume that the ordering of x_i will not change during the projection. Then the constraint function H becomes linear and the desired step can be easily calculated from the linear equation defined as the point of value 1 on the ray defined by the subgradient from the current point. In the case of unit node weights our assumption is fulfilled. To see this we must observe that the step in the function will always lead to a point with $H < 1$. Then, if we have unit node weights, $r_i \leq r_k$ for all i and k satisfying $x_i \leq x_k$, so after the projection the distance between them can only increase thus keeping the same ordering. For the case of arbitrary node weights such algorithm gives a usable approximation of the projection step length.

The whole algorithm starts with a random initialization. We assign a random position for each node such that $H < 1$ and then perform a projection to obtain a feasible starting position. We experimented also with several other initializations, but obtained significant improvements only for tree graphs. Since the optimal cut for trees can be found in linear time, we can construct a starting position that reveals it and the algorithm will only perform a few iterations to confirm that the solution does not improve. Hence to get a comprehensive picture of the algorithm behavior we decided to consider random initialization only.

After the initialization we perform iterations until convergence. To tell when the process is converged we track the minimum positional cut values obtained at each iteration. The same values will also tell us how to change the step size parameter $\texttt{stepFactor}$. If the new cut value is lower than the previous then we are making progress and we should continue with the same step size. If the value is higher than the previous then the step size is too large. If the cut does not change then we have a clue that the process has converged. Of course we do not hurry to make decisions only from one iteration. Instead we wait for a certain number of iterations controlled by the constants MAX_OSCILLATIONS and MAX_EQUAL before making the decision. Such delay also improves the convergence speed by allowing to iterate longer with a larger step size

To determine the positional cut value at each iteration proceed as follows. We consider the sorted sequence of nodes, calculate the positional cut in each interval between two consecutive nodes and take the minimum one. We can do it incrementally on the sorted sequence in time $O(n + m)$ provided the sorting, where m is the number of edges in the graph.

As suggested in one of the exercises in [5] the performance of this algorithm can be improved by taking the previous directions into account. We add the previous direction to the current reduced be some constant REDUCTION_FACTOR between 0 and 1. It models a heavy ball motion in the presence of a force in the direction of the subgradient. In our experiments such modification with REDUCTION_FACTOR = 0.95 performed substantially better.

All the steps described before can be implemented to run in time $O(n + m)$. Adding the time needed for sorting the nodes one iteration takes time $O(m + n \log n)$. The number of iterations is hard to estimate so we will provide experimental data in the next sections. The constants MAX_OSCILLATIONS and MAX_EQUAL have the most impact on the iteration count and also on the quality of the obtained cut. So we must select them carefully. After some experimentation we chose MAX_EQUAL = 200 and MAX_OSCILLATIONS = 30.

```
algorithm ratio-cut
  calculate a random feasible initial position
  acum = 0,oscillationCounter = 0,equalityCounter = 0,stepFactor = 1/14
  while (equalityCounter < MAX EQUAL)
    x = x + acum
    d = direction(x)
    x = x + d
    acum = acum * REDUCTION FACTOR + d
    if(minimum positional cut value has increased in this iteration)
      equalityCounter = 0
      oscillationCounter ++
    else if (minimum positional cut value has decreased in this iteration)
      equalityCounter = 0
    else equalityCounter ++
    if (oscillationCounter > MAX OSCILLATIONS)
      stepFactor /= 1.3
      oscillationCounter = 0
  endwhile
end
function direction (x)
  d = subgradient(x)
  step = (xmax - xmin) * stepFactor
  x1 = x + d*step
  x2 = projection of x1 on the constraint
  return x2-x
end
```

5   The Data

We evaluated the proposed algorithm on three families of graphs: random cubic graphs, random geometric graphs and random trees. We considered only graphs with unit weight nodes and edges.

Random cubic graphs are potentially hard for ratio-cut algorithms since in [13] it was shown that there is actually a O(log n) gap between the minimum ratio-cut and the maximum concurrent flow on these graphs. We generate them using the algorithm provided in [19].

Random geometric graphs are standard test suite for balanced cut problems used in several papers [11, 25]. To generate a geometric graph we place the nodes of the graph randomly in the unit square. Then we include an edge between each two nodes that are within distance δ in the graph, where δ is the minimum value such that the resulting graph is connected.

Tree graphs are seemingly easy graphs because their optimal ratio-cut can be calculated in linear time. Also it is not hard to show that only one local minimum exists for the corresponding optimization problem. Nevertheless, it is an interesting family since our experiments indicate a slow convergence of our algorithm on these graphs. Also we can compare our result with the optimal one. We generate random trees with the classical algorithm where each tree is produced with the same probability. This algorithm produces long and skinny trees, which are particularly difficult for our algorithm.

6   Experimental Results

We implemented the algorithm in C++ and evaluated its performance on a computer equipped with a Pentium III 800 MHz processor and 256 Mbytes of RAM. For each graph family we measured the running time in seconds, the number of iterations and the quality of the produced cuts. Since we did not know the exact cut values for random and geometric graphs, we evaluated how much the ratio-cut value decreases when we continue the algorithm for the same number of iterations as performed before termination. Measuring the decrease of the ratio-cut value we can estimate how far the result is from the optimal.

The algorithm was run on series of graphs of exponentially increasing size from 100 up to 204800 nodes. Ten graphs were generated for each size and the results were averaged. The average node degree for all graph families is constant. Although we cannot specify the degree explicitly for geometric graphs, due to their nature it was about 10 on all instances. The experimental results are given in figures 3 – 8. For each graph family tables show the running time, and how much the the the ratio-cut value improves when the iteration count is doubled. Let us discuss the results separately for each graph family.

6.1   Cubic Graphs

Although these graphs were suggested as hard, the algorithm performed very well on them. It took on average about 35 minutes to partition the 204800 node graphs. The running time dependence from the graph size is shown in Fig. 3.

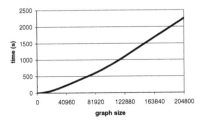

Fig. 3. Running time for cubic graphs

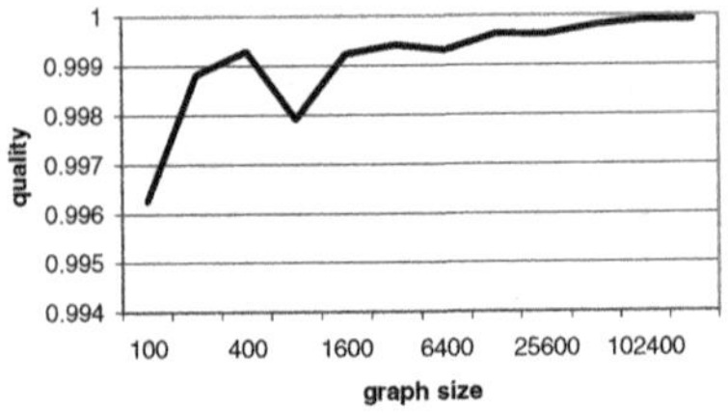

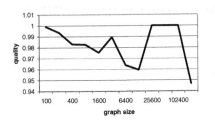

Fig. 4. Quality for cubic graphs

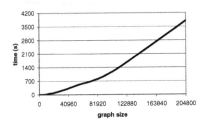

Fig. 5.   Running time for geometric graphs

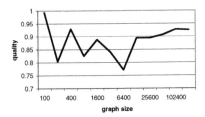

Fig. 6. Quality for geometric graphs

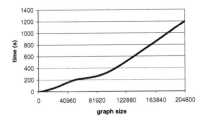

Fig. 7. Running time for tree graphs

Fig. 8. Quality for tree graphs

When we approximated the running time with a function in a form $O(n^p)$ we get the asymptotical running time about $O(n^{1.6})$ on these graphs.

The algorithm seems to find a very close to optimal cut since after doubling the iteration count the quality increased only by less than 0.4%. Even more, the quality improved for the larger graphs approaching 1 (see Fig. 4). Such behavior is not surprising since it is not hard to prove that the ratio of a cut (A, A') of a random cubic graph is $\frac{3}{n-1}$ on average independently of the sizes of A and A' (a similar proof for general random graphs is presented in [23]). Then it is unlikely that the minimum ratio cut will be much different from this average value.

6.2   Geometric Graphs

The algorithm performed very well on these graphs both in terms of speed and quality. It took on average about 1 hour to partition the 204800 node graphs. The running time dependence from the graph size is shown in Fig. 5. The asymptot-

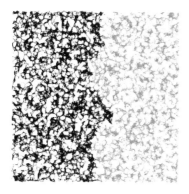

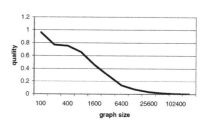

Fig. 9. The obtained cut for a 1000-node geometric graph

Fig. 10. Optimality for tree graphs

ical running time behavior on these graphs was about $O(n^{1.64})$. After doubling the iteration count the quality increased by less than 5% (see Fig. 6). Also visually the cuts seemed the best possible. Fig. 9 shows a typical cut of a 10000-node graph.

6.3   Tree Graphs

The running time for trees was better than for other families. The largest graphs were partitioned in about 20 minutes. The running time dependence from the graph size is shown in Fig. 7. The asymptotical running time on these graphs was about $O(n^{1.45})$. However the quality was poor. As shown in Fig. 10 the obtained cuts were far from the optimal and the quality decreased with increasing graph size. Also doubling the iteration count showed 10% to 20% quality improvement (see Fig. 8).

When we explored further the reason of the poor behavior we found out that convergence is much slower than for other graph families and the stopping criterion does not work correctly in this case. When we allowed the algorithm to run for a sufficiently long time, it always found the optimum solution. However we did not find a robust stopping criterion that correctly works with tree graphs and does not increase running time much for other graph families. As already mentioned a smarter initialization can be used to improve the quality of the partition if such tree or tree-like graphs are common for some application.

7   Conclusions and Open Problems

We have proposed a nondifferentiable optimization based method for solving the ratio-cut problem and presented a heuristic algorithm implementing it. We have shown that any strict local minimum is 2-optimal. The presented algorithm, however, in certain cases can find a non-strict minimum, but we can easily transform the obtained x vector into the characteristic form. Then the algorithm can be run again from this starting position and this process can be iterated until the result does not change giving a locally minimal cut, which by Theorem 5 is 2-optimal.

The obtained algorithm is simple and fast and uses amount of memory that is proportional to the size of the graph. Its running time and quality are verified experimentally. Its practical running time is about $O(n^{1.6})$ on our test data. The algorithm produce high quality cuts on random cubic and geometric graphs. On trees and other very sparse graphs the quality can be significantly improved by choosing a better starting position than a random one. We evaluated the algorithm on artificially generated data. As a further work it would be important to evaluate its performance on real-life problems. Although the algorithm performed well on most graphs, anyway it is heuristic. It is an open question whether we can find a local minimum of (2, 3, 4) in polynomial time?

Acknowledgements

The author would like to thank Paulis Kikusts and Krists Boitmanis for valuable discussion and help in preparation of the paper.

References

[1] C. J. Alpert and A. Kahng, Recent directions in netlist partitioning: a survey, tech. report, Computer Science Department, University of California at Los Angeles, 1995. 134

[2] C. J. Alpert, J. H. Huang, and A. B. Kahng. Multilevel circuit partitioning. Proc. Design Automation Conf, pp. 530 − 533, 1997. 134

[3] R. Baldick, A. B. Kahng, A. Kennings and I. L. Markov, Function Smoothing with Applications to VLSI Layout, Proc. Asia and South Pacific Design Automation Conf., Jan. 1999. 140

[4] J. W. Berry and M. K. Goldberg, Path Optimization for Graph Partitioning Problems. Technical report TR: 95-34, DIMACS, 1995.

[5] D. Bertsekas, Nonlinear Programming, Athena Scientific, 1999. 140, 142

[6] I. Bomze, M. Budinich, P. Pardalos, and M. Pelillo. The maximum clique problem. Handbook of Combinatorial Optimization, volume 4. Kluwer Academic Publishers, Boston, MA, 1999. 135

[7] K. Freivalds. A Nondifferentiable Optimization Approach to Ratio-Cut Partitioning. Manuscript 2003. Available at
http://www.gradetools.com/karlisf/papers.html. 137, 138, 139

[8] L. Hagen and A. B. Kahng, New spectral methods for ratio cut partitioning and clustering, IEEE Trans. Computer-Aided Design, 11 (1992), pp. 1074 − 1085. 134

[9] T. Hamada, C. K. Cheng, P. M. Chau. An efficient multilevel placement technique using hierarchical partitioning. IEEE Trans. on Circuits and Systems, vol. 39(6) pp. 432 − 439, June 1992. 134

[10] P. Klein, S. Plotkin, C. Stein, E. Tardos, Faster approximation algorithms for unit capacity concurrent flow problems with applications to routing and sparsest cuts, SIAM J. Computing, 3(23) (1994), pp. 466 − 488.

[11] K. Lang and S. Rao. Finding Near-Optimal Cuts: An Empirical Evaluation. 4th. ACM-SIAM Symposium on Discrete Algorithms, pp. 212 − 221, 1993. 135, 143

[12] T. Leighton, F. Makedon, S. Plotkin, C. Stein, E. Tardos, S. Tragoudas, Fast approximation algorithms for multicommodity flow problems. Journal of Computer and System Sciences, 50(2), pp. 228 − 243, April 1995. 135

[13] T. Leighton, S. Rao. Multicommodity max-flow min-cut theorems and their use in designing approximation algorithms. Journal of the ACM, vol 46 , No. 6 (Nov. 1999), pp. 787 − 832. 134, 143

[14] T. Lengauer. Combinatorial Algorithms for Integrated Circuit Layout. Stuttgart, John Wiley & Sons 1994. 134

[15] D. W. Matula, F. Shahrokhi, Sparsest Cuts and Bottlenecks in Graphs. Journal of Disc. Applied Math., vol. 27 (1990), pp. 113 − 123. 134

[16] F. Shahrokhi, D. W. Matula, On Solving Large Maximum Concurrent Flow Problems, Proceedings of ACM 1987 National Conference, pp. 205 − 209. 135

[17] F. Shahroki, D. W. Matula, The maximum concurrent flow problem. Journal of the ACM, vol. 37, pp. 318 − 334, 1990. 135

[18] N. Z. Shor, Methods of minimization of nondifferentiable functions and their applications. Kiev, "Naukova Dumka" 1979. (in Russian). 140

[19] A. Steger and N. C. Wormald, Generating random regular graphs quickly. Combinatorics, Probab. and Comput. vol. 8 (1999), pp. 377 − 396. 143

[20] M. Wang, S. K. Lim, J. Cong, M. Sarrafzadeh. Multi-way partitioning using bipartition heuristics. Proc. Asia and South Pacific Design Automation Conf., pp. 667 − 672, 2000. 134

[21] Y. C. Wei, C. K. Cheng, An Improved Two-way Partitioning Algorithm with Stable Performance, IEEE Trans. on Computer-Aided Design, 1990, pp. 1502 − 1511. 134

[22] Y. C. Wei, C. K. Cheng, A two-level two-way partitioning algorithm, Proc. Int'l. Conf. Computer- Aided Design, pp. 516 − 519, 1990. 134

[23] Y. C. Wei, C. K. Cheng, Ratio Cut Partitioning for Hierarchical Designs. IEEE Trans. on Computer- Aided Design, vol. 10, pp. 911− 921, July 1991. 134, 144

[24] C. W. Yeh, C. K. Cheng, and T. T. Y. Lin. A probabilistic multicommodity-flow solution to circuit clustering problems. IEEE International Conference on Computer-Aided Design, pp. 428 − 431, 1992. 135

[25] C. W. Yeh, C. K. Cheng, T. T. Y. Lin, An Experimental Evaluation of Partitioning Algorithms, IEEE International ASIC Conference, P14-1.1 − P14-1.4. 1991. 134, 143

Comparing Push- and Pull-Based Broadcasting

Or: Would "Microsoft Watches" Profit from a Transmitter?

Alexander Hall[1],[*] and Hanjo Täubig[2],[**]

[1] Computer Engineering and Networks Laboratory, ETH Zürich
CH-8092 Zürich, Switzerland
hall@tik.ee.ethz.ch
[2] Department of Computer Science, TU München
D-85748 Garching b. München, Germany
taeubig@in.tum.de

Abstract. The first main goal of this paper is to present Sketch-it!, a framework aiming to facilitate development and experimental evaluation of new scheduling algorithms. It comprises many helpful data-structures, a graphical interface with several components and a library with implementations of selected scheduling algorithms. Every scheduling problem covered by the classification-scheme originally proposed by Graham et al. [22] can easily be integrated into the framework.
One of the more recent enhancements of this scheme, the so called *broadcast scheduling* problem, was chosen for an extensive case study of Sketch-it!, yielding very interesting experimental results that represent the second main contribution of this paper. In broadcast scheduling many clients listen to a high bandwidth channel on which a server can transmit documents of a given set. Over time the clients request certain documents. In the *pull-based* setting each client has access to a slow bandwidth channel whereon it notifies the server about its requests. In the *push-based* setting no such channel exists. Instead it is assumed that requests for certain documents arrive randomly with probabilities known to the server. The goal in both settings is to generate broadcast schedules for these documents which minimize the average time a client has to wait until a request is answered.
We conduct experiments with several algorithms on generated data. We distinguish scenarios for which a slow feedback channel is very advantageous, and others where its benefits are negligible, answering the question posed in the title.

1 Introduction

During the past 40 years scheduling problems have received a lot of research interest, a huge theoretical background was developed. For books on the topic

---

[*] Supported by the joint Berlin/Zürich graduate program CGC, financed by ETH Zürich and German Research Foundation (DFG).
[**] Supported by DFG grant Ma 870/6-1 (SPP 1126: Algorithmics of large and complex Networks) and by DFG grant Ma 870/5-1 (Leibniz Award Ernst W. Mayr).

K. Jansen et al. (Eds.): WEA 2003, LNCS 2647, pp. 148–164, 2003.

see for instance [13, 14, 15]. While working on scheduling problems it would often be convenient to have a tool with which an algorithm could be implemented and tested quickly. "Playing" with the algorithm can help gaining intuition of how it works, or on the other hand potentially speed up the finding of counter-examples and bad cases. Quite often it is also meaningful to get hints on the performance of an algorithm by having a quick glance at its empirical behavior. Then again some heuristics can only be evaluated by conducting such experiments. But besides for testing new algorithms, a tool which is able to animate the progress of an algorithm could also prove very helpful for presentations or in teaching topics of scheduling theory.

These points stimulated the development of Sketch-it!, a framework for simulation of scheduling algorithms. To maximize the applicability, its design was closely linked to the $\alpha|\beta|\gamma$-classification-scheme, originally proposed by Graham et al. [22]. Basically all problems covered by this scheme can be tackled with the help of the framework.

In this paper we give a short overview of the framework, in order to introduce it to a broad audience. Furthermore we present experimental results in the broadcast scheduling domain, which were obtained with the help of Sketch-it!. The motivation for this is partly to demonstrate the usability of the tool, but mainly we believe that the results are of interest in their own. In the next section we motivate and define broadcast scheduling.

Motivation and Problem Statement of Broadcast Scheduling Due to the increasing availability of infrastructure that supports high-bandwidth broadcast and due to the growth of information-centric applications, broadcast scheduling is gaining practical importance [4]. The general setting of the broadcast scheduling problem is that (possibly many) clients request documents (e.g. web pages) from a server, and the server answers these requests via a high-bandwidth channel to which all clients are connected. If several clients have requested the same document, a single broadcast of this document satisfies all their requests simultaneously. One wants to determine a broadcast schedule that optimizes some objective function, usually the average response time (the time a client has to wait on average until her request is satisfied; in the scheduling literature, this is also called the average flow time).

There are two principally different settings: on the one hand *on-demand* or *pull-based* broadcasts and on the other hand *push-based* broadcasts.

In the *pull-based* setting each client has access to a low-bandwidth *feedback* channel, e.g. a modem connection, whereon it notifies the server about its requests. Two examples of such systems are @Home network [24] and DirecPC [18], which provide Internet access via cable television and via satellite, respectively.

In the *push-based* setting no feedback channel exists. The server tries to anticipate user behavior from the previously observed popularities of individual documents. A classical example are Teletext systems, where the user can select a page on her remote control and then has to wait until it appears in the periodic broadcast. A more recent application is the SPOT technology announced by Microsoft [34]. It enables special wristwatches (and in the future also other devices) to receive personalized information—like weather, events and personal messages—via a dedicated radio frequency, which is shared for all broadcasts. The watches contain no transmitter, i.e. it is not possible to add a feedback channel. Clients configure their desired contents beforehand via a Web interface.

In this paper we present the first direct comparison of the empirical performance of several well known pull-based on-line algorithms with a push-based algorithm on the same input traces. Such a comparison may help, e.g., in deciding whether or not to integrate feedback channels into a broadcast system.

We now give a more precise problem definition. In the following we restrict ourselves to the *single channel* case (at any point in time no more than one document can be broadcasted), in the literature the case of $W > 1$ channels is also considered. Furthermore we allow *arbitrary-length* documents and *preemption*, i.e. a broadcast of a document can be interrupted and continued at a later point in time. We adopt the commonly made assumption that a client can *buffer* the last portion of a document and thus can start receiving the requested document immediately at any point of it. A request is satisfied as soon as all "units" of the document were received.

Let m be the number of documents and n be the total number of requests. By $l_i \in \mathbb{N}$ we denote the length of document $i \in \{1 \ldots m\}$. R_j, where $j \in \{1 \ldots n\}$, is used both to address the j-th request and to denote its arrival time. Let D_j be the document requested by R_j and $T = R_n$ be the arrival time of the last request.

The output of an algorithm in both the pull- and the push-based setting is a schedule $S(t) \in \{0 \ldots m\}$, for $t \in \mathbb{R}^+$, giving which document is broadcasted at time t, where $S(t) = 0$ means that no document is broadcasted. For simplicity we define $S^i(t) := 1$, if $S(t) = i$, and $S^i(t) := 0$ otherwise. Let T' be the point in time when all requests R_j are answered by S, w.l.o.g. $S(t) = 0$, for $t > T'$.

Input and objective function in the two settings differ though.

Pull-Based Here the total average response time ART is given by $\frac{1}{n} \sum_{j=1}^{n} F_j$, where $F_j := C_j - R_j$ is the response/flow time of request j and C_j is its completion time. More precisely $C_j = C^{D_j}(R_j)$, with: $C^i(t) := \inf\{x| \int_t^x S^i(\theta)d\theta = l_i\}$.

$B1|r_j, pmtn|\frac{1}{n}\sum F_j$ denotes the problem of minimizing the ART in this setting. An on-line algorithm for this problem only knows of the requests with $R_j < t$, when deciding which document to broadcast at time t. It is called ρ-competitive, if it computes a schedule whose ART is at most ρ times the ART of an optimal solution.

A W-speed ρ-competitive algorithm computes a schedule on W channels whose ART is at most ρ times the ART of an optimal solution on *one* channel.

Push-Based Unlike in the pull-based setting, the algorithm does not learn the actual requests R_j. Instead it is assumed that an infinite sequence of requests is generated in a Poisson process, i.e. the request interarrival times are exponentially distributed. The algorithm knows in advance the probabilities π_i with which a request R_j, $j \in \{1 \ldots n\}$ is for document $i \in \{1 \ldots m\}$. It computes an infinite (periodic) schedule which minimizes the *expected* instead of the average response time: ERT $:= \mathbb{E}(\lim_{n\to\infty} \frac{1}{n} \sum_{j=1}^{n} F_j) = \mathbb{E}(F)$, where $\mathbb{E}(F)$ is the expected flow time of any request, given the Poisson assumption and the request probabilities π_i. The problem of minimizing the ERT is denoted by $B1|r_j, pmtn|\mathbb{E}(F)$. Note that algorithms are analyzed in an average case fashion, the output is usually not compared to an optimal solution for a given input.

An algorithm for this problem is called ρ-approximation, if it runs in polynomial time and computes a schedule whose ERT is at most ρ times the optimum ERT.

For our experiments we naturally only consider a finite sequence of requests $R_j, j \in \{1 \dots n\}$. We limit the schedule $S(t)$ computed by a push-based algorithm to the interval $[0, T']$, where T' is the point in time when all requests R_j are answered by S, and as already mentioned above we assume $S(t) = 0$, for $t > T'$. We assume that the algorithm is fair and does not starve certain documents (i.e. T' is finite).

To facilitate the comparison of the results in the different settings, we will also compute the ART in the push-based case. Also in both settings let MRT denote the *maximum response time* of a request.

Organization of this paper and Contributions In the next section we give an overview of related work. Section 2 contains a high level description of the Sketch-it! framework. Finally in Section 3 we present our experimental results, which assert the selected push-based algorithm comparatively good performance with respect to ART, independently of how many documents are present. Furthermore the quality of the schedules do not vary depending on the document sizes. Contrasting this, the MRT comparatively increases with increasing m. A nice feature of the push-based algorithm is that it is independent of the load of a system, which means it scales nicely.

Related Work

Simulation A project similar to Sketch-it! is being developed since 1999 at the Institute for Algebra and Geometry of the Otto-von-Guericke University in Magdeburg. Its name *LiSA* stands for Library of Scheduling Algorithms, but actually it is specialized in shop scheduling. It is implemented in C++. For detailed documentation see [25].

The *Cheddar Project* is a simulation tool for real time scheduling, which was developed by the LIMI/EA 2215 team at the University of Brest (France) using the Ada programming language. It comes with a graphical editor and a library of classical real time scheduling algorithms and feasibility tests. For documentation and download of the distribution see [33].

Pull-Based Broadcast In [5, 6] Aksoy and Franklin introduce pull-based broadcast scheduling and conduct experiments with a proposed on-line algorithm for the case of unit-length documents and without preemption ($B1|r_i, p_i =$

$1|\frac{1}{n}\sum F_j)$. Kalyanasundaram et al. [26], Erlebach and Hall [20] and Gandhi et al. [21] consider approximation algorithms for the corresponding off-line case. In [20] NP-hardness of the off-line setting is proved (also for $B1|r_i, pmtn|\frac{1}{n}\sum F_j$ with or without client buffering).

Acharya and Muthukrishnan [4] adopt the stretch-metric from [12] to broadcast scheduling and present experiments conducted with new on-line algorithms for arbitrary-length documents. Among other things, they show that preemption is very beneficial.

Edmonds and Pruhs [19] consider the case of arbitrary-length documents, with preemption ($B1|r_i, pmtn|\frac{1}{n}\sum F_j$) and *without* client buffering. They prove that no 1-speed o($n^{1/2}$)-competitive algorithm exists (this result extends to the case with client buffering). They furthermore present the first positive result concerning the on-line setting: an $\mathcal{O}(1)$-speed $\mathcal{O}(1)$-competitive algorithm.

In [11], $B1|r_i, pmtn| \max F_j$ is addressed. They present a PTAS (polynomial-time approximation scheme) for the off-line case and show that First-Come-First-Serve is a 2-competitive on-line algorithm.

Push-Based Broadcast Ammar and Wong [7, 8] study the case of unit-length messages broadcasted on one channel ($B1|r_i, p_i = 1|\mathbb{E}(F)$) in the context of Teletext systems. For arbitrary number of channels W, this setting is also known as broadcast disks, treated e.g. in [1, 2]. There are several results for the variation of the problem where each complete broadcast of document $i \in \{1\dots m\}$ is additionally assigned a cost c_i. The goal is to minimize the sum of the ERT and the *average broadcast cost*. Bar-Noy et al. [9] prove this is NP-hard for arbitrary c_i. They furthermore give a 9/8- and a 1.57-approximation, if $c_i = 0$ respectively c_i arbitrary, for $i \in \{1\dots m\}$. This is improved by Kenyon et al. [28]: they obtain a PTAS if the number W of channels and the costs c_i are bounded by constants.

For the case of arbitrary-length documents, without preemption see for instance [23, 38, 27]. For the case of arbitrary-length documents, with preemption Schabanel [36] presents a 2-approximation for $BW|r_i, c_i, pmtn|\mathbb{E}(F)$ based on the approach in [9] and adapts an NP-hardness proof from [27].

Acharya et al. [3] do an experimental study on interleaving push- and pull-based data broadcast for the case of unit length documents.

2 Sketch-it! – A New Scheduling Simulation Framework

The simulation framework was born in spring of 1999. It was developed under the supervision of Ernst W. Mayr, within the scope of the special research program SFB 342 of the German Research Foundation (DFG), part A7: "Efficient Parallel Algorithms and Schedules". A distribution version will be available soon at [32]. Source-code documentation can be found at [31].

Aim The original aim of the tool was to support research in the subject of scheduling that was done by the Efficient Algorithms group at the computer science department of the TUM. Such a tool should allow easy implementation of scheduling algorithms that are given by verbal description or pseudo code, and it should be flexible and open to all of the many different kinds of scheduling problems, each of which has its own special parameter settings and limitations. The goal was to prevent the user from doing the unnecessary work that always has to be done for the simulation, logging and visualization overhead. This suggested the name Sketch-it! (which was of course also chosen because of the phonetic proximity to the term 'schedule'). The researcher should be given a possibility to experiment with his own algorithms and ideas. This involves not only the implementation of algorithms, but also the creation of particular problem instances.

Another point of interest is the application of Sketch-it! as a teaching aid. This will be tested in an advanced seminar in the summer term of 2003.

Development Tools and Libraries The simulation framework is being developed in C++, using the standard compiler g++, provided by the GNU project. Since the complexity of the considered algorithms is of prime importance, we utilize the Library of Efficient Data types and Algorithms (LEDA) [30]. The visualization part of the application is implemented with Qt [35], a portable C++ development toolkit that mainly provides functionality for graphical user interfaces (GUI).

Simulation Core The core of the simulator comprises C++ classes for the different job types (single machine jobs, parallel jobs, malleable jobs, shop jobs, broadcast requests), classes for the environments (single machine, identical / uniform / unrelated parallel machines, flow / job / open shop), classes for the network topologies (set, line, mesh, hypercube) and classes to represent precedence constraints.

For a specific instance of a scheduling problem its jobs, machines and potentially precedence constraints are combined by a central class to form the *task-system*.

In the following section we describe briefly how the user can generate such a task-system and how a newly implemented or already existing algorithm can be executed with this task-system as input.

Usage

Problem Instance Generator To support testing and empirical analysis of algorithms instances can be automatically generated, one only needs to provide the parameters (environment, number of machines, number of jobs, probability distribution and expectation / variance values for stochastic problems, etc.). The tool then creates an instance of the requested problem and makes the appropriate random decisions (where necessary).

Simulation After selecting an appropriate scheduling algorithm (see below for a list of provided algorithms and how to implement a new one) the user can follow its execution either step-by-step or continuously at a certain speed. A Gantt chart which is constantly updated, shows the temporal assignment of jobs to machines. If precedence constraints are present, these can be displayed. Hereby jobs are highlighted according to their present state (i.e. released, available with respect to precedence constraints, running, finished). Other views can show the status of the machines (either in list form or in a special way, e.g. for a mesh topology). The main window also contains textual comments concerning the execution of the algorithm, e.g. about its current phase or about certain operations it is performing. A screenshot of Sketch-it!:

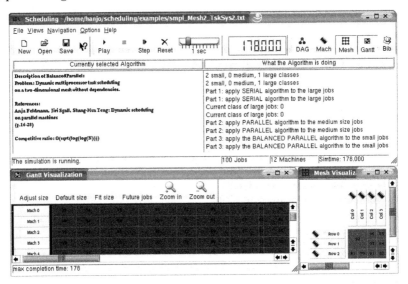

Logging Primarily to provide a standardized way to add new types of objective functions and also to ease the collection of relevant data during the execution of an algorithm, some logging capabilities are integrated in the form of so called loggers. These loggers are triggered by a carefully subdivided event system, which enables them to be notified in case selected events occur (like release of a new job, allocation of an available job, preemption of a running job, completion of a job, etc.). All standard objective functions (e.g. makespan, sum of completion times, sum of flow times, etc.) are already implemented with the help of such a logger.

Algorithms In order to implement a new scheduling algorithm, code needs to be inserted in dedicated functions for the "startup" and the "inner loop". The "startup" function is meant for initialization purposes and is called before the actual execution. The "inner loop" function is called by Sketch-it! during the execution in an on-line fashion, every time a new job/request is available/released or a job/request was completed. Note that in case the algorithm needs to make decisions at other points in time, it can return a delay at the end of the function to notify Sketch-it! when it should to be called again. Off-line algorithms have complete access to the task-system and can compute the schedule already in the "startup" function.

To give a feeling of the spectrum of possible problems which can be attacked with Sketch-it! an incomplete list of algorithms implemented to date follows: Smith's ratio rule for single machine scheduling, the algorithms of Coffman and Graham, the parallel machine algorithm of McNaughton for scheduling with preemptions, genetic algorithms for scheduling in an open shop environment, an algorithm of Ludwig and Tiwari for scheduling malleable parallel tasks, algorithms for stochastic scheduling (LEPT, SEPT, Smith weighted SEPT) and algorithms for the broadcast scheduling problems considered here. For descriptions of the multiprocessor task algorithms and of the algorithms for stochastic scheduling problems see [37] and [29].

3   Comparing Push- and Pull-Based Data-Broadcasting

In this section we present experiments which were conducted in order to compare the performance of push- to pull-based algorithms, with respect to the average response time (ART) and the maximum response time (MRT). Hereby the ART is emphasized, because it is the more widely accepted metric in the literature, and because a "good behavior" on the average seems more desirable. We compare several on-line algorithms for the pull-based setting to a push-based algorithm which is proven to be a 2-approximation for the expected response time (ERT). Note that the latter only gets the document probabilities π_i as input.

In the experiments document sizes l_i for $i \in \{1 \ldots m\}$ and requests R_j, D_j for $j \in \{1 \ldots n\}$, are generated at random. The ART/MRT of the computed schedules can vary considerably from one of these instances to the other. In order to obtain experimental data which can be compared easily for the different runs, simple lower bounds for the ART/MRT are calculated for each instance and the ART/MRT of the individual schedules are normalized by dividing with the respective lower bound. These lower bounds are described in the next section. In Section 3.2 we present the scheduling algorithms which were chosen to compete against each other. Then, in Section 3.3 we go into the details of how the instances are generated and finally discuss the results.

3.1   Simple Lower Bounds

The maximum of all message lengths is clearly a lower bound for the MRT. Let $LM := \max \{l_i \mid i \in \{1 \ldots m\}, D_j = i \text{ for some } j \in \{1 \ldots n\}\}$ denote this bound.

Deriving a lower bound LA for the ART is more interesting. First note that each request R_j, $j \in \{1 \ldots n\}$ obviously has to wait at least l_{D_j} units of time and thus $LA \geq \frac{1}{n} \sum_{j=1}^{n} l_{D_j}$.

To strengthen this trivial lower bound we consider points in time for which we know that at least one document is being broadcasted and others have to wait. To this end we define $w_i(t) := 1$, if there is a $j \in \{1 \ldots n\}$ with $D_j = i$ and $R_j \in (t - l_i, t]$, and $w_i(t) := 0$ otherwise, where $t \in [0, T_m]$ with $T_m := T + \max\{l_1 \ldots l_m\}$. In other words if $w_i(t) = 1$, there surely is at least one pending request for document i at time t. At any point in time $t \in [0, T_m]$ there are pending requests for at least $k(t) := \sum_{i=1}^{m} w_i(t)$ individual documents. Only one of these documents can be broadcasted at t and the $\geq \max\{k(t) - 1, 0\}$ requests for other documents must wait. This gives our lower bound: LA $:= \frac{1}{n} \sum_{j=1}^{n} l_{D_j} + \frac{1}{n} \int_0^{T_m} \max\{k(t) - 1, 0\} dt$.

3.2   Implemented Algorithms

Pull-Based The first two are standard greedy on-line algorithms, known from traditional scheduling problems. The third was introduced in [4] and the fourth in [19].

● **SSTF** *Shortest Service Time First*: Like the name suggests, as time advances, always broadcast the document which has a pending request that can be satisfied most quickly among all pending requests. Note that a broadcast of a long document might be preempted if a request for a short document arrives. This on-line algorithm is optimal for the equivalent standard (unicast) scheduling problem $1|r_j, pmtn|\frac{1}{n}\sum F_j$, where it is known as *shortest remaining processing time*.

To show how easy it is to implement a scheduling algorithm with Sketch-it!, to the right we exemplarily present the actual source code inserted in the "inner loop" (cf. Section 2) for SSTF.

Experiments with SSTF in [4] imply good performance with respect to ART and bad performance with respect to MRT.

```
list<SBroadcastRequ*> a =
getActiveRequ();
SBroadcastRequ *r, *p = NULL;
double s = MAXDOUBLE;
forall (r, a)
    if ( s > r→getRemProcTime() ) {
        s = r→getRemProcTime();
        p = r; }
if ( p != NULL )
    broadcastMsg( p→getMsgIndex() );
```

● **LWF** *Longest Wait First*: As time advances, always broadcast the document with the currently maximum total waiting time, i.e. the maximum total time that all pending requests for any document are waiting. In [26] it was conjectured that LWF is an $\mathcal{O}(1)$-speed $\mathcal{O}(1)$-competitive algorithm. Empirically it performs comparatively well with respect to MRT and not so well with respect to ART [4].

● **LTSF** *Longest Total Stretch First*: The current stretch of a request R_j at time $t \in [0, T']$ is the ratio of the time the request has been in the system so far $t - R_j$ (if it is still pending) respectively its response time F_j (if it is completely serviced) to the length of the requested document l_{D_j}. As time advances, LTSF always chooses to broadcast the document which has the largest total current stretch, considering all pending requests. It performs well empirically for ART; performance with respect to MRT varies [4].

● **BEQUI-EDF** *Equi-partition–Earliest Deadline First*: We only give a brief description of the algorithm, for details we refer to [19]. BEQUI-EDF runs in two stages, in the first stage Equi-partition is simulated: the broadcast bandwidth is distributed among the documents which have pending requests, where each document receives bandwidth proportional to the number of unsatisfied requests for it. Each time a document was completely broadcasted in this simulated first stage of the algorithm, a deadline is derived from the current time and the time it took to broadcast the document. In the second stage at any time the document with the earliest deadline is broadcasted.

This algorithm is special, because it is so far the only algorithm with a proven worst case performance. In [19] it is shown for any $\varepsilon > 0$ to be $(1+\varepsilon)(4+\varepsilon)$-speed $\mathcal{O}(1)$-competitive, if clients cannot buffer and where the constant $\mathcal{O}(1)$ depends on ε. In other words if we e.g. set $\varepsilon = 0.1$, the ART of a schedule computed by BEQUI-EDF for one channel is at most $\mathcal{O}(1)$ times the ART of an optimal schedule which only has access to a channel with a bandwidth of ≈ 0.22. Note that it is impossible to obtain a 1-speed competitive algorithm with ratio better than $o(n^{1/2})$, see Section 1.

Unfortunately the competitive analysis of BEQUI-EDF does not carry over to our model, where client buffering is enabled. We nevertheless think it is of interest to see how well an algorithm with provable worst case behavior (although for a slightly different model) performs compared to commonly used heuristics.

Push-Based Most results in the literature are concerned with the unit-length case, although there is some work on arbitrary-length messages, without preemption, see Section 1. The algorithm proposed in [36] (which is a 2-approximation with respect to the ERT) is the only candidate to date for the push-based setting with arbitrary-length messages and preemption.

• PUSH: The algorithm expects the probabilities π_i as input, we simply estimate these from the given requests R_j, i.e. we set $\pi_i = \frac{1}{n}|\{j \mid j \in \{1\ldots n\}, D_j = i\}|$. For randomly generated instances one has direct access to the underlying probabilities, see also next section.

The documents are split into pages of length 1 (this is possible because $l_i \in \mathbb{N}$ for $i \in \{1\ldots m\}$) and in each time-step a document is selected and its next page in cyclic order is broadcasted.

Schabanel [36] considers the case where broadcast costs c_i may be present. If these are set to zero, the analysis can be considerably simplified and yields that the pages of document $i \in \{1\ldots m\}$ should be broadcasted such that the expected distance between two consecutive broadcasts of i is $\tau_i := \mu/\sqrt{\pi_i l_i}$, where $\mu = \sum_{i=1}^{m} \sqrt{\pi_i l_i}$ is a normalizing factor. These expected distances can be achieved by a simple randomized algorithm: at each time-step choose to broadcast document $i \in \{1\ldots m\}$ with probability $1/\tau_i$. Note that $\sum_{i=1}^{m} 1/\tau_i = 1$. We implemented the derandomized greedy algorithm also given in [36].

3.3   The Experiments

Generation of Instances To be able to sample instances for broadcast scheduling problems, Sketch-it!'s generator needed to be enhanced in a straightforward way. We chose to generate the instances such that the interarrival times of requests are exponentially distributed with rate λ. This is a natural assumption which is often made in such a context, e.g. in queuing theory. For each request R_j, $j \in \{1 \ldots n\}$ we choose $D_j = i \in \{1 \ldots m\}$ at random with probability π_i. To assess realistic values for the probabilities π_i, we oriented ourselves at document popularities in the Internet. These are widely believed to behave according to Zipf's Law, thus we choose

$$\pi_i \propto i^{-1},$$

assuming that index i corresponds directly to the rank of a document, i.e. $i = 1$ is the most popular document, $i = 2$ the second most, and so on. In [10] it is stated that an exponent of -1 models reality well. PUSH can directly be given these probabilities π_i as input. Experiments showed that it makes no perceivable difference for our setup, whether PUSH obtains the π_i or their estimates as described in the previous section (the following discussion is based on the case where the π_i are estimated).

It remains to choose a realistic distribution for the document sizes. [17] contains a comprehensive study on file sizes in the Internet. It yields that it is reasonable to assume file sizes to be Pareto distributed:

$$\Pr(\text{``file size''} > x) \propto x^{-\alpha},$$

with $x \geq k$ and $\alpha \in (0, 2]$, $k > 0$, whereby it is stated in [17] that $\alpha = 1.2$ is realistic. As to not get arbitrarily large documents we chose the bounded Pareto distribution (see e.g. [16]) for the interval $[1, 100]$. Furthermore we rounded the sizes to integer values.

Conducted Experiments Of many interesting questions, we selected two:

1. "How do the scheduling algorithms perform depending on the number of documents m?" To this end we ran tests, stepwise increasing the number of documents, starting from $m = 2$ up to $m = 150$. For each fixed m we generated 100 instances with 1000–4000 requests each, as described in the previous section. From this we calculate the mean and variance of $\frac{ART}{LA}$ respectively $\frac{MRT}{LM}$ of all 100 instances for each scheduling algorithm. The empirical variance shows how predictable a scheduling algorithm is. It often is better to choose a predictable algorithm (i.e. with low variance), even if on average it performs slightly worse than others. We set the request arrival rate to $\lambda = 2$, that is in expectation 2 requests arrive per unit of time.
2. "How do the algorithms perform depending on how heavily the system is loaded?" To simulate behavior with different loads, we set $\lambda = 2^b$ for $b \in \{-6 \ldots 4\}$, instead of varying m. Everything else is done as above, m is set to 20.

Discussion of the Results We now discuss the plots in Figure 1 (appendix) row by row.

1. ART: Most strikingly, SSTF behaves very badly, PUSH behaves about the same as the others and BEQUI-EDF is second best. A closer examination of the individual algorithms follows.

- *Pull-Based*: SSTF shows the worst behavior of all algorithms. For all m it not only has the highest ART, but also a very high variance, i.e. the quality of a solution is very unpredictable and depends heavily on certain properties of the current input.

This outcome is quite surprising, because in [4] SSTF's overall empirical performance concerning ART is very good compared to LWF, LTSF and other algorithms.

A possible explanation for the observed behavior is that in some instances requests for short documents appear often enough to starve a significant number of requests for longer ones. In other words, constantly arriving requests for some short document i_1 might each time preempt the broadcast of a longer document i_2 such that more and more waiting requests for i_2 accumulate. This suggests that SSTF's performance greatly depends on how long popular documents are compared to slightly less popular ones. Note that SSTF is the only pull-based algorithm which in no way takes into account the number of requests waiting for the individual documents. This presumably explains its exceptional status in all plots.

LWF performs worst with the exception of SSTF. In [4] it has the overall worst performance among the tested algorithms with respect to the ART. LTSF empirically has the lowest ART, this confirms [4], where it also performs very well. For BEQUI-EDF the results look very promising. This is somewhat astonishing because the algorithm first simulates Equi-partition and then inserts deadlines at seemingly late points in time. A possible reason why it nevertheless performs well is given below.

- *Push-Based*: Also PUSH does comparatively well, which again is somewhat surprising: one could have expected a bigger advantage of the pull-based algorithms, especially when the number of documents in the system increases. This result is very interesting and shows that PUSH also empirically performs well on instances with exponentially distributed interarrival times. In particular it is very robust against variation of file sizes. Note: the sizes are varied independently of document popularities.

2. MRT: We do not go into details, just note that the good performance of LWF seems somehow intuitive: if requests with the longest waiting time are greedily selected, the probability that some request is starved for a long period of time is very small. On the other hand the MRT of PUSH continuously increases. This might stem from the fact that with increasing number of documents the smallest document probability π_m decreases. If such a document is requested, it might take a long time until it is completely broadcasted by PUSH.

3. ART: For the mentioned reasons SSFT is again by far worst for $\lambda > 1$. PUSH and BEQUI-EDF both are comparatively bad for low loads, but improve with higher loads. For $\lambda = 8, 16$ they even show the best performance of all algorithms.

It seems that the ART of PUSH stays at about the same level, independently of λ. This would not be surprising: if all π_i and l_i are fixed, PUSH computes a schedule which is independent of the choice of λ. Thus the expected response time is also constant (this time is obviously independent of λ for a fixed schedule). So PUSH simply benefits from a heavily loaded system because the pull-based algorithms cannot exploit their advantage of knowing which documents are currently actually requested. This they can do for lower loads, where PUSH also performs comparatively worse.

The ART of BEQUI-EDF is presumably so high in systems with low load, because for each request it has to wait quite long until it can set a corresponding deadline for EDF. The simulated Equi-partition algorithm divides the bandwidth according to the number of outstanding requests for the individual documents. Thus in systems with high loads this might implicitly give estimations of the document probabilities π_i for a certain time window. This could perhaps explain why it behaves so similarly to PUSH.

4. MRT: Except for SSTF, BEQUI-EDF performs worst for high λ. This is not astonishing because the algorithm is not at all trimmed to minimize the MRT. The MRT of PUSH again stays at about the same value, for the same reason as the ART does.

4 Conclusions

Sketch-it proved itself very useful while conducting the experiments. It was quite easy to implement the algorithms and also to enhance the generator and create the test suites.

From the experiments we conclude that the pull-based algorithm SSTF does not carry over well from traditional (unicast) scheduling. Furthermore the push-based algorithm PUSH is very robust for the case of exponentially distributed interarrival times. It performs well compared to the pull-based algorithms, independently of the number of documents and the distribution of file sizes among differently popular documents. On highly loaded systems it even outperforms pull-based algorithms, because they cannot exploit their advantage of knowing which documents the individual requests are actually for. These results are quite promising for Microsoft's SPOT technology, if one assumes that primarily information like weather and news is broadcasted (i.e. a reasonable number of documents, and the Poisson assumption is close to reality). In particular they are promising because it is probable that the system load is high: potentially a lot of users are listening to an extremely low-bandwidth channel. Moreover PULL's empirically observed independence of the load (which confirms theoretical results) makes this setting nicely scalable. On the other hand when the number of documents increases (e.g. if the possibility to send personal messages is added to the system) the ART is still reasonable (in the experiments about twice the amount of the best algorithm's output), but the MRT increases distinctly. This could mean that some users have to wait long for messages of low popularity (e.g. personal messages) and might get frustrated.

An interesting open question would be to find out how the algorithms perform on real world data (e.g. traces of Web-Servers) or generated data where the requests do not stem from a Poisson process.

Acknowledgments

The authors are much indebted to Ernst W. Mayr for stimulating the development of Sketch-it!. We would like to thank Thomas Schickinger, who played an important role in the original design and implementation of the tool. Furthermore we would like to thank Thomas Erlebach for numerous helpful discussions on the topic of broadcast scheduling and many appreciated comments concerning this paper.

References

[1] S. Acharya, R. Alonso, M. Franklin, and S. Zdonik. Broadcast disks: Data management for asymmetric communication environments. In *Proceedings of the ACM SIGMOD Conference*, pages 199–210, San Jose, CA, May 1995. 152

[2] S. Acharya, M. Franklin, and S. Zdonik. Prefetching from a broadcast disk. In *Proceedings of the 12th Int. Conference on Data Engineering (ICDE)*, pages 276–285, New Orleans, Louisiana, February 1996. 152

[3] S. Acharya, M. Franklin, and S. Zdonik. Balancing push and pull for data broadcast. In *Proceedings of the ACM SIGMOD*, Tuscon, Arizona, May 1997. 152

[4] S. Acharya and S. Muthukrishnan. Scheduling on-demand broadcasts: New metrics and algorithms. In *Proceedings of the 4th Annual ACM/IEEE International Conference on Mobile Computing and Networking (MobiCom)*, pages 43–54, Dallas, Texas, USA, October 1998. ACM Press. 149, 152, 156, 159

[5] D. Aksoy and M. Franklin. Scheduling for large-scale on-demand data broadcasting. In *Proceedings of the IEEE INFOCOM Conference*, pages 651–659, San Francisco, CA, March 1998. 151

[6] D. Aksoy and M. Franklin. RxW: A scheduling approach for large-scale on-demand data broadcast. *ACM/IEEE Transactions on Networking*, 7(6):846–860, December 1999. 151

[7] M. H. Ammar and J. W. Wong. The design of teletext broadcast cycles. *Performance Evaluation*, 5(4):235–242, Dec 1985. 152

[8] M. H. Ammar and J. W.Wong. On the optimality of cyclic transmission in teletext systems. *IEEE Transaction on Communication*, COM, 35(1):68–73, Jan 1987. 152

[9] A. Bar-Noy, R. Bhatia, J. Naor, and B. Schieber. Minimizing service and operation costs of periodic scheduling. In *Proceedings of the 9th Annual ACM-SIAM Symposium on Discrete Algorithms (SODA)*, pages 11–20, San Francisco, California, January 1998. ACM Press. 152

[10] P. Barford and M. Crovella. Generating representative web workloads for network and server performance evaluation. In *Proceedings of ACM SIGMETRICS, Measurement and Modeling of Computer Systems*, pages 151–160, Madison, Wisconsin, USA, June 1998. ACM Press. 158

[11] Y. Bartal and S. Muthukrishnan. Minimizing maximum response time in scheduling broadcasts. In *Proceedings of the 11th Annual ACM-SIAM Symposium on Discrete Algorithms (SODA)*, pages 558–559, San Francisco, California, January 2000. ACM Press. 152

[12] M. A. Bender, S. Chakrabarti, and S. Muthukrishnan. Flow and stretch metrics for scheduling continuous job streams. In *Proceedings of the 9th Annual ACM-SIAM Symposium on Discrete Algorithms (SODA)*, pages 270–279, San Francisco, California, January 1998. ACM Press. 152

[13] J. Błażewicz, K. H. Ecker, E. Pesch, G. Schmidt, and J. Węglarz. *Scheduling Computer and Manufacturing Processes*. Springer-Verlag, 1996. 149

[14] P. Brucker. *Scheduling Algorithms*. Springer-Verlag, 3rd edition, September 2001. 149

[15] P. Chrétienne, E. G. Coffman, Jr., J. K. Lenstra, and Z. Liu. *Scheduling Theory and its Applications*. John Wiley & Sons, 1995. 149

[16] M. Crovella, M. Harchol-Balter, and C. D. Murta. Task assignment in a distributed system: Improving performance by unbalancing load. In *Proceedings of ACM SIGMETRICS, Measurement and Modeling of Computer Systems*, pages 268–269, Madison, Wisconsin, USA, June 1998. ACM Press. 158

[17] M. E. Crovella and A. Bestavros. Self-similarity in world wide web traffic: Evidence and possible causes. *IEEE/ACM Transactions on Networking*, 5(6):835–846, December 1997. 158

[18] DirecPC. Website. www.direcpc.com. 149

[19] J. Edmonds and K. Pruhs. Broadcast scheduling: When fairness is fine. In *Proceedings of the 13th Annual ACM-SIAM Symposium on Discrete Algorithms (SODA)*, pages 421–430, San Francisco, California, January 2002. ACM Press. 152, 156, 157

[20] T. Erlebach and A. Hall. NP-hardness of broadcast scheduling and inapproximability of single-source unsplittable min-cost flow. In *Proceedings of the 13th Annual ACM-SIAM Symposium on Discrete Algorithms (SODA)*, pages 194–202, San Francisco, California, January 2002. ACM Press. 152

[21] R. Gandhi, S. Khuller, Y.-A. Kim, and Y.-C. Wan. Algorithms for minimizing response time in broadcast scheduling. In *Proceedings of the 9th Conference on Integer Programming and Combinatorial Optimization (IPCO)*, volume 2337 of *LNCS*, pages 425–438, Cambridge, MA, USA, May 2002. Springer-Verlag. 152

[22] R. L. Graham, E. L. Lawler, J. K. Lenstra, and A. H. G. Rinnooy Kan. Optimization and approximation in deterministic sequencing and scheduling: A survey. *Ann. Discrete Math*, 5:287–326, 1979. 148, 149

[23] S. Hameed and N. H. Vaidya. Efficient algorithms for scheduling data broadcast. *Wireless Networks*, 5(3):183–193, May 1999. 152

[24] @Home Network. Website. www.home.net. 149

[25] Institut für Algebra und Geometrie, Otto-von-Guericke-Universität Magdeburg. LiSA Homepage. lisa.math.uni-magdeburg.de. 151

[26] B. Kalyanasundaram, K. Pruhs, and M. Velauthapillai. Scheduling broadcasts in wireless networks. In *Proc. of the 8th Annual European Symposium on Algorithms (ESA)*, volume 1879 of *LNCS*, pages 290–301, Saarbrücken, Germany, September 2000. Springer-Verlag. 152, 156

[27] C. Kenyon and N. Schabanel. The data broadcast problem with non-uniform transmission times. In *Proceedings of the 10th Annual ACM-SIAM Symposium on Discrete Algorithms (SODA)*, pages 547–556, January 1999. 152

[28] C. Kenyon, N. Schabanel, and N.E. Young. Polynomial-time approximation scheme for data broadcast. In *Proceedings of the 32nd Annual ACM Symposium on Theory of Computing (STOC)*, pages 659–666, May 2000. 152

[29] B. Meier. Simulation of algorithms for stochastic scheduling problems. Diploma thesis, Technische Universität München, February 2003. 155

[30] Algorithmic Solutions Software GmbH. LEDA. www.algorithmic-solutions.com/. 153

[31] Lehrstuhl für Effiziente Algorithmen. Sketch-it! Documentation. wwwmayr.informatik.tu-muenchen.de/scheduling/ scheduling-api/html/classes.html. 152

[32] Lehrstuhl für Effiziente Algorithmen. Sketch-it! Homepage. wwwmayr.informatik.tu-muenchen.de/scheduling/. 152

[33] LIMI/EA 2215, University of Brest. The Cheddar project: a free real time scheduling simulator. http://beru.univ-brest.fr/ singhoff/cheddar/. 151

[34] Microsoft PressPass. Microsoft Presents Smart Personal Objects Technology (SPOT)-Based Wristwatches at CES. www.microsoft.com/presspass/press/2003/jan03/01-09SPOTWatchesPR.asp. 150

[35] TrollTech, Norway. Qt. www.trolltech.com/products/qt/. 153

[36] N. Schabanel. The data broadcast problem with preemption. In *Proceedings of the 17th Annual Symposium on Theoretical Aspects of Computer Science (STACS)*, volume 1770 of *LNCS*, pages 181–192, Lille, France, February 2000. Springer-Verlag. 152, 157

[37] H. Täubig. Simulation of multiprocessor scheduling problems. Diploma thesis, Technische Universität München, August 2000. 155

[38] N.H. Vaidya and S. Hameed. Scheduling data broadcast in asymmetric communication environments. *Wireless Networks*, 5(3):171–182, May 1999. 152

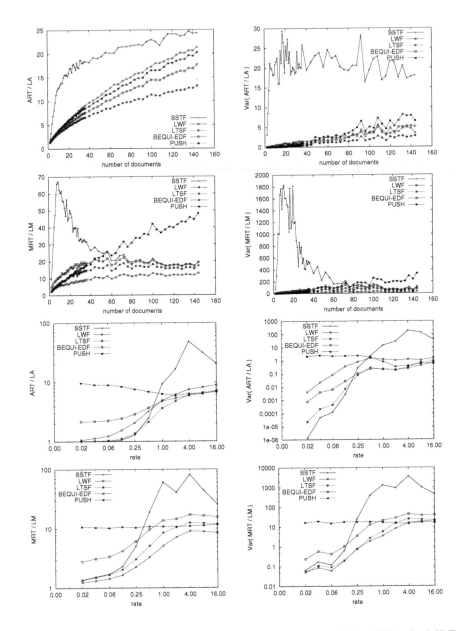

Fig. 1. Generated data. Plots of the ART, the variance of the ART, the MRT, and the variance of the MRT in dependency of the number of documents m respectively rate λ. Each quantity is normalized by the corresponding lower bound, see text for details. For each data point 100 input traces were generated, each containing 1000–4000 requests. For the top 4 plots we chose $\lambda = 2$, i.e. in expectation 2 request arrive in one unit of time. For the lower 4 plots we chose $m = 20$

Experimental Comparison of Heuristic and Approximation Algorithms for Uncapacitated Facility Location

Martin Hoefer

[1] Max-Planck-Institut für Informatik, Stuhlsatzenhausweg 85
66123 Saarbrücken, Germany
mhoefer@mpi-sb.mpg.de
[2] TU Clausthal, Marktstrasse 1
38678 Clausthal-Zellerfeld, Germany
Martin.Hoefer@tu-clausthal.de

Abstract. The uncapacitated facility location problem (UFLP) is a problem that has been studied intensively in operational research. Recently a variety of new deterministic and heuristic approximation algorithms have evolved. In this paper, we consider five of these approaches - the JMS- and the MYZ-approximation algorithms, a version of Local Search, a Tabu Search algorithm as well as a version of the Volume algorithm with randomized rounding. We compare solution quality and execution times on different standard benchmark instances. With these instances and additional material a web page was set up [26], where the material used in this study is accessible.

1 Introduction

The problem of locating facilities and connecting clients at minimum cost has been studied widely in Operations Research. In this paper we focus on the uncapacitated facility location problem (UFLP). We are given n possible facility locations and m cities. Let F denote the set of facilities and C the set of cities. Furthermore there are non-negative opening costs f_i for each facility $i \in F$ and connection costs c_{ij} for each connection between a facility i and a city j. The problem is to open a collection of facilities and connect each city to exactly one facility at minimum cost.

Instead of solving this problem to optimality, we will focus on finding approximate solutions. In the following we will present five methods, which are originating in different areas of optimization research. We will compare two approximation algorithms, two heuristics based on local search and one on LP-based approximation and rounding, which were recently developed and found to work good in practice.

1.1 Approximation Algorithms

Recently some new approximation algorithms have evolved for the metric version of the UFLP in which the connection cost function c satisfies the triangular

K. Jansen et al. (Eds.): WEA 2003, LNCS 2647, pp. 165–178, 2003.
© Springer-Verlag Berlin Heidelberg 2003

inequality. A couple of different techniques were used in these algorithms like LP-rounding ([11], [24]), greedy augmentation ([10]) or primal-dual methods ([20], [10]). In terms of computational hardness Guha and Khuller [13] showed that it is impossible to achieve an approximation guarantee of 1.463 unless $NP \in DTIME[n^{O(\log \log n)}]$. For our comparison we chose two of the newest and most promising algorithms.

JMS-Algorithm The JMS-Algorithm uses a greedy method to improve the solution. The notion of time that is involved was introduced in an earlier 3-approximation algorithm by Jain and Vazirani [20]. Later on Mahdian et al. [21] translated the primal-dual scheme into a greedy 1.861-approximation algorithm. In the third paper Jain, Mahdian and Saberi [19] presented the JMS-Algorithm (JMS), which improved the approximation bound to 1.61. However, it had a slightly worse complexity of $O(n^3)$ instead of $O(n^2 \log n)$. The following sketch of JMS is taken from [22]:

1. At first all cities are unconnected, all facilities unopened, and the budget of every city j, denoted by B_j, is initialized to 0. At every moment, each city j offers some money from its budget to each unopened facility i. The amount of this offer is equal to $max(B_j - c_{ij}, 0)$ if j is unconnected, and $max(c_{i'j} - c_{ij}, 0)$ if it is connected to some other facility i'.
2. While there is an unconnected city, increase the budget of each unconnected city at the same rate, until one of the following events occurs:
   (a) For some unopened facility i, the total offer that it receives from cities is equal to the cost of opening i. In this case, we open facility i, and for every city j (connected or unconnected) which has a non-zero offer to i, we connect j to i.
   (b) For some unconnected city j, and some facility i that is already open, the budget of j is equal to the connection cost c_{ij}. In this case, we connect j to i.

One important property of the solution of this algorithm is that it cannot be improved by simply opening an unopened facility. This is the main advantage over the previous 1.861-algorithm in [21]. In [19] experiments revealed an appealing behavior of JMS in practice.

MYZ Algorithm The MYZ algorithm could further improve the approximation factor of JMS. Mahdian, Ye and Zhang [22] applied scaling and greedy augmentation to the algorithm. For the resulting MYZ Algorithm (MYZ) the authors could prove an approximation factor of 1.52 for the metric UFLP, which is at present the best known factor for this problem for any algorithm. MYZ is outlined below. In step 4 of the algorithm C is the total connection cost of the present solution and C' the connection cost after opening a facility u.

1. Scale up all opening costs by a factor of $\delta = 1.504$
2. Solve the scaled instance with JMS
3. Scale down all opening costs by the same factor δ
4. **while** there is a unopened facility u, for which the ratio $(C - C' - f_u)/f_u$ is maximized and positive, open facility u and update solution

1.2 Heuristic and Randomized Algorithms

In terms of meta-heuristics there has not been such an intense research activity. A simulated annealing algorithm [3] was developed, which produces good results to the expense of high computation costs. Tabu search algorithms have been very successful in solving the UFLP (see [2], [23], [25]). A very elaborate genetic algorithm has been proposed by Kratica et al. over a series of papers ([16], [17], [18]). Their final version involves clever implementation techniques and finds optimal solutions for all the examined benchmarks.

Tabu Search In [23] Van Hentenryck and Michel proposed a simple Tabu Search algorithm that works very fast and outperforms the genetic algorithm in [18] in terms of solution quality, robustness and execution time. Therefore we used this algorithm for the experiments. It uses a slightly different representation of the problem. For a solution of the UFLP it is sufficient to know the set $S \subseteq F$ of opened facilities. Cities are connected to the cheapest opened facility, i.e. city j is connected to $i \in S$ with $c_{ij} = \min_{i' \in S}(c_{i'j})$. A neighborhood move from S to S' is defined as flipping the status of a facility from opened to closed ($S' = S \setminus i$) or vice versa ($S' = S \cup i$). When the status of a facility was flipped, flipping back this facility becomes prohibited (tabu-active) for a number of iterations. The number of iterations is adjusted using a standard scheme (see [23] for details). The high level algorithm can be stated as follows:

1. $S \leftarrow$ an arbitrary feasible solution
2. Set $\text{cost}(S^*) = \infty$
3. **do**
4. $bestgain$ = maximum cost savings over all possible non-tabu flips
5.   **if** $(bestgain > 0)$
6.   Apply random flip with best gain, update tabu lists and list length
7.   **else** close random facility
8.   Update S - connections of cities and datastructures
9.   **if** $(\text{cost}(S) < \text{cost}(S^*))$ do $S^* \leftarrow S$
10. **while** change of S^* in the last 500 iterations
11. **return** S^*

For every city j the algorithm uses three pieces of information: The number of the opened facility with the cheapest connection to j, the cost of this connection and the cost of the second cheapest connection to an opened facility. With this information the gains of opening and closing a facility can be updated incrementally in step 8. Thereby a direct evaluation of the objective function can be avoided. The algorithm uses priority queues to determine the second cheapest connections for each city. Due to these techniques the algorithm has an execution time of $O(m \log n)$ in each iteration.

Local Search The Local Search community has only paid limited attention to the UFLP so far. Apart from the Tabu Search algorithms there have been a few simple local search procedures proposed in [15], [10]. In this paper we use the simple version of Arya et al [4], for which the authors could prove an approximation factor of 3 on metric instances. The algorithm works with the set S of opened facilities as a solution. An operation **op** is defined as opening or closing a facility or exchanging the status of an opened and a closed facility. To improve the execution time of the algorithm we incorporated the use of incremental datastructures from the Tabu search algorithm and preferences for the simple moves as follows. We generally prefer applying the simple flips of opening and closing a facility (denoted as **ops**). As in the Tabu Search we apply one random flip of the flips resulting in best gain of the cost function. When these flips do not satisfy the acceptance condition, we pick the first exchange move found that would give enough improvement. If there is no such move left, the algorithm stops. This modified version can be stated as follows:

1. $S \leftarrow$ an arbitrary feasible solution
2. $exitloop \leftarrow false$
3. **while** $exitloop = false$
4.   **while** there is an **ops** such that $\text{cost}(\textbf{ops}(S)) \leq (1 - \frac{\epsilon}{p(n,m)}) \, \text{cost}(S)$
5.   find a random **ops*** of the **ops** with best gain
6.   do $S \leftarrow \textbf{ops}^*(S)$
7.   **if** there is an **op** such that $\text{cost}(\textbf{op}(S)) \leq (1 - \frac{\epsilon}{p(n,m)}) \, \text{cost}(S)$
8.   do $S \leftarrow \textbf{op}(S)$
9.   **else** $exitloop \leftarrow true$
10. **return** S

In our experiments the parameters were set to $\epsilon = 0.1$ and $p(n, m) = n + m$. Arya et al. suggested that the algorithm should be combined with the standard scaling techniques [10] to improve the approximation factor to 2.414. Interestingly this version performs inferior in practice. Therefore the version without scaling (denoted as LOCAL) was used for the comparison with the other algorithms. More on the unscaled and scaled versions of Local Search can be found in section 2.5 and [14].

Volume Algorithm For some of the test instances we obtained a lower bound using a version of the Volume algorithm, which was developed by Barahona in [6]. The Volume algorithm is an iterated subgradient optimization procedure, which is able to provide a primal solution and a lower bound on the optimal solution cost. To improve solution quality and speed up the computation Barahona and Chudak [7] used the rounding heuristic (RRWC) presented in [11] to find good upper bounds on the optimal dual solution cost and thereby reduced the iterations of the Volume algorithm. However, this approach has generally very high execution times in comparison to the other methods presented here. Instead we used a faster version of this algorithm which involves only a basic randomized rounding procedure and slightly different parameter settings. It will be denoted by V&RR and is available on the web page of the COIN-OR project by IBM [5]. Regarding solution quality and execution time this algorithm is generally inferior to the other algorithms. The results should only be seen as benchmark values of available optimization code. We will not go into detail describing the method, the code or the parameter settings here. The interested reader is referred to [5], [6], [7] for the specific details of the algorithm and the implementation.

2   Experiments

We tested all given algorithms on several sets of benchmark instances. First we studied the Bilde-Krarup benchmarks, which were proposed in [9]. These are non-metric small scale instances with $n \times m = 30 \times 80$ - 50×100. We chose this set because it is randomly generated, non-metric and involves the notion of increasing opening costs also present in the large scale k-median instances. Next we focused on small scale benchmarks proposed by Galvão and Raggi in [12]. These are metric instances with $n = m = 50$ - 200, which we chose because they make use of the shortest path metric and a Normal distribution to generate costs. Then we examined the performance on the cap instances from the ORLIB [8] and the M* instances, which were proposed in [18]. These are non-metric small, medium and large scale instances with $n \times m = 16 \times 50$ - 2000×2000. They have previously been used to examine the performance of many heuristic algorithms. Finally we studied metric large scale instances with $n = m = 1000$ - 3000, which were proposed in [1] and used as UFLP benchmarks for testing the performance of the Volume algorithm in [7]. So our collection of benchmark instances covers a variety of different properties: small, medium and large size; Euklidian metric, shortest path metric and non-metric costs; randomly generated costs from uniform distributions and Normal distributions.

On all instances we averaged over the performance of 20 runs for each algorithm. The experiments were done on a 866Mhz Intel Pentium III running Linux. For most problems we used CPLEX to solve the problems to optimality. The CPLEX-runs were done on a 333Mhz Sun Enterprise 10000 with UltraSPARC processors running UNIX. The execution times are about a factor of 2.5 times higher than the times for the algorithms. Here we only report average results for the different benchmark types. For more detailed results of our experiments and values for the single instances the reader is referred to [14].

With all benchmark instances, implementations of all algorithms and benchmark generators a web page was set up. All material used in this study can be accessed online at the UflLib [26].

Table 1. Parameters for the Bilde-Krarup problem classes

Type	Size $(n \times m)$	f_i	c_{ij}
B	50×100	Discrete Uniform $(1000, 10000)$	Discrete Uniform $(0,1000)$
C	50×100	Discrete Uniform $(1000, 2000)$	Discrete Uniform $(0,1000)$
Dq*	30×80	Identical, 1000*q	Discrete Uniform $(0,1000)$
Eq*	50×100	Identical, 1000*q	Discrete Uniform $(0,1000)$

* q=1,...,10

2.1 Bilde-Krarup Instances

The Bilde-Krarup instances are small scale instances of 22 different types. The costs for the different types are calculated with the parameters given in Table 1. As the exact instances are not known, we generated 10 test instances for each problem type. In Table 3 we report the results of the runs for each algorithm. In columns 'Opt' we report the percentage of runs that ended with an optimal solution. In columns 'Error' we report the average error of the final solution in percentage of the optimal solution, in columns 'Time' the average execution time in seconds. In column 'CPX' we denoted the average execution time of CPLEX to solve the instances.

The deterministic algorithms perform quite good on these instances. The average error is 2.607% at maximum although the problems are not of metric nature. MYZ performs significantly better than JMS in terms of solution quality. It can solve additional 37 problems to optimality and has a lower average error. The execution time is slightly higher because it uses JMS as a subroutine.

For the heuristic algorithms TABU provides the best results. It was able to solve problems of all classes to optimality in a high number of runs. Unfortunately it also is much slower than LOCAL, MYZ and JMS. LOCAL also performs competitive on most of these problem classes. Compared to TABU it is able to solve problems of all classes to optimality, but the overall number of instances solved is very much lower. In terms of the execution time it is much faster though. V&RR is outperformed by any of the other algorithms. It reveals the highest execution time and the worst solution quality.

2.2 Galvão-Raggi Instances

Galvão and Raggi proposed unique benchmarks for the UFLP. A graph is given with an arc density δ, which is defined as $\delta =$ connections present $/(m * n)$. Each present connection has a cost sampled from a uniform distribution in the range $[1, n]$ (except for $n = 150$, where the range is $[1, 500]$). The connection costs between a facility i and a city j are determined by the shortest path from i to j in the given graph. The opening costs f_i are assumed to come from a Normal distribution. Originally Galvão and Raggi proposed problems with $n = m = 10, 20, 30, 50, 70, 100, 150$ and 200. We will consider the 5 largest types. The density values and the parameters for the Normal distribution are listed in

Table 2.   Parameters for the Galvão-Raggi problem classes

Size	δ	Parameters for f_i	
		mean	stand. dev.
50	0.061	25.1	14.1
70	0.043	42.3	20.7
100	0.025	51.7	28.9
150	0.018	186.1	101.5
200	0.015	149.5	94.4

Table 2. The exact instances for these benchmarks are not known. As for the Bilde-Krarup benchmarks we generated 10 instances for each class. The results of our experiments are reported in Table 3. Columns 'Opt', 'Error' and 'Time' are defined as before. We also included the average execution times of CPLEX in column 'CPX'.

JMS performs slightly better than MYZ on these metric instances. Of the heuristic and randomized algorithms V&RR performs very good - even better than TABU and LOCAL - to the expense of high execution times. In fact, the times are prohibitively high as the algorithm needs much more time than CPLEX to solve the instances to optimality. LOCAL performs a little bit better than TABU, because both the execution times and the average errors are smaller. However, it is not very reliable to find optimal solutions.

2.3   ORLIB and M* Instances

The cap problems from the ORLIB are non-metric medium sized instances. The M* instances were designed to represent classes of real UFLPs. They are very challenging for mathematical programming methods because they have a large number of suboptimal solutions. In Table 4 we report the results for the different algorithms. In columns 'Opt' we again denote the percentage of runs that ended with an optimal solution. In Columns 'Error' we report the average error of the final solution in percentage of the optimal solution. For the larger benchmarks the optimal solutions are not known. Instead we used the best solutions found as a reference, which for all benchmarks were encountered by TABU. All values that do not relate to an optimal solution are denoted in brackets. In columns 'Time' we report average execution times in seconds. Furthermore in column 'CPX' we report the average execution time of CPLEX.

Again the deterministic algorithms perform very well. The maximum error for both methods was produced on the capa benchmark. Of the deterministic algorithms MYZ did perform better than JMS. It was able to solve additional 6 problems to optimality. JMS could only achieve a better performance in 4 of the 37 benchmarks. In terms of execution time MYZ becomes slightly less competitive on larger problems because the additional calculations of the greedy augmentation procedure need more time.

With a maximum average error of 0.289% TABU again is the algorithm with the best performance on these benchmarks. It is able to solve all problems to optimality - in most cases with a high frequency. Hence, our results are consistent with the values reported in [23]. However, the execution times of our code are significantly faster than the times needed by the implementation of Michel and Van Hentenryck on a similar computer (a factor of 2 and more). Compared to TABU the solution quality of LOCAL is not very competitive. It fails to find optimal solutions on 9 problems, while 7 of them are cap-benchmarks. The execution times, however, are very competitive, as it performs in most cases significantly better than TABU.

V&RR performs generally worse than the other algorithms. On some of the cap instances it achieves good solution quality. On the M* instances, however, it performs worse than all other algorithms in terms of solution quality and execution time. The execution times for the small problems exceed the times of CPLEX again. The practical use of this algorithm for small problems should therefore be avoided. For problems with $m, n \geq 100$, however, execution times of CPLEX become significantly higher.

2.4 k-Median Instances

In this section we take a look at large scale instances for the UFLP. The benchmarks considered here were originally introduced for the k-median problem in [1]. In [7] they were used as test instances for the UFLP. To construct an instance, we pick n points independent uniformly at random in the unit square. Each point is simultaneously city and facility. The connection costs are the Euklidian distances in the plane. All facility opening costs are identical. To prevent numerical problems and preserve the metric properties, we rounded up all data to 4 significant digits and then made all the data entries integer. For each set of points, we generated 3 instances. We set all opening costs to $\sqrt{n}/10$, $\sqrt{n}/100$ and $\sqrt{n}/1000$. Each opening cost defines a different instance with different properties. In [1] the authors showed that, when n is large, any enumerative method based on the lower bound of the relaxed LP would need to explore an exponential number of solutions. They also showed that the solution of the relaxed LP is, asymptotically in the number of points, about 0.998% of the optimum.

In Table 4 we report the results of our experiments for $n = 1000, 2000, 3000$. In column 'LB' we provide the lower bound on each problem calculated by V&RR. For each algorithm we report the average error and the average execution time. All errors were calculated using the lower bound in 'LB'.

On these metric benchmarks JMS again performs slightly better than MYZ. TABU is the best algorithm in terms of solution quality. LOCAL manages to find better solutions than the deterministic algorithms, but it is much slower than TABU, JMS and MYZ. The performance of V&RR is not competitive in comparison to the other algorithms. It is outperformed in terms of solution quality and execution time by all algorithms on nearly all benchmarks. Only on the larger benchmarks with small opening costs the execution times of LOCAL are equally slow. One reason for this is the use of priority queues. For the problems with smaller opening cost optimal solutions open a high number of facilities. Here the operations on the queues are getting expensive. When implemented without queues the adjustment of the datastructures when opening a facility (which is the operation used more often here) could be executed in $O(m)$. The closing operation would need $O(nm)$, which leads to inferior execution times on average. However, in this case the closing operation is most often used in the exchange step, which is called after nearly all facilities have been opened. Then most of the cities are connected to the facility located at the same site, and closing a facility affects only one city. So finding the new closest and second closest facilities can be done in $O(m)$. Thus, it is not surprising that an implementation without queues was able to improve the execution times on the large problems with $n = m > 1000$ by factors of up to 3. Nevertheless we chose to implement priority queues in our version of LOCAL as their theoretical advantage leads to shorter execution times on average.

2.5   Scaling and Local Search

In [10] a scaling technique was proposed to improve the approximation bound of local search for the metric UFLP. In the beginning all costs are scaled up by a factor of $\sqrt{2}$. Then the search is executed on the scaled instance. Of all candidates found the algorithm exits with the one having the smallest cost for the unscaled instance. With this technique the search is advised to open the most economical facilities. However, the solution space of the scaled instance might not be similar to the space of the unscaled instance. Therefore in practice it is likely that the scaled version ends with inferior solutions. It becomes obvious that this adjustment is just for lowering theoretical bounds and has limited practical use. The scaling technique was proposed for Local Search on the metric UFLP. However, it deteriorates the performance of Local Search on metric as well as non-metric instances. Please see [14] for experimental results.

3   Conclusions

The uncapacitated facility location problem was solved by 5 different algorithms from different areas of optimization research. The deterministic algorithms manage to find good solutions on the benchmarks in short execution times. Generally MYZ can improve the performance of JMS to the expense of little extra execution time. On the tested metric instances the performance of the algorithms is competitive to the heuristic and randomized algorithms tested while the execution times remain significantly shorter. Here JMS offers slightly better solutions than MYZ. The approximation algorithms reveal higher errors only on a few tested non-metric instances, but always deliver solutions that are within 5% of optimum. The presented Local Search profits from the intelligent use of datastructures. On a number of instances the execution times are able to compete with those of MYZ and JMS, but due to changing starting points the algorithm

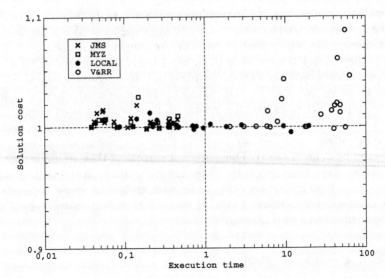

Fig. 1. Plot of solution costs and execution times in comparison to TABU

is not very robust. Scaling techniques that lead to improved approximation factors deteriorate the performance of the algorithm in practice. TABU is able to find optimal solutions in most cases. It is much faster than V&RR (and Local Search on special instances), but the execution times cannot compete with those of MYZ and JMS. The tested version of the Volume algorithm V&RR is not competitive regarding solution quality and execution times.

The preference for a method in practice depends on the properties of the problem instances and the setting. If speed is most important, JMS or MYZ should be used, especially if metric instances are to be solved. If solution quality is most important, TABU should be used. In a general setting the results indicate a preference for TABU, as it achieves best solution quality in a reasonable amount of time.

Finally we present a plot of the results in relation to TABU. The x- and y-coordinates represent values regarding execution time and solution cost, respectively. The coordinates were calculated by dividing the results of the algorithms by the results of TABU. We further adjusted some of the data by averaging over the D- and E-instances of Bilde-Krarup and the instances of the same size of k-median, respectively. There are 22 dots for each algorithm.

Only a few dots are located in the lower half of the plot, i.e. hardly any time TABU was outperformed in terms of solution cost. Moreover, there is hardly any dot in the lower left quadrangle, which indicates better performance in terms of execution time and solution cost. In the upper left quadrangle most of the dots of JMS and MYZ are located indicating faster performance with higher solution costs. In the upper right part most of the dots of V&RR are located. This means worse performance regarding execution time and solution cost. Most of the dots of LOCAL are spread closely above the line in the upper half, which is due to slightly higher solution costs, the faster performance on smaller and the slower performance on larger instances.

Acknowledgement

The author would like to thank Tobias Polzin for helpful hints and advice in the development of this study.

References

[1] S. Ahn, C. Cooper, G. Cornuéjols and A. M. Frieze. Probabilistic analysis of a relaxation for the k-median problem. *Mathematics of Operations Research*, 13:1-31, 1988. 169, 172

[2] K. S. Al-Sultan and M. A. Al-Fawzan. A tabu search approach to the uncapacitated facility location problem. *Annals of Operations Research*, 86:91-103, 1999. 167

[3] M. L. Alves and M. T. Almeida. Simulated annealing algorithm for simple plant location problems. *Rev. Invest.*, 12, 1992. 167

[4] V. Arya, N. Garg, R. Khandekar, A. Meyerson, K. Munagala and V. Pandit. Local search heuristics for k-median and facility location problems. *ACM Symposium on Theory of Computing*, pages 21-29, 2001. 168

[5] F. Barahona. An implementation of the Volume algorithm. IBM COIN-OR website, *http://oss.software.ibm.com/developerworks/opensource/coin*, 2000. 169

[6] F. Barahona and R. Anbil. The Volume algorithm: producing primal solutions with the subgradient method. Technical Report, IBM Watson Research Center, 1998. 168, 169

[7] F. Barahona and F. A. Chudak. Near-optimal solutions to large scale facility location problems. Technical Report, IBM Watson Research Center, 2000. 168, 169, 172

[8] J. E. Beasley. Obtaining Test Problems via Internet. *Journal of Global Optimization*, 8:429-433, 1996. 169

[9] O. Bilde and J. Krarup. Sharp lower bounds and efficient algorithms for the simple plant location problem. *Annals of Discrete Mathematics*, 1:79-97, 1977. 169

[10] M. Charikar and S. Guha. Improved combinatorial algorithms for the facility location and k-median problems. In *Proceedings of the 40th Annual Symposium on Foundations of Computer Science*, 1999. 166, 168, 173

[11] F. A. Chudak. Improved approximation algorithms for uncapacitated facility location. In *Proceedings of the 6th IPCO Conference*, pages 180-194, 1998. 166, 168

[12] R. D. Galvão and L. A. Raggi A method for solving to optimality uncapacitated facility location problems. *Annals of Operations Research*, 18:225-244, 1989. 169

[13] S. Guha and S. Khuller. Greedy strikes back: Improved facility location algorithms. *Journal of Algorithms*, 31:228-248, 1999. 166

[14] M. Hoefer Performance of heuristic and approximation algorithms for the uncapacitated facility location problem. Research Report MPI-I-2002-1-005, Max-Planck-Institut für Informatik, 2002. 168, 169, 173

[15] M. R. Korupolu, C. G. Plaxton and R. Rajaraman. Analysis of a local search heuristic for facility location problems. In *Proceedings of the 9th ACM-SIAM Symposium on Discrete Algorithms*, pages 1-10, 1998. 168

[16] J. Kratica, V. Filipovic, V. Sesum and D. Tosic. Solving the uncapacitated warehouse location problem using a simple genetic algorithm. In *Proceedings of the XIV International Conference on Material Handling and Warehousing*, pages 3.33-3.37, 1996. 167

[17] J. Kratica, D. Tosic and V. Filipovic. Solving the uncapacitated warehouse location problem by sga with add-heuristic. In *XV ECPD International Conference on Material Handling and Warehousing*, 1998. 167

[18] J. Kratica, D. Tosic, V. Filipovic and I. Ljubic. Solving the simple plant location problem by genetic algorithm. *RAIRO Operations Research*, 35:127-142. 2001. 167, 169

[19] K. Jain, M. Mahdian and A. Saberi. A new greedy approach for facility location problems. In *Proceedings of the 34th Symposium on Theory of Computing 2002*, forthcoming, 2002. 166

[20] K. Jain and V. V. Vazirani. Approximation algorithms for metric facility location and k-median problems using the primal-dual schema and langrangian relaxation. *Journal of the ACM*, 48:274-296, 2001. 166

[21] M. Mahdian, E. Marakakis, A. Sabieri and V. V. Vazirani. A greedy facility location algorithm analyzed using dual fitting. In *Proceedings of 5th International Workshop on Randomization and Approximation Techniques in Computer Science, Lecture Notes in Computer Science v. 2129*, pages 127-133. Springer-Verlag, 2001. 166

[22] M. Mahdian, Y. Ye and J. Zhang. Improved approximation algorithms for metric facility location problems. In *Proceedings of the 5th APPROX Conference, Lecture Notes in Computer Science v. 2462*, pages 229 - 242, 2002. 166

[23] P. Van Hentenryck and L. Michel. A simple tabu search for warehouse location. Technical Report, CS-02-05, Brown University, 2002. 167, 172

[24] M. Sviridenko. An Improved Approximation Algorithm for the Metric Uncapacitated Facility Location Problem. In *Proceedings of the 10th IPCO Conference, Lecture Notes in Computer Science v. 2337*, pages 230 - 239, 2002. 166

[25] M. Sun. A Tabu Search Heuristic Procedure for the Uncapacitated Facility Location Problem. In C. Rego and B. Alidaee (eds.) *Adaptive Memory and Evolution: Tabu Search and Scatter Search*, Kluwer Academic Publishers, forthcoming, 2002. 167

[26] UflLib. UFLP-benchmarks, optimization code and benchmark generators. |http://www.mpi-sb.mpg.de/units/ag1/projects/benchmarks/UflLib—, 2002. 165, 169

Table 3. Results for Bilde-Krarup and Galvão-Raggi instances

Type	CPX	JMS			MYZ			LOCAL			TABU			V&RR		
		Opt	Error	Time	Opt	Error	Time	Opt	Error	Time	Opt	Error	Time	Opt	Error	Time
B	6.859	50%	0.416	0.003	40%	0.588	0.003	80%	0.046	0.012	100%	0.000	0.053	70%	0.419	0.421
C	107.558	10%	1.750	0.003	30%	0.886	0.003	42%	0.848	0.014	64%	0.245	0.055	1%	4.454	0.525
D1	21.591	0%	2.445	0.001	10%	1.689	0.002	25%	1.678	0.006	54%	0.241	0.038	2%	3.719	0.239
D2	30.990	10%	1.675	0.002	20%	1.133	0.002	20%	1.758	0.006	67%	0.537	0.044	5%	3.083	0.254
D3	28.103	10%	2.607	0.002	40%	0.923	0.002	29%	0.879	0.006	89%	0.073	0.042	9%	2.245	0.235
D4	26.685	30%	0.796	0.002	30%	0.597	0.002	72%	0.530	0.006	100%	0.000	0.041	32%	1.248	0.243
D5	22.368	40%	0.647	0.002	70%	0.085	0.002	59%	0.402	0.006	94%	0.004	0.040	38%	0.995	0.246
D6	28.393	20%	1.042	0.002	30%	1.315	0.002	50%	0.882	0.006	87%	0.146	0.042	42%	0.919	0.259
D7	24.484	10%	1.771	0.002	60%	0.664	0.002	80%	0.354	0.005	100%	0.000	0.042	80%	0.214	0.259
D8	20.947	40%	1.587	0.002	40%	1.044	0.002	63%	1.000	0.006	90%	0.166	0.043	40%	1.390	0.251
D9	22.326	70%	0.846	0.002	90%	0.012	0.002	80%	0.285	0.006	100%	0.000	0.043	73%	0.496	0.259
D10	19.122	70%	0.252	0.002	80%	0.189	0.002	63%	0.760	0.006	93%	0.139	0.043	74%	0.506	0.256
E1	133.839	20%	2.265	0.003	30%	1.317	0.003	10%	1.430	0.013	57%	0.388	0.062	0%	5.712	0.516
E2	229.305	20%	1.650	0.003	40%	0.845	0.003	14%	2.712	0.013	94%	0.006	0.067	0%	4.479	0.560
E3	190.860	20%	1.610	0.003	20%	0.940	0.003	38%	0.784	0.012	63%	0.268	0.061	5%	3.419	0.553
E4	185.168	30%	1.192	0.003	30%	0.781	0.004	24%	1.577	0.013	90%	0.013	0.060	5%	2.505	0.581
E5	163.571	10%	2.560	0.003	70%	0.690	0.004	30%	2.019	0.013	100%	0.000	0.062	21%	1.924	0.546
E6	173.918	40%	1.049	0.003	50%	0.661	0.004	56%	0.969	0.013	100%	0.000	0.062	22%	1.981	0.602
E7	164.845	50%	0.759	0.004	50%	0.613	0.004	52%	0.996	0.015	100%	0.000	0.063	12%	1.802	0.586
E8	180.186	10%	1.474	0.004	40%	0.887	0.004	40%	1.043	0.014	90%	0.177	0.067	37%	1.318	0.585
E9	174.150	30%	1.232	0.004	60%	0.674	0.004	65%	0.655	0.013	100%	0.000	0.066	42%	0.896	0.592
E10	148.229	40%	0.775	0.004	60%	0.404	0.004	74%	0.948	0.013	100%	0.000	0.066	50%	0.864	0.598
50	0.200	100%	0.000	0.001	100%	0.032	0.001	90%	0.236	0.006	100%	0.000	0.026	97%	0.007	0.112
70	0.332	90%	0.038	0.003	70%	0.065	0.003	56%	0.063	0.013	90%	0.061	0.037	93%	0.001	0.238
100	0.677	90%	0.014	0.006	80%	0.099	0.007	35%	0.022	0.026	84%	0.039	0.055	90%	0.002	0.965
150	1.623	70%	0.059	0.016	60%	0.111	0.016	46%	0.020	0.062	50%	0.239	0.085	85%	0.001	3.375
200	3.355	60%	0.071	0.036	70%	0.032	0.036	40%	0.022	0.127	49%	0.131	0.133	68%	0.011	7.363

Table 4. Results for ORLIB, M and k-median instances

Type	CPX	JMS			MYZ			LOCAL			TABU			V&RR		
		Opt	Error	Time	Opt	Error	Time	Opt	Error	Time	Opt	Error	Time	Opt	Error	Time
cap7*	0.107	0%	0.401	0.001	10%	0.397	0.001	50%	0.044	0.002	100%	0.000	0.023	100%	0.000	0.047
cap10*	0.098	0%	0.499	0.001	10%	0.144	0.001	50%	0.059	0.003	100%	0.000	0.024	100%	0.000	0.073
cap13*	0.196	0%	0.493	0.002	10%	0.138	0.002	50%	0.236	0.006	100%	0.000	0.027	85%	0.017	0.159
capa-c	77.745	0%	1.990	0.152	0%	2.712	0.159	50%	0.242	0.476	82%	0.077	1.072	70%	0.039	19.457
MO*		20%	0.786	0.008	30%	0.452	0.008	63%	0.320	0.027	100%	0.000	0.067	23%	1.004	1.842
MP*		20%	0.387	0.049	30%	0.118	0.050	68%	0.086	0.166	100%	0.000	0.235	6%	1.203	11.229
MQ*		20%	0.444	0.142	30%	0.132	0.143	93%	0.041	0.523	100%	0.000	0.669	4%	1.367	25.027
MR*		[10%]	[0.381]	0.464	[10%]	[0.380]	0.476	[69%]	[0.229]	2.136	[100%]	[0.000]	1.825	0%	[2.022]	79.391
MS		[10%]	[0.000]	2.281	[10%]	[0.000]	2.323	[100%]	[0.000]	11.720	[100%]	[0.000]	6.366	0%	[1.829]	304.066
MT		0%	[0.205]	11.079	0%	[0.205]	11.241	[100%]	[0.159]	89.529	[90%]	[0.020]	31.505	0%	[1.817]	1283.285

Problem	LB	JMS		MYZ		LOCAL		TABU		V&RR	
		Error	Time	Error	Time	Error	Time	Error	Time	Error	Time
1000,10	1432737	1.085	1.758	1.486	1.810	0.457	17.282	0.500	4.713	6.972	356.955
1000,100	607591	0.920	1.758	0.902	1.799	0.358	48.320	0.327	5.166	4.713	245.225
1000,1000	220479	0.664	1.767	0.435	1.802	0.570	62.622	2.046	2.856	4.703	187.718
2000,10	2556794	1.198	8.651	1.218	8.921	0.530	85.863	0.526	16.797	9.694	1172.350
2000,100	1122455	0.996	8.516	1.581	8.684	0.547	289.785	0.529	18.803	6.729	1000.169
2000,1000	437553	0.852	8.599	0.849	8.730	0.531	682.241	0.467	88.423	3.366	912.588
3000,10	3567125	1.499	21.924	1.973	22.310	0.546	228.680	0.555	39.209	12.644	2951.249
3000,100	1600551	1.141	21.660	1.498	22.008	0.682	892.870	0.664	44.901	10.821	2677.263
3000,1000	643265	0.888	21.630	0.957	21.914	0.468	2188.568	0.375	67.246	4.309	2008.729

A Lazy Version of
Eppstein's K Shortest Paths Algorithm⋆

Víctor M. Jiménez and Andrés Marzal

DLSI, Universitat Jaume I, 12071 Castellón, Spain
{vjimenez,amarzal}@uji.es

Abstract. We consider the problem of enumerating, in order of increasing length, the K shortest paths between a given pair of nodes in a weighted digraph G with n nodes and m arcs. To solve this problem, Eppstein's algorithm first computes the shortest path tree and then builds a graph $D(G)$ representing all possible deviations from the shortest path. Building $D(G)$ takes $O(m+n \log n)$ time in the basic version of the algorithm. Once it has been built, the K shortest paths can be obtained in order of increasing length in $O(K \log K)$ time. However, experimental results show that the time required to build $D(G)$ is considerable, thereby reducing the practical interest of the algorithm. In this paper, we propose a modified version of Eppstein's algorithm in which only the parts of $D(G)$ which are necessary for the selection of the K shortest paths are built. This version maintains Eppstein's worst-case running time and entails an important improvement in practical performance, according to experimental results that are also reported here.

1 Introduction

Enumerating, in order of increasing length, the K shortest paths between two given nodes, s and t, in a digraph $G = (V, E)$ with n nodes and m arcs is a fundamental problem that has many practical applications and has been extensively studied [1, 2, 3, 5, 6, 7, 8, 9, 10, 11, 12, 13, 14, 15, 16, 17].

The algorithm with the lowest asymptotical *worst-case* time complexity that solves this problem is due to Eppstein [6]. After computing the shortest path from every node in the graph to t, Eppstein's algorithm builds a graph $D(G)$ which represents all possible one-arc deviations from the shortest path tree. Building $D(G)$ takes $O(m + n \log n)$ time in the basic version of the algorithm, and $O(m + n)$ time in a more elaborate but "rather complicated" [6] version. The graph $D(G)$ implicitly defines a path graph $P(G)$ such that the K shortest paths from its initial node to any other node represent the K shortest s-t paths in G, and it takes $O(K \log K)$ time to compute them once $D(G)$ has been built.

In [9], a different algorithm for computing the K shortest paths between two given nodes was proposed that under many circumstances runs significantly

⋆ This work has been supported by the *Generalitat Valenciana* under grant CTIDIA/2002/209 and by the Spanish *Ministerio de Ciencia y Tecnología* and *FEDER* under grant TIC2002-02684.

K. Jansen et al. (Eds.): WEA 2003, LNCS 2647, pp. 179–191, 2003.

faster in practice. This algorithm, known as the *Recursive Enumeration Algorithm* (REA), computes every new s-t path by recursively visiting at most the nodes in the previous s-t path and using a heap of candidate paths associated to each node from which the next path from s to the node is selected. After computing the shortest path from s to every node, the algorithm computes the K shortest paths in order of increasing length in $O(m + Kn \log(m/n))$ time. However, this is a worst-case bound which is only achievable in extremely rare situations. The REA can be considered as a lazy evaluation algorithm, in which the K shortest paths to intermediate nodes are only computed when they are required. In this way, a huge amount of computation effort can be saved. Experimental results, reported in [9], showed that, in practice, the REA outperforms Eppstein's algorithm [6] for different kinds of randomly generated graphs.

In this paper, we apply a similar idea to Eppstein's algorithm. The experimental results in [9], reproduced in Sect. 5, show that the time required by Eppstein's algorithm to build $D(G)$ is considerable, and the reason for its practical inefficiency. Once $D(G)$ has been built, enumerating the K shortest paths is extremely fast. Here we propose a modified version in which a recursive function builds only the parts of $D(G)$ that are necessary for the selection of the K shortest paths. In this way, the asymptotical worst-case complexity is maintained and a considerable reduction in computation effort is achieved in practice, according to experimental results with different kinds of randomly generated graphs. The new version is not only much faster than the original algorithm, but also faster than the REA in many cases.

The rest of this paper is organised as follows. After introducing some basic definitions and notation in Sect. 2 and summarizing Eppstein's algorithm in Sect. 3, the proposed modification to Eppstein's algorithm is presented in Sect. 4. Experimental results comparing this new version with the original one and with the REA are reported in Sect. 5. Finally, Sect. 6 contains conclusions and final remarks.

2   Problem Formulation and Notation

Let $G = (V, E)$ be a directed graph, where V is the set of nodes and $E \subseteq V \times V$ is the set of arcs. Given $e = (u, v) \in E$ we call $tail(e)$ to u and $head(e)$ to v. Let $\ell : E \to \mathbb{R}$ be a function mapping arcs to real-valued lengths.

Given two nodes u and v in V, a path from u to v (or u-v path) is a sequence of arcs $p = p_1 \cdot p_2 \cdot \ldots \cdot p_{|p|}$, where $tail(p_1) = u$, $head(p_{|p|}) = v$, and $head(p_i) = tail(p_{i+1})$ for $1 \leq i < |p|$. The length of p is $L(p) = \sum_{1 \leq i \leq |p|} \ell(p_i)$. In this paper we consider the problem of enumerating, in order of increasing length, the K paths from a starting node s to a terminal node t with minimum total length. It will be assumed that G does not contain negative length cycles.

For each v in V, let $d(v, t)$ be the length of the shortest v-t path. Let T be the single-destination shortest path tree containing the shortest v-t path for each v in V. Let $out(v)$ be the set of arcs with tail v that are not in the shortest path from v to t, and let $next_T(v)$ be the node w following v in T, so that the shortest path from v to t is the arc (v, w) followed by the shortest path from w to t.

For each (u, v) in E, let us define $\delta(u, v) = \ell(u, v) + d(v, t) - d(u, t)$. The value $\delta(u, v)$ is the additional cost we pay if, instead of following the shortest path from u to t, we first take the arc (u, v) and then follow the shortest path from v to t. Trivially, $\delta(u, v) \geq 0$ for each (u, v) in E, and $\delta(u, v) = 0$ for each (u, v) in T. The length of a s-t path p can be obtained by adding the value of δ for the arcs of p to $d(s, t)$ and, because $\delta(u, v) = 0$ for the arcs (u, v) of T, we only need to add the value of δ for the arcs of p that are not in T [6]. Let $sidetracks(p)$ be the subsequence of arcs of p that are not in T, let

$$L(sidetracks(p)) = \sum_{(u,v) \in sidetracks(p)} \delta(u, v), \tag{1}$$

and let S^k be the sidetracks of the kth shortest path from s to t. Since $L(p) = d(s, t) + L(sidetracks(p))$ and any path p can be made explicit from $sidetracks(p)$ and T in time proportional to $|p|$, the computation of the K shortest paths problem can be restated as the computation of T and $S^2, \ldots, S^K$ [6]. This is the aim of Eppstein's algorithm.

3   Eppstein's Algorithm

In this section we summarize, for the sake of completeness and to make clear the modification proposed in Sect. 4, the basic version of Eppstein's algorithm. We base our description on [6], where a more detailed explanation and a proof of correctness can be found.

Once T has been computed, Eppstein's algorithm builds, in time $O(m + n \log n)$, a graph $D(G)$ whose nodes are arcs in $E - T$ and which represents all possible s-t paths in G differing from T in only one arc, scored by δ. The graph $D(G)$ implicitly defines the so-called path graph $P(G)$ in which the weights of arcs are chosen so that finding the K best paths from the initial node to any destination in $P(G)$, a problem that can be solved in time $O(K \log K)$, is equivalent to finding the K best paths from s to t in G. A more detailed step by step description of this algorithm is provided below.

Step 1 Compute T and $\delta(u, v)$ for all $(u, v) \in E$. The time taken by this step depends on the kind of graph [4]. For acyclic graphs, T can be obtained in $O(m)$ time. If the graph contains cycles but the arc lengths are not negative, then Dijkstra's algorithm combined with Fibonacci heaps performs this computation in $O(m + n \log n)$ time. In the general case it can be done by means of the Bellman-Ford algorithm in $O(mn)$ time.

Step 2 For each $v \in V$, in any order such that $next_T(v)$ is processed before v:

Step 2.1 Build a heap $H_{out}(v)$ whose elements are the arcs in $out(v)$ heap-ordered by the value of δ in such a way that the root of $H_{out}(v)$, denoted $outroot(v)$, has only one child (see Fig. 1).

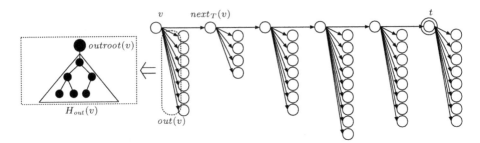

Fig. 1. The heap $H_{out}(v)$ contains all the arcs (v, w) in E except the first arc in the shortest path from v to t (represented by the horizontal arcs). The black circles represent nodes in this heap and correspond to arcs in G

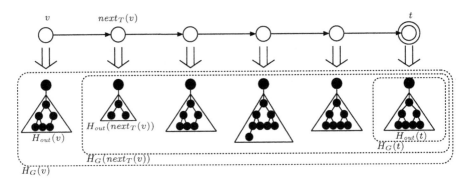

Fig. 2. $H_G(v)$ contains $H_{out}(v)$ and the elements in $H_G(next_T(v))$. $H_G(t)$ only contains $H_{out}(t)$

Step 2.2 If $v = t$, let $H_G(t)$ be the heap whose only element is $H_{out}(t)$; else, build a heap $H_G(v)$ by inserting $H_{out}(v)$ in $H_G(next_T(v))$ with score $\delta(outroot(v))$ in a persistent (non destructive) way (see Fig. 3). This insertion is guided by a balanced heap containing only $outroot(w)$ for each w in the shortest v-t path. The root of $H_G(v)$ will be denoted $h(v)$.

It is possible to build $H_{out}(v)$ in time proportional to the cardinal of $out(v)$ by first finding its root and then heapifying the rest of the elements in $out(v)$. Thus, Step 2.1 takes $O(m)$ time. Every heap $H_G(v)$ is built without modifying $H_G(next_T(v))$ in time $O(\log n)$. Therefore, the time required to perform Step 2.2 for all nodes is $O(n \log n)$, and the total time required by Step 2 is $O(m + n \log n)$.

For each v in V, the second best path from v to t differs from T in only one arc, which belongs to $out(w)$ for some node w in the shortest path from v to t. This arc is $h(v)$. In particular, $h(s)$ is S^2.

Step 3 The set of heaps $H_G(v)$ for all $v \in V$ forms a directed acyclic graph, $D(G)$, with $O(m + n \log n)$ nodes. Each node in $D(G)$ corresponds to an arc of G

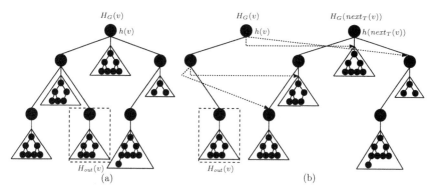

Fig. 3. (a) Heap $H_G(v)$. (b) Real representation in memory of $H_G(v)$: the nodes of $H_G(next_T(v))$ that should be modified by the insertion of $H_{out}(v)$ are replicated and linked to the corresponding subtrees in $H_G(next_T(v))$

and each arc in G is represented by at least one node in $D(G)$. For each (e, f) in which e is the parent of f in any heap $H_G(v)$, there is an arc (e, f) in $D(G)$. The graph $D(G)$ defines a different graph $P(G)$ which is the result of augmenting $D(G)$ as follows:

1. Associate a weight $\delta(f) - \delta(e)$ to each arc (e, f) in $D(G)$.
2. For each node e in $D(G)$, add an arc $(e, h(head(e)))$ with weight $\delta(h(head(e)))$. These arcs are called cross-arcs.
3. Add an initial node r and an arc from r to $h(s)$ with weight $\delta(h(s))$.

In this way, each path p in $P(G)$ from r to any node is associated to the s-t path in G whose sidetracks are the tails of the cross-arcs in p followed by the last node in p. The problem of enumerating the K shortest s-t paths in G reduces to the problem of enumerating the K shortest paths from r to any node in $P(G)$. This can be efficiently done with the help of a priority queue Q whose elements are sidetracks of s-t paths scored by the function L defined in (1). It is not necessary to explicitly build $P(G)$; instead, once $D(G)$ has been obtained, the sidetracks of the K shortest paths are obtained as follows:

Step 3.1 Initialize Q to $\{h(s)\}$.

Step 3.2 For $k = 2$ to K:

Step 3.2.1 If Q is empty, stop (no more s-t paths exist); else, extract S^k, the element in Q with the lowest score. Let e be the last sidetrack of S^k.

Step 3.2.2 Insert $S^k \cdot f$ in Q with score $L(S^k) + \delta(f)$, where f is the sidetrack $h(head(e))$.

Step 3.2.3 For each sidetrack f such that (e, f) is an arc in the graph $D(G)$, insert $prefix(S^k) \cdot f$ in Q with score $L(S^k) - \delta(e) + \delta(f)$, where $prefix(S^k)$ is the sequence S^k except its last sidetrack.

Note that all the sidetracks in Q can be efficiently represented as a prefix tree and that they are incrementally scored. Since the out-degree of each node in $D(G)$ is at most 3, Step 3 takes $O(K \log K)$ time [6].

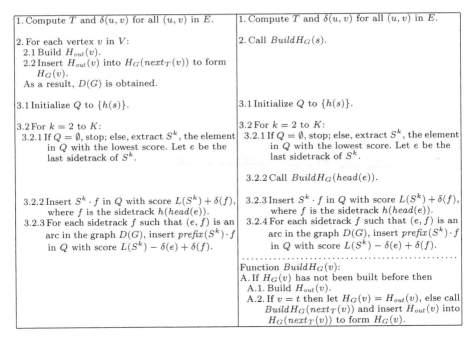

Fig. 4. Eppstein's algorithm (left) and the proposed lazy version (right), side by side

4   Lazy Version of Eppstein's Algorithm

In Step 2, Eppstein's algorithm builds $D(G)$, that is, $H_{out}(v)$ and $H_G(v)$ for every node v in V. This step takes $O(m + n \log n)$ time, which in many practical situations is a high cost, and is performed before computing S^2 in Step 3. In this section we propose a modification that reduces this cost by building only the heaps $H_{out}(v)$ and $H_G(v)$ that are really needed to compute $S^2, S^3, \ldots, S^K$.

This can be done by means of a recursive function $BuildH_G(v)$ that detects whether $H_G(v)$ has been already built. If $H_G(v)$ exists, then the function does nothing; else, it proceeds to build $H_G(v)$ by inserting $H_{out}(v)$ into $H_G(next_T(v))$ *after* recursively calling $BuildH_G(next_T(v))$. The insertion procedure is performed in the same nondestructive way as Eppstein's original algorithm in $O(\log n)$ time. The recursive calls go through the nodes of the shortest v-t path, and stop when they reach t or a node w whose $H_G(w)$ has already been built. In the base case $v = t$, $H_G(t)$ only contains $H_{out}(t)$.

In Step 3.1 of Eppstein's algorithm, the priority queue Q is initialized with $h(s)$, which is the root of $H_G(s)$. Therefore, in the new version we build $H_G(s)$ before Step 3.1 by calling $BuildH_G(s)$. This call will recursively build $H_G(w)$ for each node w in the shortest s-t path.

In Step 3.2.2 of Eppstein's algorithm, $h(head(e))$ is used, where e is the last sidetrack of S^k and $h(head(e))$ is the root of $H_G(head(e))$. In the new version, calling $BuildH_G(head(e))$ ensures that $H_G(head(e))$ is available.

Thus, instead of completely building $D(G)$ in the beginning, its construction is performed simultaneously with the generation of the K shortest paths, by carrying out a recursive traversal of paths similar to the one done by the REA. The function $BuildH_G(v)$ is called only when $H_G(v)$ is required and, therefore, the computation of $H_G(v)$ for many nodes v in V can be saved. This can be considered a lazy evaluation version of Eppstein's algorithm. The worst-case asymptotical complexity is the same as Eppstein's original algorithm because, as k increases, the lazy version will tend to compute $H_{out}(v)$ and $H_G(v)$ for every v in V, in total time $O(m + n \log n)$. However, in practice, the time saved can be considerable. The experiments presented in the next section show that the lazy version can be significantly faster, even when computing a huge number of shortest paths.

A complete description of this lazy version of Eppstein's algorithm is given in Fig. 4, side by side with the original algorithm so that the differences can be more easily appreciated.

5   Experimental Comparison

In this section, the experimental results are reported comparing, for three different kinds of random generated graphs, the lazy version of Eppstein's algorithm presented in Sect. 4 (LVEA) with the basic version of Eppstein's algorithm (EA) [6] and with the Recursive Enumeration Algorithm (REA) [9]. We reproduced the experimental conditions in [9] so that the results can be directly compared. For the sake of clarity, we used the basic version of the REA in which the sets of path candidates are implemented with heaps, although there are some cases in which a mixture of heaps and unsorted arrays may be faster [9].

All these programs were implemented in C and are publicly available at http://terra.act.uji.es/REA. Since the three algorithms share Step 1, we are interested in comparing their performance once T has been computed. Therefore, all time measurements start when Dijkstra's algorithm ends. Each point in the curves shows the average execution time for 15 random graphs generated with the same parameters, but different random seeds. The programs were compiled with the GNU C compiler (version 2.7) using the maximum optimization level. The experiments were performed on a 300 MHz Pentium-II computer with 256 megabytes of RAM, running under Linux 2.0.

Results for Graphs Generated with Martins' General Instances Generator. First, we compared the algorithms using Martins' general instances generator, the same that was used in the experiments reported in [9] and [10]. The input to this graph generator consists of four values: seed for the random number generator, number of nodes, number of arcs, and maximum arc length. The program creates a Hamiltonian cycle and then adds arcs at random. The arc lengths are uniformly distributed between 0 and 10^4.

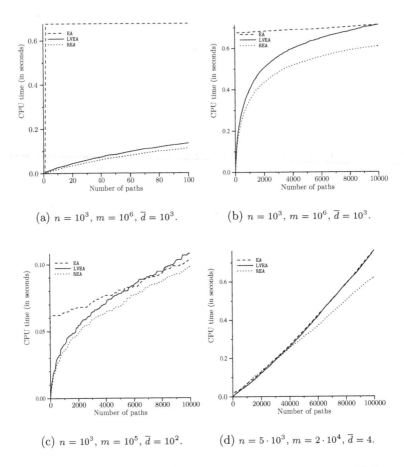

(a) $n = 10^3$, $m = 10^6$, $\bar{d} = 10^3$.   (b) $n = 10^3$, $m = 10^6$, $\bar{d} = 10^3$.

(c) $n = 10^3$, $m = 10^5$, $\bar{d} = 10^2$.   (d) $n = 5 \cdot 10^3$, $m = 2 \cdot 10^4$, $\bar{d} = 4$.

Fig. 5. Results for graphs generated with Martins' generator. CPU time as a function of the number of computed paths. (a) is an enlargement of the initial region of (b) to appreciate the differences for small values of K. ($\bar{d} = m/n$ is the average input degree.)

Figure 5 represents the CPU time required to compute up to 10^5 paths by each of the algorithms for graphs with different average degrees: high (Fig. 5b), medium (Fig. 5c), and low (Fig. 5d). Figure 5a is an enlargement of Fig. 5b for small values of K. With regard to the behaviour of EA, it can be observed that, once $D(G)$ has been built, the K shortest paths are found very efficiently. However, building this graph requires (in comparison with the other algorithms) a considerable amount of time that can be clearly identified in the figures at the starting point $K = 2$. These figures also illustrate that, in contrast, the LVEA does not invest time in the construction of $D(G)$ in the beginning, but instead builds it progressively. For large values of K, the full graph $D(G)$ is eventually

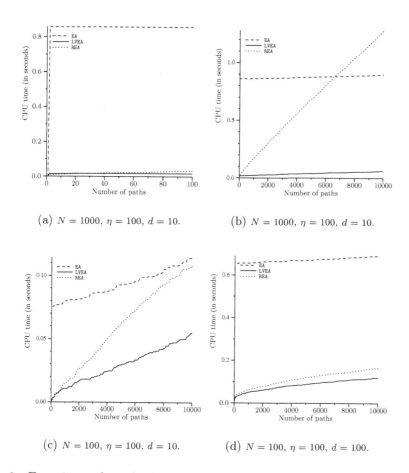

(a) $N = 1000$, $\eta = 100$, $d = 10$.

(b) $N = 1000$, $\eta = 100$, $d = 10$.

(c) $N = 100$, $\eta = 100$, $d = 10$.

(d) $N = 100$, $\eta = 100$, $d = 100$.

Fig. 6. Experimental results for multistage graphs. CPU time as a function of the number of computed paths. (a) is an enlargement of the initial region of (b). (Parameters: N, number of stages; η, number of nodes per stage; d, input degree.)

built and the running time of the **LVEA** converges to the running time of **EA**. The **REA** is still faster for these graphs.

Results for Multistage Graphs. Multistage graphs are of interest in many applications and underlie many discrete Dynamic Programming problems. A multistage graph is a graph whose set of nodes can be partitioned into N disjoint sets (stages), $V_1, V_2, \ldots, V_N$, such that every arc in E joins a node in V_i with a node in V_{i+1}, for some i such that $1 \le i < N$. We used a program that generates random multistage graphs given the number of stages, number of nodes per stage, and input degree (which is fixed to the same value for all the nodes). Arc lengths are again uniformly distributed between 0 and 10^4.

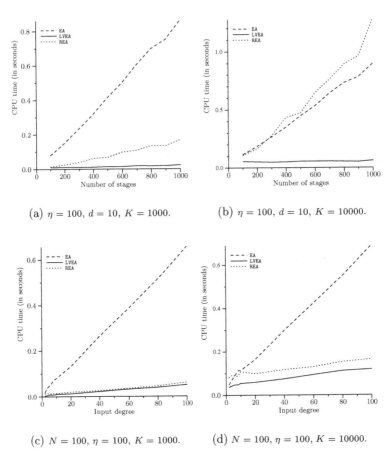

(a) $\eta = 100$, $d = 10$, $K = 1000$.   (b) $\eta = 100$, $d = 10$, $K = 10000$.

(c) $N = 100$, $\eta = 100$, $K = 1000$.   (d) $N = 100$, $\eta = 100$, $K = 10000$.

Fig. 7. Experimental results for multistage graphs. CPU time as a function of the number of stages (a and b) and the input degree (c and d). (Parameters: N, number of stages; η, number of nodes per stage; d, input degree; K, number of computed paths.)

The results are illustrated in Figs. 6 and 7. Figure 6a is an enlargement of Fig. 6b for small values of K. We can also observe here that, once $D(G)$ has been built, the K shortest paths are found by EA very efficiently, but building $D(G)$ is a highly costly operation and the main reason for the relative inefficiency of EA. The time saved by the LVEA is considerable, in this case even for large values of K. The REA does not perform as well in these graphs due to the fact that it computes the kth shortest path visiting, in the worst case, all the nodes in the $(k-1)$th shortest path, and in multistage graphs the number of nodes in any path is the number of stages. The dependency with the number of stages can be more clearly appreciated in Figs. 7a and 7b. Note how incrementing the

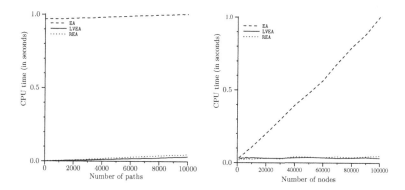

Fig. 8. Results for Delaunay triangulation graphs. CPU time as a function of (a) the number of computed paths (with 10^5 nodes) and (b) the number of nodes (to compute 10^4 paths)

value of the number of stages does not affect the **LVEA** (which benefits from the efficient representation of alternative paths as sidetracks), while it clearly affects **EA** (which needs to build heaps for a larger number of nodes).

Finally, Figs. 7c and 7d represent the dependency of the running time with the input degree to compute 10^3 and 10^4 paths, respectively. In these figures the input degree ranges from 2 to 100, and this includes the evolution from Fig. 6c to Fig. 6d, as well as the behaviour for smaller values of the input degree. It can be seen that the time required by the **REA** and the **LVEA** only increases slightly when the input degree increases, while **EA** is clearly more affected by this parameter, so that the difference between the algorithms increases as the input degree increases.

Results for Delaunay Triangulation Graphs. Delaunay triangulations are a particular kind of graph that we have chosen because they share with many real-world graphs, such as road maps, the fact that shortest paths tend to be in a certain region of the graph, in this case, close to the straight line connecting the origin and destination points. Nodes in other regions of the graph do not participate in the K shortest paths. Both the **REA** and the **LVEA** can take advantage of this fact and avoid building heaps associated to them. We used a graph generator that uniformly distributes a given number of points in a square, computes their Delaunay triangulation and assigns the Euclidean distance between the two joined points (nodes) to each arc. The initial and final nodes are located on opposite vertices of the square.

The results are illustrated by Fig. 8. In this case, the lazy version of Eppstein's algorithm is much faster than the original algorithm (which is far from being competitive) and slightly faster than the **REA**.

6   Conclusions

The algorithm proposed by Eppstein to compute the K shortest paths between two given nodes in a graph is outstanding because of its low asymptotical worst-case complexity [6]. However, this algorithm includes a initial stage to build a graph $D(G)$ from which the K shortest paths are then very efficiently computed, and the time required by this initial stage is considerable in practice, as the experimental results in Sect. 5 illustrate. For this reason, a simpler solution, the so-called *Recursive Enumeration Algorithm* (REA) [9] is significantly faster in many cases. In this paper, we have combined ideas from both algorithms to propose a lazy version of Eppstein's algorithm that avoids building $D(G)$ completely when only a part of it is necessary. In this way, we maintain Eppstein's worst-case complexity and achieve a considerable reduction in computation effort in practice, according to experimental results with different kinds of randomly generated graphs: the new version is not only faster than the original algorithm, but also faster than the REA in many cases. It might, therefore, be a useful practical alternative. Our implementation of these algorithms is publicly available at http://terra.act.uji.es/REA.

References

[1] J. Azevedo, M. Costa, J. Madeira, and E. Martins. An algorithm for the ranking of shortest paths. *European J. Op. Res.*, 69:97–106, 1993. 179

[2] J. Azevedo, J. Madeira, E. Martins, and F. Pires. A computational improvement for a shortest paths ranking algorithm. *European J. Op. Res.*, 73:188–191, 1994. 179

[3] R. Bellman and R. Kalaba. On kth best policies. *J. SIAM*, 8(4):582–588, 1960. 179

[4] T. Cormen, C. E.Leiserson, R. Rivest, and C. Stein. *Introduction to Algorithms*. The MIT Press, Cambridge, MA, 2nd edition, 2001. 181

[5] S. Dreyfus. An appraisal of some shortest-path algorithms. *Op. Res.*, 17:395–412, 1969. 179

[6] D. Eppstein. Finding the k shortest paths. *SIAM J. Computing*, 28(2):652–673, 1999. 179, 180, 181, 183, 185, 190

[7] B. L. Fox. Calculating kth shortest paths. *INFOR - Canad. J. Op. Res. and Inform. Proces.*, 11(1):66–70, 1973. 179

[8] B. L. Fox. Data structures and computer science techniques in operations research. *Op. Res.*, 26(5):686–717, 1978. 179

[9] V. M. Jiménez and A. Marzal. Computing the k shortest paths: a new algorithm and an experimental comparison. *Lecture Notes in Computer Science*, 1668:15–29, 1999. (http://terra.act.uji.es/REA/papers/wae99.ps.gz). 179, 180, 185, 190

[10] E. Martins and J. Santos. A new shortest paths ranking algorithm. Technical report, Departamento de Matemática, Universidade de Coimbra, 1996. (http://www.mat.uc.pt/~eqvm). 179, 185

[11] E. Q. V. Martins. An algorithm for ranking paths that may contain cycles. *European J. Op. Res.*, 18:123–130, 1984. 179

[12] K. Mehlhorn and M. Ziegelman. Resource constrained shortest paths. *Lecture Notes in Computer Science*, 1879:326–337, 2000. 179

[13] S. P. Miaou and S. M. Chin. Computing K-shortest paths for nuclear spent fuel highway transportation. *European J. Op. Res.*, 53:64–80, 1991. 179

[14] D. R. Shier. Iterative methods for determining the k shortest paths in a network. *Networks*, 6:205–229, 1976. 179

[15] D. R. Shier. On algorithms for finding the k shortest paths in a network. *Networks*, 9:195–214, 1979. 179

[16] C. C. Skicism and B. L. Golden. Computing k-shortest path lengths in euclidean networks. *Networks*, 17:341–352, 1987. 179

[17] C. C. Skicism and B. L. Golden. Solving k-shortest and constrained shortest path problems efficiently. *Annals of Op. Res.*, 20:249–282, 1989. 179

Linear Algorithm for 3-Coloring
of Locally Connected Graphs

Martin Kochol

MÚ SAV, Štefánikova 49, 814 73 Bratislava 1, Slovakia
kochol@savba.sk

Abstract. The problem to decide whether a graph is 3-colorable is NP-complete. We show that if G is a locally connected graph (neighborhood of each vertex induces a connected graph), then there exists a linear algorithm which either finds a 3-coloring of G, or indicates that such coloring does not exist.

By Garey, Johnson and Stockmayer [2], it is an NP-complete problem to decide whether a graph is 3-colorable. This problem remains NP-complete for much smaller classes of graphs, as shown for example in [1, 2, 3, 5]. On the other hand, by König [4], a graph is 2-colorable if and only if it is bipartite. Thus there exists a linear algorithm which either finds a 2-coloring of a graph, or decides that such coloring does not exist. In this paper we show that similar algorithm exists for 3-coloring of locally connected graphs.

We deal with finite graphs without multiple edges and loops. If G is a graph, then $V(G)$ and $E(G)$ denote the sets of vertices and edges of G, respectively. For every $U \subseteq V(G)$ and $A \subseteq E(G)$, denote by $G[U]$ and $G[A]$ the subgraphs of G induced by U and A, respectively (note that $G[U]$ and $G[A]$ are the maximal and minimal subgraphs of G satisfying $V(G[U]) = U$ and $E(G[A]) = A$). For every $v \in V(G)$, denote by $N(v)$ the set of neighboring vertices of v. We say that G is *locally connected* if $G[N(v)]$ is connected for every $v \in V(G)$. G is *uniquely k-colorable* if there exists a unique partition of $V(G)$ into k independent sets.

We start with an auxiliary statement.

Lemma 1. *Suppose there is a quadruple (G, U, u, φ) such that G is a graph with m edges, U is an independent set of vertices from G, $u \in U$, $U \cup N(u) = V(G)$, φ is a mapping from U to $\{1, 2, 3\}$, $\varphi(u) \neq \varphi(x)$ for every $x \in U \setminus \{u\}$ which is not an isolated vertex of G, and each component of $G[N(u)]$ contains a vertex adjacent with a vertex from $U \setminus \{u\}$. Then there exists an algorithm running in time $O(m)$ which either finds a mapping ψ from $N(v)$ to $\{1, 2, 3\}$ such that ψ and φ form a 3-coloring of G or indicates that such ψ does not exist. Furthermore, if ψ exists, then it is unique.*

Proof. We construct sets $C \subseteq V(G)$ such that $U \subseteq C$ and a mapping φ_C from $C \setminus U$ to $\{1, 2, 3\}$ such that φ and φ_C form a 3-coloring of $G[C]$. Furthermore, φ_C is the unique mapping with this property. In the same time we color the edges from $G - u$ such that the edges having exactly two ends from C are colored by color 5, the edges having exactly one end in C are colored by color 4 and the rest of edges are colored by color 3. The edges of G having one end u are colored by color 2 if the second end is from C and by color 1 if the second end is not from C. The edge coloring is only an auxiliary process, which helps to count the number of steps. In each stage, we always increase the color of an edge if we use this edge during the process. Hence the algorithm has running time $O(m)$.

Step 1: We set $C := U$, $\varphi_C := \emptyset$, color the edges having one end in $U \setminus \{u\}$ and $\{u\}$ by colors 4 and 1, respectively, and the rest of edges by color 3. Go to Step 2.

Step 2: Check whether there exists an edge colored by 4. If there is no such edge, then, since each component of $G[N(u)]$ contains a neighbor of $U \setminus \{u\}$, we get that C contains all vertices from $N(v)$, i.e., $C = V(G)$ and thus $\psi = \varphi_C$ has the desired properties. If there is an edge colored by 4, take its end $v \in N(u) \setminus C$ and consider the set C_v of vertices from C which are joined with v by an edge having color 4. If the vertices from C_v are colored by more than one color, ψ does not exist because the vertices from C_v are not isolated in G, whence, by the assumptions of lemma, $\varphi(x) \neq \varphi(u)$ for every $x \in C_v$. If all vertices from C_v are colored by the same color α, then setting $\varphi'(v) \in \{1,2,3\}$, $\varphi'(v) \neq \alpha, \varphi(u)$ and $\varphi'(x) = \varphi_C(x)$ for every $x \in C \setminus U$ we get a mapping φ' from $C \cup \{v\}$ to $\{1,2,3\}$ such that φ and φ' form a 3-coloring of $G[C \cup \{v\}]$. Furthermore, φ' is the unique mapping with this property (because φ_C is so and $\alpha \neq \varphi(u)$). Thus we can set $C := C \cup \{v\}$, $\varphi_C := \varphi'$, increase the numbers of colors of the edges incident with v by 1, and repeat Step 2. $\square$

Note that ψ from Lemma 1 can be constructed in $O(m)$ time no matter on the cardinality of U (i.e, the number of isolated vertices of G).

Theorem 1. *Let G be a locally connected graph and $m = \max\{|V(G)|, |E(G)|\}$. Then there exists an algorithm running in $O(m)$ time which either finds a 3-coloring of G, or indicates that G is not 3-colorable. Furthermore, if G is 3-colorable, then every component of G is uniquely 3-colorable.*

Proof. We construct sets $C \subseteq V(G)$, $T \subseteq C$ and a 3-coloring φ_C of $G[C]$ such that

(1) $G[C]$ is uniquely 3-colorable;
(2) if $v \in T$, then $N(v) \subseteq C$;
(3) if $v \in C$, then $N(v) \cap T \neq \emptyset$.

In the same time we decompose $E(G)$ into sets $E_1, \ldots, E_5$ such that

(a) $E_1 = E(G - C)$;
(b) $E_2 = E(G) \setminus (E(G[C]) \cup E_1)$;
(c) $E_3 = E(G[C \setminus T])$;

(d) $E_4 = E(G[C]) \setminus (E_3 \cup E(G[T]))$;
(e) $E_5 = E(G[T])$.

Notice that E_1, E_3, E_5 are the sets of edges having both ends in $V(G) \setminus C$, $C \setminus T$, T, respectively, and E_2, E_4 are the sets of edges having exactly one end in C, T, respectively. For simplicity, we assume that the edges from E_i are colored by color i ($i = 1, \ldots, 5$). We start the algorithm with $C = T = \emptyset$, $E_1 = E(G)$, $E_2 = E_3 = E_4 = E_5 = \emptyset$ (i.e., all edges of G are colored by 1). Similarly as in Lemma 1, the edge-coloring helps to count he number of steps in the algorithm.

Step 1: If $V(G) = \emptyset$, we are ready. Otherwise choose $v \in V(G)$. If $N(v) = \emptyset$ color v by color 3, set $G := G - v$, and go to Step 1. If $N(v) \neq \emptyset$, take $v_1 \in N(v)$ and a bijective mapping $\varphi : \{v, v_1\} \rightarrow \{2, 3\}$, and apply the algorithm from Lemma 1 for the quadruple $(G[N(v) \cup \{v\}] - vv_1, \{v, v_1\}, v, \varphi)$, which satisfies the assumptions of this lemma. We get either a (unique) 3-coloring ψ of $G[N(v) \cup \{v\}]$, or $G[N(v) \cup \{v\}]$ and G are not 3-colorable. In the first case set $T := \{v\}$, $C := N(v) \cup \{v\}$, $\varphi_C := \psi$, (which satisfy $(1) - (3)$), color the edges of $G[C]$ following the rules (a) – (d), and go to Step 2.

Step 2: Check whether $C = T$. If $C = T$, then by $(1) - (3)$, $G[C]$ is a (uniquely) 3-colorable component of G with 3-coloring φ_C. Thus we set $G := G - C$ and go to Step 1. If $C \neq T$, then go to Step 3.

Step 3: Choose $v \in C \setminus T$. Now we need to check whether $C' = C \cup N(v)$ and $T' = T \cup \{v\}$ satisfy $(1) - (3)$. First take edges $e_1 = vv_1, \ldots, e_n = vv_n$, $n = |N(v)|$, incident with v.

For $i = 1, \ldots, n$, do the following. If e_i has colors 2, 3 or 4, replace it by colors 4, 4 or 5, respectively. If $v_i \in C$, then v_i is already colored. If $v_i \notin C$, then consider the edges $e_{i,1} = v_i v_{i,1}, \ldots, e_{i,n_i} = v_i v_{i,n_i}$ incident with v_i and different from e_i. If $e_{i,j}$ has colors 1 or 2, replace it by colors 2 or 3, respectively. (Note that if $e_{i,j}$ joins two vertices from $N(v) \setminus C$, then we increase its color twice, from 1 to 2 with respect to one end and then from 2 to 3 with respect to the second end.)

Let H be the graph arising from $G[N(v) \cup \{v\}]$ after deleting all edges belonging to $G[C]$ and adding the (isolated) vertices from $C \setminus (N(v) \cup \{v\})$. Then $E(H)$ is the set of edges of G for which we have increased a color from $\{1, 2\}$ to a color from $\{3, 4\}$ and the quadruple (H, C, v, φ_C) satisfies the assumptions of Lemma 1. Apply the algorithm from Lemma 1 for this quadruple. After $O(|E(H)|)$ steps, either we extend the 3-coloring φ_C to a 3-coloring $\varphi_{C'}$ of $G[C']$, or we show that $G[C']$ and G are not 3-colorable. If $G[C']$ is 3-colorable, then (by Lemma 1 and since C and T satisfy $(1) - (3)$ and $G[N(v)]$ is connected) also C' and T' satisfy $(1) - (3)$. Set $C := C'$, $T := T'$, $\varphi_C := \varphi_{C'}$. The new colors of the edges of G satisfy conditions (a) – (e) for the new sets C and T. Go to Step 2.

We finish the algorithm when we get $V(G) = \emptyset$ or when we show that G is not 3-colorable. Furthermore, every component of G is uniquely 3-colorable if G is 3-colorable. In Step 3, the number of processes is a multiple of the number of edges for which we have increased the color. The same holds for Step 1 if the chosen vertex v is not isolated. In Step 2, we only delete (or separate) the vertices which are already colored. Similarly in Step 1 if the chosen vertex v is isolated. Thus the algorithm has running time $O(m)$. $\square$

Theorem 1 cannot be generalized to k-colorability for $k > 3$ unless P = NP. To show this, construct for any connected graph G a locally connected graph G_n, $n \geq 1$, adding to G a copy of K_n and joining every vertex from G with every vertex from K_n. Then G_n is $n + r$-colorable iff G is r-colorable. Hence, by [2], the problem to decide whether a locally connected graph is k-colorable is NP-complete for every fix $k > 3$.

References

[1] Ehrlich, G., Even, S., Tarjan, R. E.: Intersection graphs of curves in the plane, J. Combin. Theory Ser. B **21** (1976) 8–20. 191
[2] Garey, M. R., Johnson D. S., Stockmeyer, L.: Some simplified NP-complete graph problems, Theor. Comput. Sci. **1** (1976) 237–267. 191, 194
[3] Kochol, M., Lozin, V., Randerath, B.: The 3-colorability problem on graphs with maximum degree 4, Preprint 2/2003, Mathematical Institute, Slovak Academy of Sciences, Bratislava, Slovakia [http://www.mat.savba.sk/preprints]. 191
[4] König, D.: Theorie der endlichen und unendliched Graphen, Akademische Verlagsgesellschaft M. B. H., Leipzig, 1936. 191
[5] Maffray, F., Preissmann, M.: On the NP-completeness of the k-colorability problem for triangle-free graphs, Discrete Math. **162** (1996) 313–317. 191

A Clustering Algorithm for Interval Graph Test on Noisy Data

Wei-Fu Lu[1] and Wen-Lian Hsu[2]

[1] Institute of Computer and Information Science
National Chiao Tung University, Hsin-Chu, Taiwan, ROC
gis84812@cis.nctu.edu.tw
[2] Institute of Information Science
Academia Sinica, Taipei, Taiwan, ROC
hsu@iis.sinica.edu.tw

Abstract. An interval graph is the intersection graph of a collection of intervals. One important application of interval graph is physical mapping in genome research, that is, to reassemble the clones to determine the relative position of fragments of DNA along the genome. The linear time algorithm by Booth and Lueker (1976) for this problem has a serious drawback: the data must be error-free. However, laboratory work is never flawless. We devised a new iterative clustering algorithm for this problem, which can accommodate noisy data and produce a likely interval model realizing the original graph.

1   Introduction

An interval graph is the intersection graph of a collection of intervals. This class of graphs has a wide range of applications. An important application of interval graphs is the construction of physical maps for the genome research. Physical maps are critical in hunting for specific genes of interest, and also useful for further physical examination of DNA required for other genome project. The term "physical mapping" means the determination of the relative position of fragments of DNA along the genome by physicochemical and biochemical methods. The construction of physical maps is generally accomplished as follows. Long DNA sequences are broken to smaller fragments, and then each fragment is reproduced into the so-called clones. After deciding some fingerprints for each clone, two clones are considered overlapping if their fingerprints are sufficiently similar. Finally, information on pairwise overlapping determines the relative positions of clones, thus completing the construction of physical maps [3, 4, 6, 9, 17, 19, 20].

The error free version of the mapping problem can be modeled as an interval graph recognition problem: given a graph $G = (V, E)$, finding a family of intervals such that each interval corresponding to one vertex of the graph, and two vertices are adjacent if and only if their corresponding intervals are overlapping [2, 10, 11, 12]. However, data collected from laboratories unavoidably contain errors, such as false positives (FPs, two overlapping clones are actually

K. Jansen et al. (Eds.): WEA 2003, LNCS 2647, pp. 195–208, 2003.

non-overlapping) and false negatives (FNs, two non-overlapping clones are actually overlapping). Because a single error might cause the clone assembly to fail, traditional recognition algorithms can hardly be applied on noisy data directly. Moreover, no straightforward extension of traditional algorithm can overcome the drawbacks.

Four typical models have been proposed for dealing with errors. The definitions are as follows: 1) *interval graph completion problem*: assume the input data only contain FNs and minimize the number of edges whose addition makes the graph an interval [13, 15, 18]. 2) *interval graph deletion problem*: assume there are only FPs in the input data, and minimize the number of edges whose deletion makes the graph an interval graph [7]. 3) *interval sandwich problem*: assume that some pairs of clones are definite overlaps, some are definite non-overlaps, and the rest are unknown, then construct an interval graph under these overlapping constraints [8, 14]. 4) *intervalizing k-color graph problem*: assume that clones are created from k copies of DNA molecule, and some pairs of clones are definite overlaps. The objective is to generate a k-colorable interval graph with the overlapping conditions [1, 5, 7, 8]. However, the above models suffer from the following two unpleasant phenomena: 1. all of the above models have been shown to be NP-hard [5, 7, 8, 22], and it would be difficult to define an associated "single objective optimization problem" for approximation due to the errors could be intertwined together; 2. even if one can find the best solution, this solution might not make any biological sense.

To cope with this dilemma, consider the nature of error treatment. Generally, data collected in real life contain a small percentage of errors. Suppose the error percentage is 5% with carefully control. The challenge is thus to discover the 95% correct information versus the 5% incorrect information automatically. We designed an algorithm to deal with errors based on local structure matching. The idea is very similar to the one employed in [16]. Our philosophy is that, in order to determine whether certain overlapping information is valid or noisy, we check the neighborhood data to see if it conforms "approximately" to a particular local structure dictated by the problem. The probability that an isolated piece of spurious information has a well-behaved neighborhood structure is nil. More precisely, in our analysis, if there is enough valid information in the input data, then a certain monotone structure of the overlapping information on the neighborhood will emerge, allowing us to weed out most errors. We do not set any "global" objective to optimize. Rather, our algorithm tries to maintain a certain "local" monotone structure, namely, to minimize the deviation from the local monotone structure as much as possible.

The kind of error-tolerant behavior considered here are similar in nature to algorithms for voice recognition or character recognition problems. Thus, it would be difficult to "guarantee" that the clustering algorithm always produces a desirable solution (such as one that is a fixed percentage away from the so-called "optimal solution"); the result should be justified through benchmark data and real life experiences. Our experimental results show that, when the error percentage is small, our clustering algorithm is robust enough to discover certain errors and to correct them automatically most of the time.

The remaining sections are organized as follows. Section 2 gives the basic definitions of some notations. An interval graph test based on [16] is discussed in Section 3, which forms the basis of our clustering algorithm. Section 4, the main part of this paper, illustrates how to deal with errors in the input data. The experimental results are shown in Section 5. Section 6 contains some conclusion remarks.

2   Basic Definitions

In this paper all graphs are assumed to be undirected, simple, and finite. For a graph $G = (V, E)$, denote its number of vertices by n and its number of edges by m. Given a vertex u in G, define $N[u]$ to be the set of vertices including u and those vertices adjacent to u in G; define $N(u)$ to be $N[u] - u$. For some subset M of V, define $N(M)$ be the set of those vertices that are not in M but adjacent to some vertices in M. Thus, we have $N(N[u]) = \{x | x$ is not in $N[u]$ but adjacent to some vertices in $N[u]\}$, which is the second-tier neighborhood in a breadth-first-search from u. This kind of neighborhood plays a crucial role on our clustering analysis. We define relations between two adjacent vertices using the above set of neighbors. Two adjacent vertices u, v in G are said to be *strictly adjacent (STA)*, if none of $N[u]$ and $N[v]$ is contained in the other. We denote the set consists of those vertices strictly adjacent to u by $STA(u)$. A vertex u is said to be *contained* in another vertex v, if $N[u]$ is contained in $N[v]$.

Each interval graph has a corresponding interval model in which two intervals overlap if and only if their corresponding vertices are adjacent[1]. However, the corresponding interval model is usually far from unique, because of variations of the endpoint orderings. To obtain the unique interval model representation, consider the following block structure of endpoints: Denote the right (resp. left) endpoint of an interval u by $R(u)$ (resp. $L(u)$). In an interval model, define a maximal contiguous set of right (resp. left) endpoints as an *R-block* (resp. *L-block*). Thus, the endpoints can be grouped as an alternating left-right block sequence. Since an endpoint block is a set, the endpoint orderings within the block are ignored. The overlapping relationship remains unchanged if one permutes the endpoint order within each block. Denote the right block containing $R(u)$ by $B_R(u)$, the left block containing $L(u)$ by $B_L(u)$, and the set of block subsequence from $B_L(u)$ to $B_R(u)$ by $[B_L(u), B_R(u)]$. An endpoint $R(w)$ (resp. $L(w)$) is said to *be contained in an interval* u if $B_L(w)$ (resp. $B_R(w)$) is contained in $[B_L(u), B_R(u)]$.

Let G be an interval graph. Consider an interval model for G. For an interval u, the neighborhood of u can be partitioned into $A(u)$, $B(u)$, $C(u)$, $D(u)$, where $A(u)$ consists of those intervals that strictly overlap u from left side; $B(u)$ consists of those intervals that strictly overlap u from right side; $C(u)$ consists

---

[1] For convenience, we shall not distinguish between these two terms, "vertex" and its corresponding "interval".

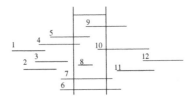

Fig. 1.   An example of $A(u)$, B(u), $C(u)$,D(u),$LL(u)$, and $RR(u)$

of those vertices that properly contain u; $D(u)$ consists of those vertices that are properly contained in u. We call these sets $A(u)$, $B(u)$, $C(u)$, $D(u)$, the left neighborhood, the right neighborhood, the outer neighborhood, and the inner neighborhood. On the other hand, the second-tier neighborhood of u can be partitioned into $LL(u)$ and $RR(u)$, where $LL(u)$ consists of those intervals that are completely to the left of u and overlap some neighbors of u; and $RR(u)$ consists of those intervals that are completely to the right of u and overlap some neighbors of u. We call $LL(u)$ the left second-tier neighborhood and $RR(u)$ the right second-tier neighborhood.An example of $A(u)$, $B(u)$, $C(u)$, $D(u)$, and $RR(u)$ is shown in Figure 1, where $A(u) = \{4,5\}$, $B(u) = \{9,10\}$, $C(u) = \{6,7\}$, $D(u) = \{8\}$, $LL(u) = \{1,2,3\}$, and $RR(u) = \{11,12\}$.

3   An Interval Graph Test

To our best knowledge, no straightforward extension of existing linear time algorithms can handle errors. The idea of [16], however, can be modified to yield a clustering version that can deal with noisy data. In this section, we describe a quadratic time interval graph test, which adopts some techniques similar to [16]. Notably, the time complexity is not a major concern for algorithms on noisy data.

The basic idea of this algorithm is very simple: The vertices are processed one by one according to an ascending order of their degree. For each vertex u, we decide the unique left-right block sequence that records the relative positions of endpoints within u, based on a robust local structure on its neighbors. If the unique left-right block sequence within u intersects other existing left-right block sequences, all the left-right block sequences are further merged into a new left-right block sequence. Finally, if graph G is an interval graph, after all vertices have been processed, we will obtain the unique left-right block sequence that realize graph G; otherwise, the algorithm will terminate in some iteration due to the failure of left-right block sequence construction.

For each vertex u in G, our algorithm performs three main steps: 1) neighborhood classification, 2) block sequence determination, and 3) vertex replacement. The first step, neighborhood classification, classifies vertices adjacent to u into $A(u)$, $B(u)$, $C(u)$ and $D(u)$. Since the block sequence within u relates to a robust local structure on $A(u)$ and $B(u)$, this classification is significant for our interval graph test. The second step, block sequence determination, decides the unique left-right block sequence within u according to a monotone structure on $A(u)$ and $B(u)$, and merge this block sequence with another existing block sequence, if necessary. The last step, vertex replacement, generates a "special vertex" u^s which is adjacent to all neighbors of u and special vertices strictly adjacent to u. We shall associate u^s with the corresponding left-right block sequence of u constructed in the second step. Remove vertices whose endpoints are both contained in the block sequence of u^s, and delete all edges between $A(u^s)$ and $B(u^s)$, since information about those deleted edges and vertices is no longer needed. After vertex replacement, the graph is further reduced.

The main iteration of our interval graph test is described in Algorithm 1. The following definitions are needed to describe the algorithm.

Definition 1. *A collection of sets is said to be monotone if every two sets S_i, S_j in the collection are comparable, that is, either $S_i \supseteq S_j$ or $S_j \subseteq S_i$.*

Definition 2. *A interval u is said to be compatible with a left-right block sequence LB_1, RB_1, LB_2, RB_2, $\ldots$, LB_d, RB_d if the left (resp. right) endpoints within u are contained in LB_1, LB_2, $\ldots$, LB_d (resp. RB_1, RB_2, $\ldots$, RB_d), and let RB_{j_1} (resp. LB_{j_2}) be the leftmost R-block (resp. rightmost L-block) having nonempty intersection with endpoints within u, all blocks in between (but excluding) RB_{j_1} and RB_{j_2} are contained in $N(u)$.*

Algorithm 1 *The Interval-graph-test: Processing an original vertex*
1. *Neighborhood Classification:*
    1.1 *Construct the following set: $C(u) \leftarrow \{w|N(w) \supset N(u)\}$, $D(u) \leftarrow \{w|N(w) \subseteq N(u)\}$ and $STA(u) \leftarrow N(u) - C(u) - D(u)$.*
    1.2 *Partition $STA(u)$ into $A(u)$ and $B(u)$:*
        (1) *Let u^* be a vertex in $STA(u)$ with the largest $|N(u^*) \cap (N(STA(u)) - N(u))|$.*
        (2) *Let $LL(u) \leftarrow \{w|w \in N(u*) \cap (N(STA(u)) - N(u))\}$, and $RR(u) \leftarrow N(STA(u)) - N(u) - LL(u)$.*
        (3) *Let $A(u) \leftarrow STA(u) \cap N(LL(u))$ and $B(u) \leftarrow STA(u) - A(u)$.*
    1.3 *Let u_{SL} be the special interval such that $u_{SL} \in A(u)$, and u_{SR} be the special interval such that $u_{SR} \in B(u)$.*
2. *Block sequence determining:*
    2.1 *Find the collection of sets $\{N(w) \cap B(u)|w \in A(u)\}$.*
    2.2 *Check the following:*
        (1) *The collection $\{N(w) \cap B(u)|w \in A(u)\}$ is monotone such that the right endpoints of intervals in $A(u)$ and the left endpoints of interval in $B(u)$ can be uniquely partitioned with $R(u_{SL})$ located on the first R-block and $L(u_{SR})$ located on the last L-block.*
        (2) *Every interval in $D(u)$ is compatible with the block sequence determined by the above two sets and the remaining intervals in $D(u)$.*
    2.3 *If there is any violation, G is not an interval graph and the test is terminated*
3. *Vertex replacement:*
    3.1 *Create new special interval us with $N(u^s) \leftarrow N(u_{SL}) \cap N(u) \cap N(u_{SR})$.*
    3.2 *Suppose that x is a vertex with its right endpoint in us but not its left endpoint, and y is a vertex with its left endpoint in us but not its eight endpoint. Remove edge (x, y) if it exists.*
    3.3 *Remove u, u_{SL} and u_{SR} and vertices whose left endpoints and right endpoints are both contained in u^s.*

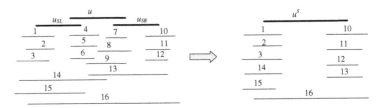

Fig. 2. An example of the *Interval-graph-test*

An example of the *Interval-graph-test* is shown in Figure 2. The left half of Figure 2 is the interval graph at the beginning of the iteration that interval u is processed. Intervals u_{SL} and u_{SR} are the two special intervals strictly overlapping u. The corresponding block sequence of u_{SL} is $\{L(u_{SL})\}$, $\{R(1)\}$, $\{L(4), L(5)\}$, $\{R(2), R(3)\}$, $\{L(6), L(u)\}$, $\{R(u_{SL})\}$, and the corresponding block sequence of u_{SL} is $\{L(u_{SR})\}$, $\{R(7), R(8)\}$, $\{L(10), L(11)\}$, $\{R(9), R(u)\}$, $\{L(12)\}$, $\{R(u_{SL})\}$. In neighborhood classification, the neighborhood of u is classified into $A(u) = \{14, 15\}$, $B(u) = \{13\}$ and $C(u) = \{16\}$, and $D(u) = \{4, 5, 6, 7, 8, 9\}$. Based on the monotone property of $A(u)$ and $B(u)$, as well as the compatible property of $D(u)$ and the block sequence decided by $A(u)$ and $B(u)$, we can obtain the block sequence within u, say $\{L(u)\}$, $\{R(u_{SL})\}$, $\{L(9), L(13)\}$, $\{R(6)\}$, $\{L(8)\}$, $\{R(4), R(5)\}$, $\{L(7)\}$, $\{R(14)\}$, $\{L(u_{SR})\}$, $\{R(u)\}$. In the vertex replacement step, do the following:

1. Create a new special interval u^s with $N(u^s) = N(u_{SL}) \cup N(u) \cup N(u_{SR}) = \{1, 2, 3, 4, 5, 6, 7, 8, 9, 10, 11, 12, 13, 14, 15, 16\}$, and associates u^s with the block sequence that records the relative positions of endpoints within u^s.
2. Delete all edges connecting vertices in $\{4, 5, 6, 14, 15\}$ and $\{7, 8, 9, 13\}$.
3. Remove intervals u_{SL}, u_{SR}, and intervals contained in u^s (namely, intervals 4, 5, 6, 7, 8, and 9).

At the end of this iteration, the corresponding interval graph becomes the right half of Figure 2.

We can prove that our algorithm decides whether a graph G is an interval graph or not correctly based on the following lemmas and theorems (the details are omitted).

Lemma 1. *If graph G is an interval graph, then the collections $\{N(w) \cap B(u) | w \in A(u)\}$ and $\{N(w) \cap A(u) | w \in B(u)\}$ are both monotone.*

Lemma 2. *If $\{N(w) \cap B(u) | w \in A(u)\}$ is monotone, then the right endpoints in $A(u)$ and the left endpoints in $B(u)$ can be partitioned into $LB_2, \ldots, LB_n$ and $RB_1, RB_2, \ldots, RB_{n-1}$ respectively, such that the $LB_1, RB_1, \ldots, LB_n, RB_n$ is the left-right block sequence within u, where $LB_1 = \{L(u)\}$ and $RB_n = \{R(u)\}$.*

Theorem 1. *A graph is an interval graph iff the following conditions hold for each iteration of the interval-graph-test algorithm:*

1. *The collection of set $\{N(w) \cap B(u) | w \in A(u)\}$ is monotone and the right endpoints of $A(u)$ and the left endpoints of $B(u)$ can be uniquely partitioned with $R(u_{SL})$ located on the first right block and $L(u_{SR})$ located on the last left block.*
2. *Every interval in $D(u)$ is compatible with the block sequence determined by the above two sets and the remaining intervals in $D(u)$.*

If the given graph G is an interval graph, then the proposed algorithm will yield an interval model of graph G, otherwise, the algorithm will terminate in step 2 of the *Interval-graph-test*.

4   Treating the Errors

In this section, we present the clustering version of interval graph testing. The method to perform neighbor classification on noisy data will be discussed in Section 4.1. Section 4.2 illustrates the block sequence determining while taking FNs and FPs into account. The complete clustering version of interval graph test (under noise) is summarized in Section 4.3.

4.1   The Error-Tolerant Neighborhood Classification

If the input data contain errors, it is more intricate for neighborhood classification. However, based on clustering analysis on the neighborhood and second-tier neighborhood of u, we are able to classify neighbors of interval u roughly into four sets $A(u)$, $B(u)$, $C(u)$, and $D(u)$. Our strategy is to classify the second-tier neighbors of u, $N(N[u])$, into $LL(u)$ and $RR(u)$ first, and then classify the neighbors of $u$into $A(u)$, $B(u)$, $C(u)$ and $D(u)$ based on $LL(u)$ and $RR(u)$. Let $OV(w, v) = |N[w] \cap N[v]|$ denote the overlap function between two intervals w and v. The overlap function is used to measure the degree of overlapping for each pair of intervals in $N(N[u])$. The clustering of $LL(u)$ and $RR(u)$ uses a greedy strategy based on the overlap function. The classification of $LL(u)$ and $RR(u)$ is described in Algorithm 2 below.

Algorithm 2 *The LL-RR-classification Algorithm*
1. *For each interval in $N(N[u])$, associate it with a cluster that consists of that interval.*
2. *Calculate $OV(w, v)$ for each pair of intervals in $N(N[u])$.*
3. *Select a pair of intervals such that these two intervals are in different clusters and attain the highest $OV(u, v)$ value. Merge the corresponding clusters of w and v into one cluster.*
4. *Reiterate Step 3 until all there are two clusters.*
5. *Let one cluster be $LL(u)$ and the other be $RR(u)$.*

An example for explaining neighborhood classification is shown in Figure 3. In this case, the input data are noisy, but we can only depict part of errors in this figure. The solid lines and the dotted lines represent intervals overlapping u and those not overlapping u in the input data, respectively. All intervals depicted are located at the original "correct" position. Thus, 4 overlaps u originally but does not overlap u in the input data due to FN, and 10 does not overlap u originally

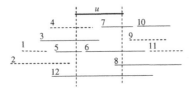

Fig. 3. An example for neighborhood classification

but overlap u in the input data due to FP. Furthermore, assume that there are FPs between interval 1 and 11 and between 2 and 11. At the first iteration, merge the corresponding clusters of 1 and 2 into one cluster, since $OV(1,2) = 4$ is the highest. At the second and the third iterations, merge the corresponding clusters of 2 and 4, and the corresponding clusters of 9 and 11, respectively. Finally, we have that $LL(u) = \{1,2,4\}$ and $RR(u) = \{9,11\}$.

We classify $SAT(u)$ into $A(u)$ and $B(u)$ using the simple heuristic rule: intervals in $A(u)$ should not overlap any interval in $RR(u)$, and intervals in $B(u)$ should not overlap any interval in $LL(u)$. Thus, the overlapping relation between $A(u)$ and $RR(u)$, and between $B(u)$ and $LL(u)$ could be considered as FPs. For each interval w in $SAT(u)$, classifying w into $A(u)$, if the number of FPs due to classify w into $A(u)$ less than the number of FPs due to classify w into $B(u)$, we conclude that w is in $A(u)$, on the contrary, we conclude that w is in $B(u)$. Such a classification scheme is summarized in Algorithm 3.

Algorithm 3 *The A-B-classification Algorithm*
1. *Calculate the error functions of w as follows:*
   $E_A(w) \leftarrow |\{(w,v)|v \in N(w) \cap RR(u)\}|$
   $E_B(w) \leftarrow |\{(w,v)|v \in N(w) \cap LL(u)\}|.$
2. *Classify w into $A(u)$, $B(u)$:*
   If $E_A(w) < E_B(w)$ then classifying w into $A(u)$
   else then classifying w into $B(u)$

In the example of Figure 3, $E_A(3) = 0$, $E_B(3) = 3$, $E_A(5) = 0$, $E_B(5) = 2$, $E_A(6) = 1$, $E_B(6) = 1$, $E_A(7) = 1$, $E_B(7) = 0$, $E_A(8) = 2$, $E_B(8) = 0$, $E_A(10) = 2$, $E_B(10) = 0$. Thus, $A(u) = \{3,5\}$, $B(u) = \{6,7,8,10\}$.

The above sets $A(u)$, $B(u)$, $LL(u)$ and $RR(u)$ could still be misclassified due to those FPs and FNs related to interval $u$itself. To prevent this kind of errors (or to minimize its effect), we shall reclassify intervals currently in $LL(u) \cup A(u)$ into new $LL(u)$ and $A(u)$ as follows (The reclassification of $RR(u) \cup B(u)$ into new $RR(u)$ and $B(u)$ can be done similarly). Denote $LL(u) \cup A(u)$ by L-part(u), and $RR(u) \cup B(u)$ by R-part(u). To reclassify intervals currently in $LL(u) \cup A(u)$ into new $LL(u)$ and $A(u)$, it suffices to determine the location of $L(u)$.

Once $L(u)$ is located, then those intervals of L-part(u) whose right endpoints are to the right (resp. left) of $L(u)$ are considered neighbors, $A(u)$ (resp. non-neighbors, $LL(u)$), of u. We shall locate the right endpoint of intervals in L-part(u) first, and then decide the position of $L(u)$. To do that, we need to determine the relative positions among right endpoints of intervals in L-part(u).

Interestingly enough, R-$part(u)$ will play an important role in this process based on the following simple lemma.

Lemma 3. *Let S and T be two sets of intervals. If the right (resp. left) endpoint of every interval in T is to the right (resp. left) of the right (resp. left) endpoint of every interval in S, then the right (resp. left) endpoint of interval w in S with the largest $|N(w) \cap T|$ value is the rightmost right endpoint (resp. leftmost left endpoint) among all right (resp. left) endpoints of intervals in S.*

Based on Lemma 3, we shall order right endpoints of intervals in L-$part(u)$ from right to left iteratively as follows. Initially, set S to be L-$part(u)$ and T to be R-$part(u)$. Note that S and T will be changed at each iteration. We shall maintain that the right endpoint of every interval in T is to the right of the right endpoints of intervals in S. Thus, we shall make the right endpoint of interval $w_{|L\text{-}part(u)|}$ in S with the largest $|N(w_{|L\text{-}part(u)|}) \cap T|$ value to be the rightmost right endpoint of intervals in S ($= L$-$part(u)$). Next, delete interval $w_{|L\text{-}part(u)|}$ from S and add $w_{|L\text{-}part(u)|}$ into T. Now, make the right endpoint of interval $w_{|L\text{-}part(u)|-1|}$ in the resultant S with the largest $|N(w_2) \cap T|$ value to be the rightmost right endpoint of intervals in the remaining S ($= L$-$part(u)$-$\{w_{|L\text{-}part(u)|}\}$). Thus, $R(w_{|L\text{-}part(u)|-1|})$ is to the left of $R(w_{|L\text{-}part(u)|})$, but to the right of all right endpoints of other intervals in L-$part(u)$. Then, delete interval $w_{|L\text{-}part(u)|-1|}$ from S and add $w_{|L\text{-}part(u)|-1|}$ into T. Reiterate the above process until all right endpoints of intervals in L-$part(u)$ have been ordered.

After that, call the ordered right endpoints of intervals in L-$part(u)$ from left to right as $R(w_1)$, $R(w_2)$, ..., $R(w_{|L\text{-}part(u)|})$. But, in some cases due to noise, although $|N(x) \cap T| > |N(y) \cap T|$ (which would entail that $R(x)$ is to the right of $R(y)$), $R(x)$ might be, in fact, to the left of $R(y)$. However, if the error rate is quite small, (say, no more than 5%), we can expect that $R(x)$ will be ordered to the right of $R(y)$ with high probability. Thus, we can obtain the approximate ordering of the right endpoints of intervals in L-$part(u)$. Similarly, we can also order left endpoints of intervals in R-$part(u)$ form left to right as $L(v_1)$, $L(v_2)$, ..., $L(v_{|R\text{-}part(u)|})$.

To decide the position of $L(u)$, we calculate the "cost" of $L(u)$ for each position that $L(u)$ could be placed. If $L(u)$ is placed between $R(w_i)$ and $R(w_{i+1})$, u must overlap all intervals w_j with $j > i$, otherwise (u, w_j) is a FN. Moreover, intervals w_j and w_k such that $j, k > i$ must overlap each other, otherwise (w_j, w_k) is a FN. On the other hand, intervals w_j with $j \leq i$ should not overlap u, otherwise (w_i, u) is a FP. Let $ErrL(u, i)$ be the total number of FNs and FPs, if $L(u)$ is placed between $R(w_i)$ and $R(w_{i+1})$. Thus, $ErrL(u, i) = |\{(u, w_j)|(u, w_j) \notin E(G) \text{ and } j > i\}| + |\{(w_j, w_k)|(w_j, w_k) \notin E(G) \text{ and } j, k > i\}| + |\{(w_i, u)|(w_i, u) \in E(G) \text{ and } j \leq i\}|$. Note that $ErrL(u, 0)$ is defined as the total number of errors that place $L(u)$ to the left of all the right endpoints of intervals in L-$part(u)$. We conclude that $L(u)$ should be placed between $R(w_i)$ and $R(w_{i+1})$, if $ErrL(u, i)$ is the minimum among all of the $ErrL$ values. Similar strategy can be used to decide the position of $R(u)$. The heuristic to distinguish neighbors and non-neighbors of u is described in Algorithm 4.

Algorithm 4 *The Neighborhood-decision Algorithm.*

1. *Order the right endpoints of intervals in L-part(u) from left to right as $R(w_1)$, $R(w_2)$, ..., $R(w_{|L-part(u)|})$ as follows :*
    1.1 *Let S be L-part(u), T be R-part(u), and i be $|L-part(u)|$.*
    1.2 *Let w^* be an interval in S with the largest $|N(w^*) \cap T|$, and denote $L(w^*)$ by $L(w_i)$.*
    1.3 *Delete w_i from S and add w_i into T.*
    1.4 *Decrease i by 1.*
    1.5 *Reiterate Step 1.2 to Step 1.4, until S is empty.*
2. *For $0 \leq i \leq |L-part(u)|$, let $ErrL(u, i) = |\{(u, w_j)|(u, w_j) \notin E(G) \text{ and } j > i\}| + |\{(w_j, w_k)|(w_j, w_k) \notin E(G) \text{ and } j, k > i\}| + |\{(w_i, u)|(w_i, u) \in E(G) \text{ and } j \leq i\}|.*
3. *If $ErrL(u, t)$ is the minimum among all $ErrL$'s, we conclude that $L(u)$ should be placed between $R(w_t)$ and $R(w_{t+1})$. Let $A(u) = \{w_i|i > t\}$ and $LL(u) = \{w_j|j \leq t\}$.*
4. *Order the left endpoints of intervals in R-part(u) from left to right as $L(v_1)$, $L(v_2)$, ..., $L(v_{|R-part(u)|})$ as follows:*
    4.1 *Let S be R-part(u), T be L-part(u), and i be 1.*
    4.2 *Let v^* be an interval in S with the largest $|N(w*) \cap T|$, and denote $L(v^*)$ by $L(w_i)$.*
    4.3 *Delete v_i from S and add v_i into T.*
    4.4 *Increase i by 1.*
    4.5 *Reiterate Step 4.2 to Step 4.4, until S is empty.*
5. *For $0 \leq i \leq |R-part(u)|$, let $ErrR(u, i) = |\{(u, v_j)|(u, v_j) \notin E(G) \text{ and } j \leq i\}| + |\{(v_j, v_k)|(v_j, v_k) \notin E(G) \text{ and } j, k \leq i\}| + |\{(v_i, u)|(v_i, u) \in E(G) \text{ and } j > i\}|.*
6. *If $ErrR(u, t)$ is the minimum among all $ErrR$'s, we conclude that $R(u)$ should be placed between $L(v_t)$ and $L(v_{t+1})$, and let $B(u) = \{v_i|i \leq t\}$ and $RR(u) = \{v_j|j > t\}$.*

For example, in Figure 3, we can order right endpoints of intervals in L-part(u) from left to right as $R(1)$, $R(2)$, $R(5)$, $R(4)$, $R(3)$, and order left endpoints of intervals in R-part(u) from left to right as $L(7)$, $L(6)$, $L(8)$, $L(9)$, $L(10)$, $L(11)$. Furthermore, $ErrL(u, 2) = 1$ is the minimum among all $ErrL$'s, and $ErrR(u, 3) = 1$ is the minimum among all $ErrR$'s. Thus, we conclude that $L(u)$ should be located between $R(5)$ and $R(2)$, and $R(u)$ should be located between $L(8)$ and $L(9)$. Hence, L-part(u) and R-part(u) could be reclassified into $LL(u) = \{1, 2\}$, $A(u) = \{3, 4, 5\}$, $B(u) = \{9, 10, 11\}$, $RR(u) = \{6, 7, 8\}$, and the FPs and FNs relative to u itself have been corrected.

4.2   Deciding Endpoint Block Sequence under the Influence of FNs and FPs

In this section we determine the left-right block sequence within u on noisy data. The monotone collection $\{N(w) \cap N(u)|w \in A(u)\}$ provides a very strong structural property for interval graphs. This property is stable enough for us to obtain a "good" left-right block sequence within interval u. In case the above collection of sets does not satisfy the monotone property, one could remove some elements and/or add some elements into the sets to make it satisfy the monotone property. We denote the removed elements as *removals* and the added elements as *fill-ins*. The removals and fill-ins could be considered as FPs and FNs, respectively. Note that it is a relative matter to decide removals and fill-ins, and there is a trade-off in determining FPs and FNs. Suppose we suspect an edge to be a FP. There are two possibilities. One is that we simply remove this edge. The other is that we let it stay, which would possibly create some FN(s) we need to fill in to preserve the monotone property. Our strategy is to detect and remove potential FPs first, and then deal with the FNs. Note that the minimum fill-in problem is NP-complete [22] and a polynomial approximation for the problem has been proposed in [18].

We use the FP-Screening algorithm in Algorithm 5 to determine a FP. Let w_1, w_2 , $\ldots$, $w_{|A(u)|}$ be a list in $A(u)$ ordered according to their ascending $|N(w) \cap B(u)|$ values. If $\{N(w) \cap B(u)|w \in A(u)\}$ is monotone, we should have $N(w_i) \cap B(u) \subseteq N(w_j) \cap B(u)$ for all $i < j$. Since data is noisy, this might not hold for all $i < j$, but it should hold with high probability due to low error rate. So for each $v \in N(w_i) \cap B(u)$, if $|\{j|j > i$ and $v \in N(wj) \cap B(u)\}| \geq 3$, the entry (w_i, v) is considered a FP. The threshold is set to be 3 since the probability that there are more than three FPs in the same interval is relatively low.

Algorithm 5 *The FP-screening Algorithm*
1. *Sort intervals in $A(u)$ into a list $\{w_1, w_2, ïK, w_{|A(u)|}\}$ according to their ascending $|N(w) \cap B(u)|$ values.*
2. *For each $w \in A(u)$, if $|\{j|i < j$ and $v \in N(w_i) \cap B(u)$ and $v \notin N(w_j) \cap B(u)\}|$, the pair of intervals (w_i, v) is considered a FP. Remove edge (w_i, v).*

After the FPs are determined and removed, we determine fill-ins that make the collection $\{N(w) \cap B(u)|w \in A(u)\}$ monotone using the following greedy strategy shown in Algorithm 6. Initially, consider all intervals in $A(u)$ unselected. For each unselected interval w in $A(u)$, define its "fill-in cost" to be the minimum number of edges whose addition will satisfy $N(w) \cap B(u) \subseteq N(w') \cap B(u)$ for every unselected interval w' in $A(u)$, namely, define $fill\text{-}in(w) = |\{(w', v)|v \in N(w) \cap B(u)$ and $v \notin N(w') \cap B(u)$ for all $w' \in A(u)$, w' is unselected, and $w' \neq w\}|$. Each time, select the interval, say w^*, with the minimum "fill-in cost" among unselected intervals in $A(u)$. Once w^* has been selected, adding all edges counted in $fill\text{-}in(w^*)$ and mark w^* a selected interval. Reiterate the above process until all intervals in $A(u)$ are selected.

Algorithm 6 *The FN-screening Algorithm*
1. *Select an unselected interval w^* in $A(u)$ with the minimum $|fill\text{-}in(w^*)|$ value.*
2. *Consider each element (w', v) counted in $|fill\text{-}in(w^*)|$ as a FN and add edge (w', v).*
3. *Mark w^* a selected interval.*
4. *Reiterate Step 1 to Step 3 until all intervals in $A(u)$ have been selected.*

4.3   The Clustering Version of Interval Graph Test

Finally, we summarize the algorithms of these section in Algorithm 7 below. The intervals are processed according to an ascending order of their degrees.

Algorithm 7 *The Interval-graph-clustering-test*
1. *Neighbor Classification:*
   1.1 *Let $C(u) \leftarrow \{w|N(w) \supset N(u)\}$, $D(u) \leftarrow \{w|N(w) \subseteq N(u)\}$ and $STA(u) \leftarrow N(u) - C(u) - D(u)$.*
   1.2 *Construct $LL(u)$ and $RR(u)$ using the LL-RR-classification algorithm.*
   1.3 *Partition $STA(u)$ into $A(u)$, $B(u)$ using the A-B-classification algorithm.*
   1.4 *Distinguish the neighbors and non-neighbor of u using the Neighborhood-decision algorithm.*
   1.5 *Let u_{SL} be the special interval in $A(u)$, and u_{SR} be the special interval in $B(u)$.*
2. *Block sequence determination*
   2.1 *Screen out FPs using the FP-screening algorithm.*

  2.2 *Fill in FNs using the FN-screening algorithm.*
  2.3 *Construct the left-right block sequence within u using the collection $\{N(w) \cap B(u) | w \in A(u)\}$ and intervals in $D(u)$.*
3. *Vertex replacement:*
  3.1 *Create a new special interval u^s with $N(u^s) \leftarrow N(u_{SL}) \cup N(u) \cup N(u_{SR})$.*
  3.2 *Suppose that x is a vertex only with its right endpoint in u^s and y is a vertex only with its left endpoint in u^s. Remove edge (x, y) if it exists.*
  3.3 *Remove u, u_{SL} and u_{SR} and vertices whose left endpoints and right endpoint are both contained in u^s.*

5   Experimental Results

We conduct experiments based on synthetic data. We start with a fixed interval model and, in each experiment, randomly generate errors on the edge connections, then feed the resultant graph to our algorithm to get a left-right block ordering. Three fixed graphs of sizes 100, 200, and 400 are used. These graphs are generated randomly under the constraint that the number of endpoints an interval contains (roughly corresponds to its "coverage") ranges from 5 to 15. The combined error rates of FPs and FNs are set to be 3%, 5% and 10%, respectively. Within each error percentage, set the ratio of the number of FPs and that of FNs to be 1 to 4, namely, every generated FP accompanies 4 FNs. For various combination of graph size and error rate, we repeat the experiment 50 times using different random seeds. The results are evaluated by comparing the resultant interval ordering from that of the original ordering, based on the measurement defined below.

Regard the position of an interval as the position of the "left endpoint" of the interval. For an interval u, let d_1 be the number of intervals ordered to the left of u but whose indices are greater than u and d_2, the number of intervals ordered to the right of u whose indices are less than u. Let the displacement $d(u)$ of interval u be the larger of d_1 and d_2. The displacement $d(u)$ gives an approximate measure of the distance of interval u from its "correct" position. It should be noted that defining an exact measure is difficult here since many other intervals have to be moved simultaneously in order to place a particular interval "correctly". We use the following criterion for measuring the total deviation of the resultant ordering from the original one: If the displacement of an interval u is more than 4, we say u is a jump interval, which means that the position of u is quite far from its ordinary position. For example, in Figure 4, $d(2) = 6$ (there are 6 intervals ordered to the left of interval 2 whose indices are greater than 2), $d(6) = 1$, and $d(8) = 6$ (there are 6 intervals ordered to the right of interval 8 whose indices are less than 8). Thus, interval 2 and interval 8 are jump intervals.

We measure the performance of our algorithm by counting the number of jump intervals in the resultant interval model. Table 1 lists the statistics for the number of jump intervals. As one can see, when the error rate less than 5%, the number of jump intervals is less than 10, and even when the error rate increase to 10%, the number of jump intervals remains less than 15 in most runs, indicating that the final interval ordering produced by the algorithm is a good approximation for the original.

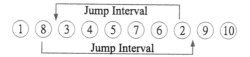

Fig. 4. An example of jump intervals

Table 1. Statistics for the jump intervals

Number of Jump Intervals	100 vertices			200 vertices			400 vertices		
	3%	5%	10%	3%	5%	10%	3%	5%	10%
0~5	48	34	20	38	28	4	48	40	27
6~10	1	4	17	12	17	11	1	5	13
11~15	0	0	9	0	4	23	0	3	6
16~20	1	2	1	0	1	7	1	1	3
21~25	0	0	2	0	0	1	0	0	0
26~30	0	0	1	0	0	0	0	1	0
31~35	0	0	0	0	0	1	0	0	0

6   Concluding Remark

In this paper we propose a clustering algorithm for interval graph test on noisy
data. The physical mapping problem in human genome research can be modeled
as an interval graph recognition problem, if the overlap information is error-free.
However, data collected from laboratories unavoidably contain errors. Traditional recognition algorithms can hardly be applied directly on noisy data, and
related models for the imperfection are shown to be NP-hard. In our algorithm,
for two typical error types FPs and FNs, we check the neighborhood data to see
whether they conform "approximately" to a particular local structure dictated
by interval graphs to determine whether overlapping information are valid or
noisy. The experimental results show that, when the error percentage is small,
the clustering algorithm is robust enough to discover certain errors and to correct
them automatically most of the time.

References

[1] H. L. Bodlaender, B. de Fluiter. On Intervalizing k-Colored Graphs for DNA Physical Mapping. *Discrete Applied Math.*, 71, 55-77, 1996. 196
[2] K. S. Booth and G. S. Lueker, Testing for the Consecutive Ones Property, Interval
Graphs, and Graph Planarity Using PQ-tree Algorithms, *J. Comput Syst. Sci.*,
13, 335-379, 1976. 195
[3] A. V. Carrano, P. J. de John, E. Branscomb, T. Slezak, B. W. Watkins. Constructing chromosome and region-specific cosmid map of the human genome, *Genome*,
31, 1059-1065, 1989. 195

[4] A. Coulson, J. Sulston, S. Brenner, J. Karn. Towatrd a physical map of the genome of the nematode, Caenorhabditis Elegans. *Proc. Natl. Acad. Sci. USA*, 83, 7821-7825, 1987. 195

[5] M. R. Fellows, M. T. Hallett, H. T. Wareham. DNA Physical Mapping: Three Ways Difficult. *LNCS* 726, 260-271, 1993. 196

[6] R. M. Gemmil, J. F. Coyle-Morris, F. D. Jr. McPeek, L. F. Wara-Uribe, F. Hecht. Construction of long-range restriction maps in human DNA using pulsed filed gel electrophoresis. *Gene Anal. Technol.*, 4, 119-131, 1987. 195

[7] P. W. Goldberg. M. C. Golumbic, H. Kaplan and R. Shamir. Four strikes against physical mapping of DNA. *J. Comput. Biol.*, 2, 139-152, 1995. 196

[8] M. C. Golumbic, H. Kaplan, R. Shamir. On the complexity of DNA physical mapping. *Advances in Applied Mathematics*, 15, 251-261, 1994. 196

[9] E. D. Green and M. V. Olson. Chromosomal region of the cystic fibrosis gene in yeast artificial chromosomes: a model for human genome mapping, *Science*, 250, 94-98, 1990. 195

[10] W. L. Hsu, A simple test for interval graphs, *LNCS* 657, 11-16, 1992. 195

[11] W. L. Hsu and T. H. Ma, Substitution Decomposition on Chordal Graphs and Applications, *LNCS* 557, 52-60, 1991. 195

[12] N. Korte and R. H. Mohring, An Incremental Linear-Time Algorithm for Recognizing Interval Graphs, *SIAM J. Comput.*, 18, 68-81, 1989. 195

[13] H. Kaplan, R. Shamir. Pathwidth, bandwidth, and completion problems to proper interval graphs with small cliques. *SIAM J. Comput.*, 25(3), 540-561,1996. 196

[14] H. Kaplan, R. Shamir. Bounded Degree Interval Sandwich Problems. *Algorithmica*, 24, 96-104, 1999. 196

[15] H. Kaplan, R. Shamir, R. E. Tarjan. Tractability of Parameterized Completion Problems on Chordal, Strongly Chordal, and Proper Interval Graphs. *SIAM J.Comput.*, 28(5), 1906-1922, 1999. 196

[16] W. F. Lu and W. L. Hsu. A test for the Consecutive Ones Property on Noisy Data. *To appear in Journal of Computational Biology.* 196, 197, 198

[17] F. Michiels, A. G. Craig, G. Zehetner, G. P. Smith, and H. Lehrach. Molecular approaches to genome analysis: A strategy for the construction of ordered overlapping clone libraries. *Comput. App. Biosci.*, 3(3),203-210, 1987. 195

[18] A. Natanzon, R. Shamir, R. Sharan. A Polynomial Approximation Algorithm for the Minimum Fill-In Problem. *STOC* 1998, 41-47, 1998. 196, 205

[19] M. V. Olson, E. Dutchik, M. Y. Graham, G. M. Brodeur, C. Helms, M. Frank, M. MacCollin, R. Acheinman, T. Frand. Random-clone strategy for genomic restriction mapping in yeast, *Proc. Natl. Acad. Sci. USA*, 83, 7826-7830, 1989. 195

[20] M. V. Olson, L. Hood, C. Cantor, and D. Botstein, A common language for physical mapping of the human genome, *Science*, 234, 1434-1435, 1985. 195

[21] M. Stoer and F. Wagner. A simple min-cut algorithm. *J. ACM*, 44(4), 585-591, 1997.

[22] M. Yannakakis. Computing the Minimum Fill-In is NP-Complete, *SIAM J. Alg. Disc. Meth*, 2, 77-79, 1981. 196, 205

Core Instances for Testing: A Case Study*
(Extended Abstract)

Monaldo Mastrolilli and Leonora Bianchi

IDSIA
Strada Cantonale Galleria 2, CH-6928 Manno, Switzerland
{monaldo,leonora}@idsia.ch
http://www.idsia.ch

Abstract. Data generation for computational testing of optimization algorithms is a key topic in experimental algorithmics. Recently, concern has arisen that many published computational experiments are inadequate respect to the way test instances are generated. In this paper we suggest a new research direction that might be useful to cope with the possible limitations of data generation. The basic idea is to select a finite set of instances which 'represent' the whole set of instances. We propose a measure of the representativeness of an instance, which we call ε-*representativeness*: for a minimization problem, an instance x_ε is ε-representative of another instance x if a $(1 + \varepsilon)$-approximate solution to x can be obtained by solving x_ε. Focusing on a strongly NP-hard single machine scheduling problem, we show how to map the infinite set of all instances into a finite set of ε-representative core instances. We propose to use this finite set of ε-representative core instances to test heuristics.

1 Introduction

Motivation. The literature addressing computational testing of optimization algorithms has a long history and, until recently, computational testing has been performed primarily through empirical studies. In the operations research community, concern has arisen that many published computational experiments are inadequate, especially respect to the way test instances are generated (see [7], [9], [10], [1], [15] and [11]). Among the limitations of computational tests is that they often involve simply devising a set of supposedly 'typical' instances, running the algorithm and its competitors on them, and comparing the results. The usefulness and importance of such experiments strongly depends on how 'typical' the sample instances actually are. Moreover, the number of such comparisons is quite limited. As a consequence, it is probably unwise to make any hard and fast generalizations by using this kind of computational tests.

---

* Supported by Swiss National Science Foundation project 20-63733.00/1, "Resource Allocation and Scheduling in Flexible Manufacturing Systems", and by the "Metaheuristics Network", grant HPRN-CT-1999-00106.

K. Jansen et al. (Eds.): WEA 2003, LNCS 2647, pp. 209–222, 2003.

Our Contribution. In this paper we suggest a new research direction that might be useful to cope with the possible limitations of data generation. The basic idea is to select a finite set of typical instances which 'represent' the whole set of instances. The key phrase here is 'represent'. We concentrate on the concept of *representativeness* of a set of problem instances. We define a measure of the representativeness of an instance that we call ε-*representativeness*. Given a minimization problem[1] $\mathcal{P}$, an instance x_ε is ε-representative of another instance x if a $(1 + \delta)$-approximate solution for x_ε can be transformed in polynomial time into a $(1 + \delta + \varepsilon)$-approximate solution for x, for $\varepsilon, \delta \geq 0$. In other words, if a 'good' solution for x_ε can be mapped into a 'good' solution for x.

In this paper we restrict to a strongly NP-hard single machine scheduling problem, which in the notation of Graham et al. [6], is noted $1|r_j, q_j|C_{\max}$. For this problem we propose a methodology for experimental testing of algorithms which is based on the concept of ε-representativeness. The test methodology works as follows. For any $\varepsilon > 0$, we map the infinite set of all instances I into a finite set of instances I_ε which are ε-representative of the instances in I. We call I_ε *core set*, and the instances belonging to it *core instances*. More precisely, we exhibit a linear time algorithm f that transforms any given instance x from I into an instance x_ε that is ε-representative of instance x, and that belongs to a defined finite set I_ε. Moreover, a $(1 + \delta)$-approximate solution for x_ε can be transformed in linear time (by algorithm g) into a $(1+\delta+\varepsilon)$-approximate solution for x, for any $\delta \geq 0$. Assume now that you have a heuristic H_ε that has been tested on the finite set of core instances I_ε. If H_ε is an (experimentally) good heuristic H_ε (i.e. an algorithm that returns $(1 + \delta)$-approximate solutions) for this finite set I_ε of core instances, then it can be used to get a good algorithm for the infinite instances of problem $1|r_j, q_j|C_{\max}$ (i.e. $(1 + \delta + \varepsilon)$-approximate algorithm). In Figure 1 we give a pictorial view of the $(1+\delta+\varepsilon)$-approximation algorithm that works as follows: for any given instance x from I, first transform x into x_ε, then apply H_ε on x_ε, and transform back the obtained solution for x_ε into $(1+\delta+\varepsilon)$-approximate solution for x. It follows that by testing the heuristic on the core instances, we can obtain an approximation algorithm whose performance ratio depends on the experimental results and on the representativeness ε of the instances used for testing. We remark that this approach can be seen as a 'blackbox' testing and used to test any kind of heuristics (such as simulated annealing, genetic algorithms, tabu search, ant colony optimization algorithms, etc.).

Unfortunately, we could only find a core instance set whose cardinality is exponential in $1/\varepsilon$. A natural question addressed in this paper is to understand if we can hope to reduce considerably the size of the core set, i.e., to be polynomial in $1/\varepsilon$. For strongly NP-hard problems, we exhibit a first negative results which states that the size of I_ε cannot be a polynomial in $1/\varepsilon$, unless P=NP. (We do not know if weakly NP-hard problems admit a polynomial in $1/\varepsilon$ core set size.) This means that for 'small' values of ε the number of core instances may actually be too large to be fully tested. This limitation suggests a second natural question, that is, what can we say about the performances of an algorithm by

---

[1] A similar definition can be given for maximization problems.

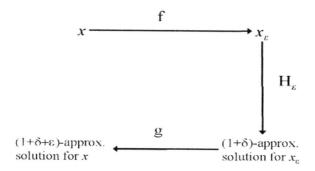

Fig. 1. The resulting $(1 + \delta + \varepsilon)$-approximation algorithm

performing a 'small' (polynomial in $1/\varepsilon$) number of tests? A first positive answer to this question can be obtained by using Monte Carlo method for sampling on the core set. We show that within a 'small' number of tests we can check if an algorithm performs 'well' (or better than another algorithm) for most of the core instances with high probability.

We believe that this paper presents a new research direction and shows only preliminary results. Many questions remain to be explored to understand if it can be applied in a direct way for experimental analysis of algorithms/heuristics.

Techniques. Interestingly, the techniques used to reduce I to I_ε are a combination of the 'standard' techniques to obtain polynomial time approximation schemes (PTAS) (see [17] for a survey). In the literature there are several polynomial time approximation schemes that make use of the relative error ε to reduce the number of potentially 'interesting' solutions from exponential to polynomial. We exploit the error ε to reduce the number of 'interesting' instances. These strong similarities let us hope that analogous results can be obtained for several other problems that admit a PTAS. Indeed, we claim that, by using the ideas introduced in this paper and the techniques described in [5], the presented approach for testing can also be used for other scheduling problems, such as the identical and unrelated parallel machine scheduling problem with costs, the flow-shop and the job-shop scheduling problems with fixed number of machines and of operations per job. More generally, we believe that this approach can be extended to several other problems belonging to the class EPTAS[2].

Organization of the Paper. The remainder of the paper is organized as follows. Section 2 focus on the case study of the single machine scheduling problem, for which, in Section 2.1 we derive the core instance set. Section 3 shows the

---

[2] EPTAS is the class of optimization problems which admit approximation schemes running in time $f(1/\varepsilon)n^c$, where f is an arbitrary function, n is the input size and c is a constant independent of ε (see [3]).

theoretical lower bound on the core set cardinality. Section 4 shows how to sample and how to test a subset of core instances by means of Monte Carlo sampling. Experimental results of the proposed testing methodology will be given in the full version of this paper. Conclusions and future research issues are in Section 5.

2   Case Study: Scheduling on a Single Machine

We shall study the following problem. There is a set of n jobs J_j $(j = 1, ..., n)$. Each job J_j must be processed without interruption for $p_j \geq 0$ time units on a single machine, which can process at most one job at a time. Each job has a release date $r_j \geq 0$, which is the time when it first becomes available for processing. After completing its processing on the machine, a job requires an additional delivery time $q_j \geq 0$. If s_j $(\geq r_j)$ denotes the time J_j starts processing, job J_j is delivered at time $s_j + p_j + q_j$. Delivery is a *non-bottleneck* activity, in that all jobs may be simultaneously delivered. Our objective is to minimize, over all possible schedules, the *maximum delivery time*, i.e., $\max_j s_j + p_j + q_j$. The problem as stated is strongly NP-hard [12], and in the notation of Graham et al. [6] it is noted as $1|r_j, q_j|C_{\max}$.

The first phase of our test method requires mapping the set of all instances to the set of core instances. This is performed in Subsection 2.1.

2.1   Core Reduction

We start defining set I_ε of core instances, then we prove that the infinite set I of all instances can be mapped to the finite set I_ε, for every $\varepsilon > 0$. In the rest of the paper we assume, for simplicity of notation, that $1/\varepsilon$ is an integer. Let $\nu = (1 + \frac{1}{\varepsilon})^2 + \frac{2}{\varepsilon} = O(1/\varepsilon^2)$. The set of integers $0, 1, ..., k$ is denoted $[k]$.

Definition 1. *An instance x_ε belongs to I_ε if and only if the following holds:*

1. *x_ε has ν jobs;*
2. *for every job j of x_ε, $r_j = h \cdot \varepsilon$ and $q_j = i \cdot \varepsilon$ for some $h, i \in [1/\varepsilon]$;*
3. *for every job j of x_ε, $p_j = i\frac{\varepsilon}{\nu}$ for some $i \in [\nu/\varepsilon]$;*

The size of I_ε is bounded by $(1/\varepsilon)^{O(1/\varepsilon^2)}$. Indeed, let us say that two jobs are of the same *type* if they have the same processing, release and delivery time. Clearly, the number of different types is bounded by $\tau = O(1/\varepsilon^5)$. Let us denote these τ types by $t_1, ..., t_\tau$. Let u_i denote the total number of jobs of type t_i, for $i = 1, ..., \tau$. A core instance is any vector $(u_1, u_2, ..., u_\tau)$ with $\sum u_i = \nu$. Therefore, we have at most $\tau^\nu = (1/\varepsilon)^{O(1/\varepsilon^2)}$ core instances.

The following theorem proves that the infinite set I of all instances can be mapped to I_ε. Our approach uses several transformations of the given input instance x which may potentially increase the objective function value by a factor of $1 + O(\varepsilon)$. Therefore we can perform a constant number of transformations while still staying within $1 + O(\varepsilon)$ of the original optimum. At the end, the resulting transformed instance x_ε belongs to I_ε and any $(1+\delta)$-approximate solution for x_ε can be transformed in linear time into a $(1+\delta+O(\varepsilon))$-approximate solution for x. Throughout this paper, when we describe this type of transformation, we shall say it produces $1 + O(\varepsilon)$ *loss*. (The proof of the following theorem can be found in appendix.)

Theorem 1. *For any fixed $\varepsilon > 0$, with $1 + 6\varepsilon$ loss, any given instance $x \in I$ can be transformed into an instance $x_\varepsilon \in I_\varepsilon$ in $O(n)$ time, where the constant hidden in the $O(n)$ running time is reasonably small and does not depend on the error ε.*

Note that, although the size of I_ε may be decreased, we prove in Section 3 that $|I_\varepsilon|$ cannot be polynomial in $1/\varepsilon$, unless P=NP.

3   On the Core Set Size

In 1977, Berman and Hartmanis [2] investigated the density of NP-complete problems. The *density* of a language A is a function $C_A : \mathbb{N} \longrightarrow \mathbb{N}$ defined by $C_A(n) = |\{x \in A : |x| \leq n\}|$. Recall that a language A is *sparse* if there exists a polynomial q such that for every $n \in \mathbb{N}$, $C_A(n) \leq q(n)$. In [2], Berman and Hartmanis proposed the density conjecture, that no sparse language can be NP-complete. The density conjecture was proved to be equivalent to P$\neq$NP by Mahaney [13].

In this section we give a general condition that assures that a problem $\mathcal{P}$ cannot have a core set size which is polynomial also in $1/\varepsilon$. Theorem 2 applies to the problem addressed in this paper. The proof of the following theorem uses the result proved by Mahaney [13].

For any problem $\mathcal{P}$ and for any instance x of $\mathcal{P}$, let $\max(x)$ denote the value of the largest number incurring in $x \in I$, where I denotes the set of instances of problem $\mathcal{P}$. For simplicity, let us focus on minimization problems. Let $SOL(x)$ be a function that associates to any input instance $x \in I$ the set of feasible solutions of x. Let m be a function that associates with any instance $x \in I$ and with any feasible solution $y \in SOL(x)$ a positive rational $m(x, y)$ that denotes the measure of solution y. The value of any optimal solution of x will be denoted as $m^*(x)$. The following definition formalize the notion of *core reducibility*, that maps instances of I to instances of I_ε, but it also maps back good solutions for I_ε to good solutions for I.

Definition 2. *For any fixed $\varepsilon > 0$, problem $\mathcal{P}$ is core reducible, if two functions f and g, and a finite set I_ε of instances exist such that, for any instance $x \in I$:*

1. *$f(x, \varepsilon) \in I_\varepsilon$;*
2. *For any $y \in SOL(f(x, \varepsilon))$, $g(x, y, \varepsilon) \in SOL(x)$;*
3. *f and g are computable in polynomial time with respect to both the length of the instance and $1/\varepsilon$;*

4. *There exists a polynomial time computable solution $s \in SOL(f(x,\varepsilon))$ such that $m(x, g(x, s, \varepsilon)) \leq (1 + \varepsilon) \cdot m^*(x)$.*

We are now ready to give the following fundamental result. (The proof of the following theorem can be found in appendix.)

Theorem 2. *Let $\mathcal{P}$ be a strongly NP-hard problem that admits a polynomial p such that $m^*(x) \leq p(|x|, \max(x))$, for every input $x \in I$. If $\mathcal{P}$ is core reducible, then the density of I_ε cannot be polynomial in $1/\varepsilon$, for any $\varepsilon > 0$, unless P=NP.*

The two main requirements of Theorem 2 are that $\mathcal{P}$ have solution values that are not too large and that $\mathcal{P}$ is NP-complete in the strong sense. These hold for many problems of interest, including, for example, the bin packing problem, the graph coloring problem, the maximum independent set problem and the minimum vertex cover problem. Theorem 2 also applies to the addressed scheduling problem. Thus we have a general method of considerable power for ruling out the possibility of a core set size polynomial in $1/\varepsilon$.

4   Sampling on the Core Set

In Subsection 2.1 we have presented a core set for problem $1|r_j, q_j|C_{\max}$ whose cardinality is exponential in $1/\varepsilon$. Moreover, in Section 3 we have shown that it is unlikely that a 'small' core set exists. Thus, for 'small' values of the ε parameter, the core set may be too big to be fully tested. A way to deal with this problem is to select a subset of the core set which is small enough to be fully tested. In this case the problem is to select an appropriate subset for which any measurement done on the subset is a good estimator of a measurement on the whole core set. In this section we use Monte Carlo sampling to select by random sampling an appropriate subset of the core set, in order to do algorithm testing.

4.1   Testing by Monte Carlo Sampling

Let $\mathcal{P}$ be a core reducible problem and I_ε be the core set of this problem for every fixed $\varepsilon > 0$. Let A be an algorithm for $\mathcal{P}$ which we want to evaluate by experimental testing. Let us define a test to make on A, that is, let us define a Boolean function $f : I_\varepsilon \to \{0, 1\}$ over I_ε. We define the set $I_\varepsilon^1 = \{x \in I_\varepsilon | f(x) = 1\}$ as the pre-image of 1. The function f may be thought of as an indicator of success ($f(x) = 1$) or failure ($f(x) = 0$) of the algorithm A over the core set I_ε. For example, given a core instance $x \in I_\varepsilon$, the algorithm A has success when it finds a solution which is within a certain distance from the optimum, while A fails in the opposite case. The problem is to estimate the size of I_ε^1 respect to the size of I_ε, that is, the fraction $|I_\varepsilon^1|/|I_\varepsilon|$ of 'successes' in our test of the algorithm.

An obvious approach to estimating $|I_\varepsilon^1|/|I_\varepsilon|$ is to use the classical Monte Carlo method (see, e.g. [16]). This involves choosing N independent samples from I_ε, say $x_1, ..., x_N$, and using the value of f on these samples to estimate the probability that a random choice will lie in I_ε^1. More formally, define the random variables $X_1, ..., X_N$ as follows:

$$X_i = \begin{cases} 1 \text{ if } f(x_i) = 1 \\ 0 \text{ otherwise.} \end{cases}$$

By this definition, $X_i = 1$ if and only if $x_i \in I_\varepsilon^1$. Finally, we define the estimator random variable

$$Z = |I_\varepsilon| \sum_{i=1}^{N} \frac{X_i}{N} .$$

It is easy to verify that $\mathbf{E}[Z] = |I_\varepsilon^1|$. We might hope that with high probability the value of Z is a good approximation to $|I_\varepsilon^1|$. In particular, we want to have a small error with respect to $|I_\varepsilon|$, when using Z as an estimator for $|I_\varepsilon^1|$. Of course, the probability that the approximation is good depends upon the choice of N. The following theorem (see also [16]) relates the value of N to η and δ.

Theorem 3. *For $\eta \in (0,1]$ and $\delta \in (0,1]$, the Monte Carlo method computes an estimator Z such that*

$$\Pr\left[\left|\frac{|I_\varepsilon^1|}{|I_\varepsilon|} - \frac{Z}{|I_\varepsilon|}\right| < \eta\right] \geq 1 - \delta,$$

provided

$$N \geq \frac{4}{\eta^2} \ln \frac{2}{\delta} .$$

In practice, after running A on the sample, we count the number of times A has been successful. Let X denote this value (X is linked to the estimator Z by $Z = |I_\varepsilon|\frac{X}{N}$). Theorem 3 implies that, provided the sample is big enough, $|I_\varepsilon^1|$ lies in the interval $((\frac{X}{N} - \eta)|I_\varepsilon|, (\frac{X}{N} + \eta)|I_\varepsilon|)$, with probability at least $1 - \delta$. Therefore, with probability at least $1 - \delta$, algorithm A is successful for at least the $(\frac{X}{N} - \eta) \cdot 100\%$ of core instances.

4.2   Comparison of Two Algorithms

The goal of this section is to describe a method to compare algorithms. Let $\mathcal{P}$ be a core reducible problem and I_ε be the core set of this problem for every fixed $\varepsilon > 0$. Let A and B be two algorithms for $\mathcal{P}$ which we want to compare. Let us define a test to make on A and B, that is, let us state when an algorithm is 'better', 'worse' or 'equivalent' to another algorithm. For example, given a core instance $x \in I_\varepsilon$, the algorithm A is better than B on x if it returns a better solution than B.

In the following we are interested into computing how many times A is better, worse or equivalent to B. For every given algorithm A and B the set I_ε can be seen as partitioned into three subsets $I_\varepsilon^A, I_\varepsilon^B$ and $I_\varepsilon^{A=B}$ of instances, where I_ε^A is the set of instances for which A is better than B, I_ε^B the opposite and $I_\varepsilon^{A=B}$ is the set of instances for which A and B are equivalent. We are interested in the size of these subsets and in the following we describe how to approximate these values.

Let η and δ be two parameters, and let N be the size of a sample from the core set. After running A and B on the same sample, we count the number of times A is better, worse or equal to B. Let α, β and γ denote, respectively, these values. Then, we define $Z_\varepsilon^A = |I_\varepsilon|\frac{\alpha}{N}$, $Z_\varepsilon^B = |I_\varepsilon|\frac{\beta}{N}$ and $Z_\varepsilon^{A=B} = |I_\varepsilon|\frac{\gamma}{N}$ as the estimators of $|I_\varepsilon^A|, |I_\varepsilon^B|$ and $|I_\varepsilon^{A=B}|$, respectively. Again it is easy to verify that $\mathbf{E}[Z_\varepsilon^A] = |I_\varepsilon^A|$, $\mathbf{E}[Z_\varepsilon^B] = |I_\varepsilon^B|$ and $\mathbf{E}[Z_\varepsilon^{A=B}] = |I_\varepsilon^{A=B}|$. By theorem 3, we can say that, provided the sample is big enough (i.e. $N \geq \frac{4}{\eta^2}\ln\frac{2}{\delta}$), $|I_\varepsilon^A|, |I_\varepsilon^B|$ and

$|I_\varepsilon^{A=B}|$ lie in intervals $\left((\frac{\alpha}{N}-\eta)|I_\varepsilon|, (\frac{\alpha}{N}+\eta)|I_\varepsilon|\right)$, $\left((\frac{\beta}{N}-\eta)|I_\varepsilon|, (\frac{\beta}{N}+\eta)|I_\varepsilon|\right)$ and $\left((\frac{\gamma}{N}-\eta)|I_\varepsilon|, (\frac{\gamma}{N}+\eta)|I_\varepsilon|\right)$, respectively, with probability at least $1-\delta$.

4.3   Sampling from the Core Set of $1|r_j, q_j|C_{\max}$

The methods described in Subsection 4.1 and 4.2 can be applied to problem $1|r_j, q_j|C_{\max}$ if we can choose, uniformly at random, N independent samples from the core set I_ε of Definition 1. This can be done without generating the (exponentially large) core set I_ε. Indeed, we can choose, uniformly at random, an instance from I_ε as follows. Recall that $\tau = O(1/\varepsilon^5)$ is the number of different job types, and that ν is the number of jobs of any instance from I_ε (see Subsection 2.1). Choose independently and uniformly at random ν numbers from $[\tau]$. Each number defines the type of the ν jobs of the instance. It is easy to check that any instance from I_ε has the same probability to be selected. The number of random bits used is polynomial in $1/\varepsilon$.

We observe that every polynomial algorithm for problem $1|r_j, q_j|C_{\max}$ can be tested in time polynomial in $1/\eta$, $\log(1/\delta)$ and $1/\varepsilon$. Indeed, the required number of experiments is a polynomial function of $1/\eta$ and $\log(1/\delta)$, and the size of every core instance is polynomial in $1/\varepsilon$.

5   Future Work and Research

Focusing on a strongly NP-hard single machine scheduling problem, we have shown how to map the infinite set of all instances into a finite set of ε-representative core instances. The ε-representative core instances can be used to test heuristics. In the full version of this paper we will provide experimental applications of the described testing methodology. We remark that this approach can be seen as a 'black-box' testing and used to test any kind of heuristics (such as simulated annealing, genetic algorithms, tabu search, ant colony optimization algorithms, etc.).

We believe that this paper is just an initial step in the exploration of such idea for testing and so we state some open problems and possible directions.

We have seen that the number of core instances cannot be polynomial in $1/\varepsilon$ when the problem is strongly NP-hard. Does the latter hold also for a weakly NP-hard problem?

By sampling uniformly and testing heuristics on the sampled instances, we can understand if they perform 'well' for most of the core instances. Other ways of sampling should be investigated to understand on which instances heuristics perform poorly. With this aim, the statistical technique called *importance sampling* (see [16] p. 312) might be fruitful.

Finally, we believe that there are possibilities to connect this work to parameterized complexity [4].

References

[1] R. Barr, B. Golden, J. Kelly, M. Resende, and W. S. Jr. Designing and reporting on computational experiments with heuristic methods. *Journal of Heuristics*, 1:9–32, 1995. 209

[2] L. Berman and J. Hartmanis. On isomorphisms and density of NP and other complete sets. *SIAM Journal on Computing*, 6(2):305–322, June 1977. 213

[3] M. Cesati and L. Trevisan. On the efficiency of polynomial time approximation schemes. *Information Processing Letters*, 64(47):165–171, 1997. 211

[4] R. Downey and M. Fellows. *Parameterized Complexity*. Springer, 1998. 217

[5] A. Fishkin, K. Jansen, and M. Mastrolilli. Grouping techniques for scheduling problems: simpler and faster. In *9th Annual European Symposium on Algorithms (ESA'01)*, volume LNCS 2161, pages 206–217, 2001. 211

[6] R. Graham, E. Lawler, J. Lenstra, and A. R. Kan. Optimization and approximation in deterministic sequencing and scheduling: A survey. In *Annals of Discrete Mathematics*, volume 5, pages 287–326. North–Holland, 1979. 210, 212

[7] H. Greenberg. Computational testing: why, how and how much. *ORSA Journal of Computing*, 2:94–97, 1990. 209

[8] L. Hall and D. Shmoys. Approximation algorithms for constrained scheduling problems. In *Proceedings of the 30th IEEE Symposium on Foundations of Computer Science*, pages 134–139, 1989. 219

[9] J. Hooker. Needed: An empirical science of algorithms. *Operations Research*, 42:201–212, 1994. 209

[10] J. Hooker. Testing heuristics: We have it all wrong. *Journal of Heuristics*, 1:33–42, 1995. 209

[11] D. Johnson. A theoretician's guide to the experimental analysis of algorithms. In G. Johnson and McGeoch, editors, *to appear in Proceedings of the 5th and 6th DIMACS Implementation Challenges*. American Mathematical Society, 2002. 209

[12] J. Lenstra, A. R. Kan, and P. Brucker. Complexity of machine scheduling problems. *Annals of Operations Research*, 1:343–362, 1977. 212

[13] S. Mahaney. Sparse complete sets for *NP*: Solution of a conjecture of Berman and Hartmanis. *Journal of Computer and System Sciences*, 25(2):130–143, Oct. 1982. 213, 221

[14] M. Mastrolilli. Grouping techniques for one machine scheduling subject to precedence constraints. In *Proceedings of the 21st Foundations of Software Technology and Theoretical Computer Science*, volume LNCS 2245, pages 268–279, 2001. 219

[15] C. McGeoch. Toward an experimental method for algorithm simulation. *INFORMS Journal of Computing*, 8:1–15, 1996. 209

[16] R. Motwani and P. Raghavan. *Randomized Algorithms*. Cambridge University Press, Cambridge, 1995. 214, 215, 217, 221

[17] P. Schuurman and G. Woeginger. Approximation schemes - a tutorial. Technical report, 2001. 211

A   Appendix

A.1   Proof of Theorem 1

We start describing procedure Core-Reduction which maps the infinite set I of all instances to a finite set I_ε of core instances. Then, we analyze this algorithm and show that by applying Core-Reduction, any given instance x from I can be transformed with $(1 + 6\varepsilon)$ loss into a core instance belonging to I_ε, for every $\varepsilon > 0$.

The procedure Core-Reduction consists of scaling the input data and 'merging' jobs to form an approximate version of the original instance that belongs to a finite set of different instances. The procedure Core-Reduction is as follows.

1. *Checking number of jobs.* Let $\nu = (1+\frac{1}{\varepsilon})^2+\frac{2}{\varepsilon} = O(1/\varepsilon^2)$. If the input instance has no more than ν jobs, go to point (3), otherwise go to point (2);

2. *Grouping small jobs together.* Let $P = \sum_j p_j$, $r_{max} = \max_j r_j$ and $q_{max} = \max_j q_j$. Round down every release date (delivery time) to the nearest value among the following $\rho = 1 + \frac{1}{\varepsilon} = O(1/\varepsilon)$ values: $0, \varepsilon r_{max}, 2\varepsilon r_{max}, ..., r_{max}$ $(0, \varepsilon q_{max}, 2\varepsilon q_{max}, ..., q_{max})$. For simplicity of notation, let us use r_j and q_j to denote the resulting rounded release and delivery times of any job j. Divide jobs into *large* and *small* jobs, according to the length of their processing times, $L = \{j : p_j > \varepsilon P\}$ and $S = \{j : p_j \leq \varepsilon P\}$. Let us say that two small jobs, j and h, belong to the same *class* iff $r_j = r_h$ and $q_j = q_h$. The number of distinct job classes is bounded by $\kappa = (1+\frac{1}{\varepsilon})^2 = O(1/\varepsilon^2)$. Let us partition the set of small jobs into at most κ classes $K_1, K_2, ..., K_\kappa$. Let j and h be two small jobs from K_i such that $p_j + p_h \leq \varepsilon P$. We 'glue' together these two jobs to form a composed job with the same release date and delivery time as j (and h), and in which the processing time is equal to the sum of their processing times. We repeat this process until at most one job from K_i has processing time not greater than $\varepsilon P/2$. At the end, all jobs in class K_i, have processing times not greater than εP. The same procedure is performed for every class. At the end of the process, the total number of composed jobs is no larger than $\frac{P}{\varepsilon P/2} + \kappa = \nu$.

3. *Normalizing.* Let p_{max}, r_{max} and q_{max} denote the maximum processing, release and delivery time, respectively, of the instance after steps (1) and (2). Let $LB = \max\{p_{max}, q_{max}, r_{max}\}$. Normalize the instance by dividing every release, delivery and processing time by LB.

4. *Rounding release and delivery times.* Round down every release date and delivery time to the nearest value among the following $\rho = O(1/\varepsilon)$ values: $0, \varepsilon, 2\varepsilon, ..., 1$.

5. *Rounding processing times.* Round down every processing time to the nearest value among the following $\pi = O(1/\varepsilon^3)$ values: $0, \varepsilon/\nu, 2\varepsilon/\nu, ..., 1$.

In the following we analyze Core-Reduction and show that the transformation is with $(1 + 6\varepsilon)$ loss. First we observe that if the number of jobs of the input instance is not less than ν then, by following the ideas in [14], small jobs can be merged together as described in step (2) with $1 + 3\varepsilon$ loss (see Lemma 1 for a proof of this).

Note that step (3) can be performed with no loss, the normalized instance is equivalent to the one after steps (1) and (2). Observe that after step (3) every processing, release and delivery time is not greater than 1. Now, let us consider the resulting instance x after step (1), (2) and (3), and instance x' obtained from x by rounding release, processing, and delivery times as in steps (4) and (5) of Core-Reduction. In steps (4) and (5), by rounding down the values we cannot obtain a modified instance with an optimal value larger than the the the optimal value of instance x. Following the ideas in [8], every feasible solution for the modified instance x' can be transformed into a feasible solution for the original instance just by adding ε to each job's starting time, and re-introducing the original delivery and processing times. It is easy to see that the solution value may increase by at most 3ε (observe that there are at most ν jobs and each processing time may increase by at most ε/ν).

Let us now focus on the time complexity of Core-Reduction. It is easy to check that steps $(1), (3), (4)$ and (5) of the procedure requires linear time. Suppose now that step (2) must be performed, that is, $n \geq \nu$. The process of partitioning small jobs into κ job classes, as described in step (2) of the procedure, can be implemented to take $O(n + \kappa)$ time. Since in this case $n \geq \kappa$, step (2) also takes $O(n)$ time.

Lemma 1. *With $1 + 3\varepsilon$ loss, the number of jobs can be reduced to be at most* $\min\{n, O(1/\varepsilon^2)\}$.

Proof. In the following we show that by using the instance with grouped small jobs (as described in step (2) of Core-Reduction) we can get an optimal solution with $1 + 3\varepsilon$ loss.

Following the ideas in [8], every feasible solution for the modified instance with rounded release and delivery times can be transformed into a feasible solution for the original instance just by adding $\varepsilon r_{\max}$ to each job's starting time, and re-introducing the original delivery times. It is easy to see that the solution value may increase by at most $\varepsilon(r_{\max} + q_{\max}) \leq 2\varepsilon opt$, where opt is the optimal value of the original instance (see [8]). Let us consider an optimal schedule y^* for the instance x with rounded release and delivery times. Modify x to obtain instance x^* in which the processing times of all jobs remain unchanged, while release and delivery times are set as follows

$$\begin{aligned}
\tilde{r}_j = r_j, \tilde{q}_j = q_j & \qquad \text{for } j \in S, \\
\tilde{r}_j = s_j^*, \tilde{q}_j = opt(x) - p_j - s_j^* & \text{ for } j \in L,
\end{aligned} \tag{1}$$

where s_j^* denote the starting time of job j in y^*. Clearly $opt(x) = opt(x^*)$.

Merge small jobs as described before and denote by $\tilde{x}$ the resulting modified instance. By showing that there exists a schedule $\tilde{y}$ for $\tilde{x}$ such that $m(\tilde{x}, \tilde{y}) \leq (1 + \varepsilon)opt(x^*)$ the claim follows.

Consider the following algorithm, known as *Extended Jackson's Rule*: whenever the machine is free and one of more jobs is available for processing, schedule an available job with largest delivery time. Apply the *Extended Jackson's Rule* to the modified instance $\tilde{x}$. Let use $\tilde{y}$ to denote the resulting schedule. Let us define a critical job J_c as one that finishes last in y, i.e., its delivery is completed last. Associated with a critical job J_c there is a critical sequence consisting of those jobs tracing backward from J_c to the first idle time in the schedule. Let us fix a critical job J_c and denote the last job J_b in the critical sequence with $\tilde{q}_c > \tilde{q}_b$ as interference job for the critical sequence. Let B denote the set of jobs processed after J_b in the critical sequence. By the way that J_b was chosen, clearly $\tilde{q}_j \geq \tilde{q}_c$ for all jobs $J_j \in B$. We claim that if there is no interference job then $m(\tilde{x}, \tilde{y}) = opt(x^*)$, otherwise $m(\tilde{x}, \tilde{y}) \leq opt(x^*) + p_b$. It is easy to show that $m(\tilde{x}, \tilde{y}) \leq p_b + \min_{j \in B} \tilde{r}_j + \sum_{j \in B} p_j + \min_{j \in B} \tilde{q}_j$ and $opt(x^*) \geq \min_{j \in B} \tilde{r}_j + \sum_{j \in B} p_j + \min_{j \in B} \tilde{q}_j$. By the way that the interference job J_b was chosen, we have $\tilde{q}_c = \min_{j \in B} \tilde{q}_j$. Let U denote the set B of jobs obtained by "unsticking" small jobs. We can bound the length of an optimal schedule for x^* as $opt(x^*) \geq \min_{j \in U} \tilde{r}_j + \sum_{j \in U} p_j + \min_{j \in U} \tilde{q}_j$. It is easy to see that $\min_{j \in B} \tilde{r}_j + \sum_{j \in B} p_j + \min_{j \in B} \tilde{q}_j = \min_{j \in U} \tilde{r}_j + \sum_{j \in U} p_j + \min_{j \in U} \tilde{q}_j$. Indeed, we have glued only small jobs having the same release and delivery times, and therefore $\min_{j \in B} \tilde{q}_j = \min_{j \in U} \tilde{q}_j$ and $\min_{j \in B} \tilde{r}_j = \min_{j \in U} \tilde{r}_j$. Therefore, $m(\tilde{x}, \tilde{y}) \leq p_b + opt(x^*)$ and if there is no interference job then $m(\tilde{x}, \tilde{y}) = opt(x^*)$.

Then if there is no interference job or J_b is a small job we have $m(\tilde{x}, \tilde{y}) \leq (1 + \varepsilon) \cdot opt(x^*)$, by definition of small jobs ("glued" or not). Now, consider the last case in which the interference job J_b is a large job. By the definition of the extended Jackson's rule, no job $j \in B$ could have been available at the time when J_b is processed, since otherwise such a job j would have taken priority over job J_b. Thus, $\tilde{r}_c > \tilde{r}_b$ and $\tilde{q}_c > \tilde{q}_b$, but J_b cannot be a large job since by construction if $\tilde{r}_c > \tilde{r}_b$ then $\tilde{q}_b \geq \tilde{q}_c$.

A.2   Proof of Theorem 2

Let us assume that there exists a polynomial q_1 such that the density C_{I_ε} of I_ε is bounded by $C_{I_\varepsilon}(n) \leq q_1(n, 1/\varepsilon)$, for every $n \in \mathbb{N}$ and for any $\varepsilon > 0$. Since $\mathcal{P}$ is strongly NP-hard, there exists a polynomial q_2 such that problem $\tilde{\mathcal{P}}$ obtained by restricting $\mathcal{P}$ to only those instances x for which $\max(x) \leq q_2(|x|)$, is NP-hard. We show that the assumption is equivalent to P=NP by reducing problem $\tilde{\mathcal{P}}$ to a problem $\mathcal{P}'$ whose underlying language is sparse and NP-complete.

For every $(x, y) \in I \times SOL$, the measure function $m(x, y)$ of $\mathcal{P}$ is defined to have values in $\mathbb{Q}$. It is however possible to transform any such optimization problem into an equivalent one such that $m(x, y)$ is a positive integer. We assume this in the following.

Consider set $I(n) = \{x \in I : |x| = n\}$, for any $n \in \mathbb{N}$. For every $x \in I(n)$, if we set the error to $\varepsilon(n) = 1/(p(n, q_2(n)) + 1)$, then any $y^* \in SOL^*(f(x, \varepsilon(n))$ can be transformed into the optimal solution for x by means of function g. Indeed, since $\tilde{\mathcal{P}}$ is core reducible we have $m(x, g(x, y^*, \varepsilon(n))) \leq (1+\varepsilon(n)) \cdot m^*(x) < m^*(x)+1$, where the last inequality is due to the fact that $m^*(x) \leq p(|x|, \max(x)) \leq p(n, q_2(n))$. From the integrality constraint on the measure function m, it follows that $m(x, g(x, y^*, \varepsilon(n))) = m^*(x)$, that is, $g(x, y^*, \varepsilon(n))$ is an optimal solution for x. The time complexity to compute $f(x, \varepsilon(n))$ and $g(x, y^*, \varepsilon(n))$ is polynomial in $|x|$. Let us partition set $I(n)$ into equivalence classes, such that $x, z \in I(n)$ belong to the same class iff $f(x, \varepsilon(n)) = f(z, \varepsilon(n))$. Consider set $I'(n)$ which contains for every equivalence class the corresponding core instance, i.e., $I'(n) = \{f(x, \varepsilon(n)) : x \in I(n)\}$. Let q_3 be the polynomial limiting the computation time of $f(x, \varepsilon(n))$, for $x \in I(n)$. Then, for any $x \in I(n)$, $|f(x, \varepsilon(n))| \leq q_3(n)$. If $|f(x, \varepsilon(n))| < n$ we increase the length of $f(x, \varepsilon(n))$ up to n, such that the modified instance is equivalent, except for the length, to the former (for instance, we can do that by adding a dummy symbol several times till length n is touched). For simplicity, let us denote this modified length-increased set by $I'(n)$ again. It is easy to see that $I(n)$ polynomially reduces to $I'(n)$, since a polynomial time algorithm for $I'(n)$ implies the existence of a polynomial time algorithm for $I(n)$. According to the assumption, the cardinality of $I'(n)$ is bounded by a polynomial $q_4(n)$, where $q_4(n) = q_1(n, p(n, q_2(n)) + 1)$. Since for every $n \in \mathbb{N}$, $I(n)$ polynomially reduces to $I'(n)$, then $I = \cup_{n\in\mathbb{N}}I(n)$ polynomially reduces to $I' = \cup_{n\in\mathbb{N}}I'(n)$, and the density $C_{I'}$ of I' is bounded by a polynomial, indeed $C_{I'}(n) = |\{x \in I' : |x| \leq n\}| \leq |\cup_{i\leq n} I'(i)| = \sum_{i\leq n} |I'(i)|$. Hence, problem $\tilde{\mathcal{P}}$ can be reduced in polynomial time to a problem $\mathcal{P}'$ with a sparse set of instances. It is easy to see that $\mathcal{P}'$ is NP-hard, and the underlying language is sparse and NP-complete. This is equivalent to P=NP [13].

A.3   Proof of Theorem 3

The claim follows by proving that

$$\Pr\left[Z \leq |I_\varepsilon^1| - \eta|I_\varepsilon|\right] \leq \frac{\delta}{2} \tag{2}$$

and

$$\Pr\left[Z \geq |I_\varepsilon^1| + \eta|I_\varepsilon|\right] \leq \frac{\delta}{2}, \tag{3}$$

for $N \geq \frac{4}{\eta^2} \ln \frac{2}{\delta}$.

Define $X = \sum_{i=1}^{N} X_i$, $\rho = |I_\varepsilon^1|/|I_\varepsilon|$ and the estimator $Z = |I_\varepsilon|X/N$. Then,

$$\Pr\left[Z \leq |I_\varepsilon^1| - \eta|I_\varepsilon|\right] = \Pr\left[X \leq N\frac{|I_\varepsilon^1|}{|I_\varepsilon|} - N\eta\right] = \Pr\left[X \leq N\rho\left(1 - \frac{\eta}{\rho}\right)\right].$$

Since $\mathbf{E}[X] = N\rho$, by a straightforward application of the Chernoff bound (see, e.g. [16]) to the rightmost term we obtain

$$\Pr\left[Z \leq |I_\varepsilon^1| - \eta|I_\varepsilon|\right] \leq \exp^{-N\frac{\eta^2}{4\rho}} \leq \exp^{-N\frac{\eta^2}{4}}.$$

For the given lower bound on N we obtain (2). Similarly, we can prove (3). It is easy to see that inequalities (2) and (3) imply

$$\Pr\left[\left|\frac{|I_\varepsilon^1|}{|I_\varepsilon|} - \frac{Z}{|I_\varepsilon|}\right| < \eta\right] \geq 1 - \delta .$$

The Reliable Algorithmic Software Challenge RASC

Kurt Mehlhorn

Max-Planck-Istitut für Informatik
66123 Saarbrücken, Germany
`www.mpi-sb.mpg.de/~mehlhorn`

Abstract. Algorithms are the heart of computer science. They make systems work. The theory of algorithms, i.e., their design and their analysis, is a highly developed part of theoretical computer science [7].
In comparison, algorithmic software is in its infancy. For many fundamental algorithmic tasks no reliable implementations are available due to a lack of understanding of the principles underlying reliable algorithmic software, some examples are given during the talk. The challenge is
- to work out the principles underlying reliable algorithmic software and
- to create a comprehensive collection of reliable algorithmic software components.

I describe what I consider a major challenge in algorithmics, and then outline some venues of attack.

References

[1] E. Berberich, A. Eigenwillig, M. Hemmer, S. Hert, K. Mehlhorn, and E. Schömer. A computational basis for conic arcs and boolean operations on conic polygons. to appear in ESA 2002, 2002.

[2] C. Burnikel, S. Funke, K. Mehlhorn, S. Schirra, and S. Schmitt. A separation bound for real algebraic expressions. In *ESA 2001*, Lecture Notes in Computer Science, pages 254-265, 2001.

[3] CGAL (Computational Geometry Algorithms Library). `www.cgal.org`.

[4] M. Dhiflaoui, S. Funke, C. Kwappik, K. Mehlhorn, M. Seel, E. Schömer, R. Schulte, and D. Weber. Certifying and repairing solutions to large LPs, How good are LP-solvers? to appear in SODA 2003, 2003.

[5] D. Kratsch, R. McConnell, K. Mehlhorn, and J. P. Spinrad. Certifying algorithms for recognizing interval graphs and permutation graphs. SODA 2003 to appear, 2003.

[6] K. Mehlhorn, S. Näher, T. Schilz, S. Schirra, M. Seel, R. Seidel, and C. Uhrig. Checking geometric programs or verification of geometric structures. In *Proceedings of the 12th Annual Symposium on Computational Geometry (SCG'96)*, pages 159-165, 1996.

[7] T. Ottmann and P. Widmayer. *Algorithmen und Datenstrukturen*. Spektrum Akademischer Verlag, 1996. 222

[8] C. K. Yap. *Fundamental Problems in Algorithmic Algebra*. Oxford University Press, 1999.

K. Jansen et al. (Eds.): WEA 2003, LNCS 2647, p. 222, 2003.
© Springer-Verlag Berlin Heidelberg 2003

A New Class of Greedy Heuristics for Job Shop Scheduling Problems

Marco Pranzo[1], Carlo Meloni[2], and Dario Pacciarelli[1]

[1] Dipartimento di Informatica e Automazione, Università degli Studi di Roma Tre
Via della Vasca Navale 79, 00146, Roma, Italy
mpranzo@dia.uniroma3.it
[2] Dipartimento di Elettrotecnica ed Elettronica, Politecnico di Bari
via E. Orabona 4, 70125, Bari, Italy

Abstract. In this paper we introduce a new class of greedy heuristics for general job shop scheduling problems. In particular we deal with the classical job shop, i.e. with unlimited capacity buffer, and job shop problems with blocking and no-wait. The proposed algorithm family is a simple randomized greedy family based on a general formulation of the job shop problem. We report on an extensive study of the proposed algorithms, and comparisons with other greedy algorithms are presented.

1 Introduction

A scheduling problem arises when a set of competing jobs requires some processings on a set of machines. From the modeling point of view most works on job shop scheduling problems are based on the disjunctive graph formulation of Roy and Sussman [23]. In this paper we deal with the *alternative graph* [12] model, a generalization of the disjunctive graph capable of modeling several constraints arising in real-world applications. In particular we deal with the classical job shop scheduling problem, i.e. with unlimited buffer, and with job shop scheduling problems with blocking and no-wait in process. As solution technique we introduce a new class of randomized greedy heuristics based on the alternative graph formulation.

The paper is organized as follows. In Section 2 a brief survey on greedy algorithms is reported, in Section 3 we introduce the notation and the alternative graph formulation. In Section 4 we discuss the arc greedy heuristic algorithms and we report on our computational results. Some conclusions follow.

2 Literature Review

In this section we review literature on greedy heuristics, these algorithms are *constructive*, and aim to build a solution by repeatedly enlarging a consistent partial solution.

Over the years, a great number of Priority Dispatching Rules (PDR) has been proposed (see, for example [5], [7], [10], [20]). These very fast one-pass algorithms

K. Jansen et al. (Eds.): WEA 2003, LNCS 2647, pp. 223–236, 2003.

select an operation from a subset according to some simple heuristic criterion, and schedule it at the end of the partial sequence. These procedures can be either deterministic or randomized, usually in the randomized heuristics the best over several run is considered. It is therefore common to consider the best dispatching solution over a set of different rules, as, for example, in [11], [19]. However, as reported by several authors [11], [21], the behavior of these dispatching rules is quite erratic, and no rule clearly outperforms the others.

Other greedy techniques are based on insertion algorithms, which aim to build processing order of the operations on each machine. At each step an operation is selected and inserted into the partial solution. It is worth noting that the operation could be inserted in any point of the sequence, and not only at the end as in the PDRs. The insertion algorithm were first applied to the flow shop scheduling by Nawaz et al [17] and to job shop by Werner and Winkler [25]. Moreover both the INSA (INSertion Algorithm) of Nowicki and Smutnicki [18], and the Bidir algorithm of Dell'Amico and Trubian [8] are insertion algorithms and are used as initial solution of a tabu search algorithm.

The Shifting Bottleneck Procedure (SB) is a tailored heuristic algorithm for solving the job shop scheduling problem. This algorithm was first proposed in 1988 by Adams, Balas and Zawack [1]. The Shifting Bottleneck procedure is a constructive algorithm that iteratively schedules a single machine problem, using the Carlier [6] approach. Balas, Lenstra and Vazacopoulos [4] consider as one-machine relaxation for the SB procedure the one-machine problem with delayed precedence constraints and solve it to optimality. At each step m single machine relaxed problems are solved, and the critical (bottleneck) machine is first identified and then scheduled. Every time a machine is scheduled, all the previous scheduled machines are kept fixed. As a stand alone heuristic, the Shifting Bottleneck procedure performs better than the simple Priority Dispatching Rules heuristics but it requires more computational times, better results are obtained by reiterating the SB procedure.

The Beam Search technique is an approach to improve the performance of single heuristics. The beam search is a heuristic search strategy [15] that consists of a limited breadth first visit of the branching tree. At each step of the search process only the best β candidates are maintained, and all the other candidates are discarded. The parameter β is called *beam width*, and it influences the performance of the algorithm. The beam search procedure has been applied to the job shop scheduling by Sabuncuoglu and Bayiz [24] superimposing it to priority dispatching rules.

3   The Alternative Graph Formulation

In this section we introduce the alternative graph formulation [12], [13]. The job shop scheduling problem is the problem of allocating machines to competing jobs over time, subject to the constraint that each machine can handle at most one job at a time. The sequence of machines for each job is prescribed and the processing of a job on a machine is called an *operation* and cannot be interrupted.

An operation requires a fixed, non-preemptive processing time. Each job consists of a sequence of operations that have to be processed in a specified order, and it is known in advance. The goal is to minimize the makespan, i.e. the completion time of the last operation.

We focus on the sequencing of operations rather than on the sequencing of jobs. We have therefore a set of operations $\{o_0, o_1, \ldots, o_n\}$ which have to be performed on m machines $\{m_1, m_2, \ldots, m_m\}$. Each operation o_i requires a specified amount of processing time p_i on a specified machine $M(i)$, and cannot be interrupted from its starting time t_i to its completion time $c_i = t_i + p_i$. o_0 and o_n are dummy operations, with zero processing time, that we call "start" and "finish" respectively. Each machine can process only one operation at a time.

There is a set of precedence relations among operations. A *precedence relation* (i, j) is a constraint on the starting time of operation o_j, with respect to t_i. More precisely, the starting times of the successor o_j must be greater or equal to the starting time of the predecessor o_i plus a given *delay* f_{ij}, which in our model can be positive, null or negative. Finally, we assume that o_0 precedes $o_1, \ldots, o_n$, and o_n follows $o_0, \ldots, o_{n-1}$. Precedence relations are divided into two sets: *fixed* and *alternative*. Alternative precedence relations are partitioned into pairs. They usually represent the constraints that each machine can process only one operation at a time.

A *schedule* is an assignment of starting times $t_0, t_1, \ldots, t_n$ to operations $o_0, o_1, \ldots, o_n$ respectively, such that all fixed precedence relations, and exactly one for each pair of the alternative precedence relations, are satisfied. Without loss of generality we assume $t_0 = 0$. The goal is to minimize the starting time of operation o_n. This problem can be therefore formulated as a particular *disjunctive program*, i.e. a linear program with logical conditions involving operations "and" ($\wedge$, conjunction) and "or" ($\vee$, disjunction), as in [3].

Problem 1

$$\min t_n - t_0$$
$$\text{s.t.} \quad t_j - t_i \geq f_{ij} \qquad\qquad\qquad (i, j) \in F$$
$$(t_j - t_i \geq a_{ij}) \vee (t_k - t_h \geq a_{hk}) \ ((i, j), (h, k)) \in A$$

Associating a node to each operation, Problem 1 can be usefully represented by the triple $\mathcal{G} = (N, F, A)$ that we call *alternative graph*. The alternative graph is as follows. There is a set of nodes N, a set of directed arcs F and a set of pairs of directed arcs A. Arcs in the set F are *fixed* and f_{ij} is the length of arc $(i, j) \in F$. Arcs in the set A are *alternative*. If $((i, j), (h, k)) \in A$, we say that (i, j) and (h, k) are *paired* and that (i, j) is the alternative of (h, k). Let a_{ij} be the length of the alternative arc (i, j). In our model the arc length can be positive, null or negative.

A *selection* S is a set of arcs obtained from A by choosing at most one arc from each pair. The selection is *complete* if exactly one arc from each pair is chosen. Given a selection S let $\mathcal{G}(S)$ indicate the graph $(N, F \cup S)$. A selection S is *consistent* if the graph $\mathcal{G}(S)$ has no positive length cycles. With this notation each schedule is associated with a complete consistent selection on the corresponding alternative graph. By definition, the *makespan* of a consistent selection S is the length of the longest path from node 0 to node n in $\mathcal{G}(S)$, and we call it *critical path*. The makespan of a schedule is therefore the makespan of the associated complete consistent selection. Moreover the critical path is defined even if the current selection S is a partial and consistent selection. Given a selection S, we denote the value of a longest path from i to j in $\mathcal{G}(S)$ by $l^S(i, j)$ or simply $l(i, j)$. We use the convention that if there is no path from i to j, then $l(i, j) = -\infty$.

The alternative graph [12], [13] is a generalization of the disjunctive graph of Roy and Sussman [23]. In fact, in the disjunctive graph the pairs of alternative arcs (called disjunctive arcs) are all in the form $((i, j), (j, i))$, where i and j are two operations to be processed on the same machine. With the alternative graph formulation it is possible to model in a precise way, and to solve effectively, a number of complex practical scheduling issues for which there were no successful methodologies so far, see for instance [14], [22]. In this paper we refer mainly to four job shop scheduling problems, namely the ideal (i.e. with infinite capacity buffer), the blocking with swap allowed (i.e. with zero buffer assumption), the blocking with no swap (i.e. with zero buffer and without swaps) and the no-wait (i.e. jobs are not allowed to idle). See the paper of Mascis and Pacciarelli [13] for more details on the alternative graph formulation of the considered constraints.

4   Arc Greedy Heuristics

In this section we introduce the *Arc Greedy Heuristics* (AGH), a family of greedy randomized heuristic algorithms. These heuristics are a randomized extension of the heuristics introduced in Mascis and Pacciarelli [12]. And all the heuristics in the AGH family have the same algorithmic structure and differ only for the evaluation criterion applied to select at each step the next arc in A. These heuristics are based on the idea of repeatedly extending a feasible selection. In particular, at each time, the algorithms select one unselected pair from the set A. The step is repeated until a feasible solution is built or a positive length cycle is detected. Given an alternative graph $\mathcal{G} = (N, F, A)$, we define L_{ij} as the length of the critical path passing through the arc (i, j), and thus the following quantity:

$$L_{ij} = l(0, i) + a_{ij} + l(j, n). \tag{1}$$

Clearly L_{ij} can be considered as a myopic evaluation, and a rough lower bound, of the resulting makespan in the partially selected alternative graph obtained by selecting the alternative arc (i, j). At each step of the algorithm and for each unselected pair $(u, v) = ((i, j), (h, k))$, the values L_{ij} and L_{hk} are computed.

We define the heuristic criterion function $HC((u, v))$ as an operator that applied to an unselected pair $(u, v) \in A$ returns the integer value associated with that possible choice. Note that, in what follows, we refer to the heuristic meaning

the Arc Greedy Heuristic algorithm not depending on the specific heuristic criterion. Besides we refer to the heuristic criterion meaning both the function $HC(\cdot)$
and the Arc Greedy Heuristic algorithm with that specified heuristic criterion.

The main difference between an Arc Greedy Heuristic and a priority dispatching rule or an insertion algorithm is that the latter algorithms at each step
schedule an unscheduled operation. In terms of alternative graph scheduling an
operation o_i means to direct all the outgoing alternative arcs of the node i.
Besides an Arc Greedy Heuristic selects at each time only an unselected pair
in A. Therefore an Arc Greedy Heuristics requires a higher number of steps to
build a solution, and thus a higher computation time, but it has more freedom
of choice and can yield better results.

4.1 Algorithmic Description

Now we describe in more details the proposed greedy algorithm, and the different
heuristic criteria functions implemented. In Figure 1, the general sketch of the
AGH is shown.

At each step the Arc Greedy Heuristic chooses the next unselected pair
$(i, j) \in A$ according to a *pseudo-random proportional* rule [9]. With probability $0 \le q_0 \le 1$ a deterministic choice is made and the best arc is selected, that
is $(u, v) = \arg\min HC_{(a,b) \in A}((a, b))$. Instead with probability $1 - q_0$ the algorithm operates a random choice with uniform probability in a cardinality based
Restricted Candidate List (RCL). The Restricted Candidate List contains the
best n elements according to the heuristic criterion. The size of the Restricted
Candidate List is $\lceil \alpha \mid A \mid \rceil$, where $0 \le \alpha \le 1$ is an input parameter of the
algorithm and $\mid A \mid$ is the number of unselected pairs in the current alternative
graph. Clearly the size of the RCL, i.e. the number of candidates in the RCL,

---

Algorithm Arc Greedy Heuristic
Input:
    Alternative graph $\mathcal{G} = (N, F, A)$
    Heuristic Criterion $HC(\cdot)$
    int q_0, α
begin
    $S := \emptyset$
    while $A \neq \emptyset$ **do**
        with probability q_0 choose the pair $(u, v) \in A$ minimizing $HC(\cdot)$
        with probability $1 - q_0$ choose randomly a pair $(u, v) \in A \cap RCL(\alpha)$
        Select one arc belonging to (u, v)
        Update $S := S \cup \{u\}$ and $A := A - \{(u, v)\}$
        if $\mathcal{G}(S)$ is unfeasible **then** STOP, failed in finding a feasible solution
    end
end.

Fig. 1. The general sketch of the AGH algorithm

is dependent on a parameter α and on the number of unselected pairs in $\mathcal{G}(S)$. Since the number of unselected pairs decreases at every step of the algorithm, at the beginning the RCL contains more elements than at the end of the run of the algorithm.

Fixing the input parameters to $q_0 = 1$ and $\alpha = 0$, the AGH has a deterministic behavior, i.e. it always chooses the best unselected pair in A; whereas with $q_0 = 0$ and $\alpha = 1$ a completely unbiased random choice is made. Note that if a complete random choice is made, then the heuristic criterion adopted has no influence on the performance of the algorithm, i.e. all the Arc Greedy Heuristic algorithms behave in the same way.

Each criterion defines a function $HC((u, v))$ that takes as input an alternative pair belonging to the set A of alternative pairs and gives as output an integer value. In the following we describe in details the different heuristic criteria proposed.

Once the pair $(u, v) = ((i, j), (h, k)) \in A$ is chosen, the algorithm has to decide how to select it, two alternatives arise (i, j) or (h, k), the choice is made depending on which heuristic criterion is adopted. We distinguish two different orderings for the values given by the heuristic criteria: increasing and decreasing. For increasing ordering we mean that the lower values of the heuristic criterion are more attractive for the algorithm and thus are selected first (we call this strategy "Select"), whereas in the decreasing ordering the higher values of the heuristic criterion are sorted in the first positions (and we refer to this strategy as "Avoid").

When selecting one arc in the chosen alternative pair (u, v) it is possible that the resulting selection is not consistent, i.e. the graph $\mathcal{G}(S)$ has a postive length cycle and thus the schedule is unfeasible. In this case the AGH algorithm tries to select the other alternative arc in the pair. If the resulting selection is again not consistent, then the procedure stops failing to find a feasible solution. If, on the contrary, the resulting selection is consistent the algorithm continues the run. Nevertheless it has to be pointed out that in the case of the ideal job shop scheduling problem all the Arc Greedy Heuristics are always able to find a feasible schedule.

In Table 1 we summarize the classification of our heuristics. The name of the heuristic criterion results in the table, the ordering type of the criterion values is reported in the rows, whereas in the columns we report the operator applied to L_{ij} and L_{hk}. The first letter of the name is "A"void ("S"elect) and reflects the decreasing (increasing) ordering adopted. In Select heuristics the algorithm tries to make the best choice, whereas in the Avoid heuristics the algorithm

Table 1. Classification of the heuristics

Ordering	$\max\{L_{ij}, L_{hk}\}$	$\min\{L_{ij}, L_{hk}\}$	$L_{ij} + L_{hk}$	$\|L_{ij} - L_{hk}\|$
Decreasing	AMCC	ALCP	AMSP	AMBP
Increasing	SMCP	SLCP	SMSP	SMBP

tries to avoid the worst choice. The next two letters represent the operator applied to L_{ij} and L_{hk}. "M"ost "C"ritical means the maximum whereas "L"ess "C"ritical stands for the minimum, "M"ost "S"imilar is the sum, and finally "M"ost "B"alanced is the absolute value of the difference between L_{ij} and L_{hk}. The last letter always stands for "P"air, with the only exception of the AMCC criterion where the last letter stands for "C"ompletion time. We decided to use the name AMCC instead of using AMCP, in order to maintain the same names as in [12].

For example, the AMCC (Avoid Most Critical Completion time) criterion gives as output $\max\{L_{ij}, L_{hk}\}$ and the SLCP (Select Less Critical Pair) returns $\min\{L_{ij}, L_{hk}\}$. Once that the pair $((i,j),(h,k)) \in A$ is chosen, the algorithm has to decide how to select it. If the heuristic criterion uses a decreasing ordering (Avoid type) then the arc (i,j) is selected with $L_{ij} < L_{hk}$. In case of an increasing ordering (Select type) of the heuristic criterion the arc (i,j) is selected to satisfy $L_{ij} > L_{hk}$.

5   Experimental Results

The performances of each algorithm are analyzed on a set of benchmark instances from literature. We tested the performance of each algorithm on 54 instances from the literature on the job shop scheduling problem.

We consider three problems (ABZ5-7) generated by Adams, Balas & Zawack [1]; ten 10 × 10 problems (ORB01-10) generated by Applegate and Cook [2] and 40 problems of different sizes (LA01-40) generated by Lawrence [11]. Besides we consider the famous instance (FT10) proposed by Fisher and Thompson in a book edited by Muth and Thompson [16]. The size of the benchmark instances varies from 10 to 30 jobs and from 5 to 20 machines.

For all the considered benchmark problems the number of alternative pairs is $|A| = m * n * (n-1)/2$, where m is the number of the machines and n is the number of the jobs. Note that the expression holds because each job has to perform exactly one operation on each machine. In Table 2 the size in terms of jobs, machines and alternative pairs of the problems is given.

Table 2. Sizes of the problems

♯ Jobs	♯ Machines	♯ Alternative Pairs
10	5	225
15	5	525
20	5	950
10	10	450
15	10	1050
20	10	1900
30	10	4350
15	15	1575
15	20	2100

It has to be noticed that all the AGH algorithms are not able to find often a feasible solution to the more constrained problems, namely the blocking no swap, the blocking with swap allowed and no-wait. For this reason in this section only the results for the IJSP are presented. The issue of finding feasible solutions in constrained cases has to be tackled by metaheuristic algorithms, and it will be investigated in future research.

For each algorithm we considered the following configurations of parameters q_0 and α. We have defined two sets Q and A of the possible value of the input parameters $q_0 \in Q = \{0, 0.25, 0.50, 0.80, 0.90, 0.95\}$ and $\alpha \in A = \{0.05, 0.10, 0.20, 0.50, 0.75, 1\}$. At this set of configurations we added the deterministic configuration $q_0 = 1$ and $\alpha = 0$.

In Table 3, for all the different heuristic criteria, we report on a comparison between the deterministic version of the heuristic criteria and the results obtained by the best configuration over the 54 test instances. The average relative error (Error(%)) is obtained as $100 * (AGH - Opt)/Opt$, where AGH is the value of the solution found by the heuristic and Opt is the optimal value of the instance. The first column reports on the heuristic criterion adopted. In the next two columns the results of the deterministic configuration are shown. In columns 4 and 5 we show the best configuration and the average relative error over twenty runs respectively, whereas in the last two columns the results of the best over twenty repetitions are shown. It can be noticed that the deterministic criteria are more effective, at least in the best performing heuristic criteria, over the single run of the AGH. Moreover if we consider the best over several runs, then the best configurations are not the deterministic configurations anymore.

In Figure 2 we plot the average computational times required of three different AMCC configurations. The times refer to a single run on a Pentium III 900 MHz processor. We consider the deterministic configuration ($q_0 = 1$ and $\alpha = 1$), the best configuration ($q_0 = 0.80$ and $\alpha = 0.20$) and the random configuration ($q_0 = 0$ and $\alpha = 1$). The computational times do not depends on the heuris-

Table 3. Comparison between the deterministic configuration and the best configuration (average on a single run and best over twenty runs)

Heuristic Criterion	Deterministic		Single Run		Twenty Runs	
	(q_0, α)	Error (%)	(q_0, α)	Error (%)	(q_0, α)	Error (%)
AMCC	(1.00, 0.00)	8.34	(0.80, 0.10)	8.91	(0.80, 0.10)	4.15
AMSP	(1.00, 0.00)	14.24	(0.90, 0.05)	14.91	(0.90, 0.05)	7.94
AMBP	(1.00, 0.00)	14.85	(0.95, 1.00)	15.41	(0.95, 1.00)	8.04
SLCP	(1.00, 0.00)	21.45	(0.50, 1.00)	20.25	(0.50, 1.00)	11.49
SMCP	(1.00, 0.00)	30.49	(0.00, 1.00)	21.10	(0.00, 1.00)	11.94
SMBP	(1.00, 0.00)	41.24	(0.00, 0.75)	21.13	(0.00, 0.75)	11.95
ALCP	(1.00, 0.00)	27.59	(0.00, 0.20)	21.88	(0.00, 0.20)	12.02
SMSP	(1.00, 0.00)	27.17	(0.00, 1.00)	21.25	(0.00, 1.00)	12.50

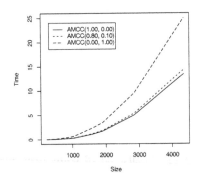

Fig. 2. Computational time of a single run for AMCC heuristics

tic criterion adopted, but rather they depend on the configuration, in fact the deterministic configuration is almost twice faster than the random configuration.

In order to analyze the influence of the two parameters (q_0 and α) on the quality of the achieved solutions, we introduce the *randomness* of a configuration R as $R = 100*(\alpha+1-q_0)$. Clearly $R = 0$ for the deterministic configuration, i.e. $q_0 = 1$ and $\alpha = 0$, whereas the maximum value of the randomness is obtained by q_0 and $\alpha = 1$ which corresponds to a completely unbiased random choice. In Figure 3 we show the best results obtained over twenty repetitions by varying the randomness of the Arc Greedy Heuristic. We are aware that the choice of summing q_0 and α is an arbitrary choice but there is no significant difference in the achieved results of two summed configurations.

In order to analyze the influence of the two parameters (q_0 and α) on the quality of the achieved solutions, we introduce the *randomness* of a configuration R as $R = 100*(\alpha+1-q_0)$. Clearly $R = 0$ for the deterministic configuration, i.e. $q_0 = 1$ and $\alpha = 0$, whereas the maximum value of the randomness is obtained by q_0 and $\alpha = 1$ which corresponds to a completely unbiased random choice. In Figure 3 we show the best results obtained over twenty repetitions by varying the randomness of the Arc Greedy Heuristic. We are aware that the choice of summing q_0 and α is an arbitrary choice but there is no significant difference in the achieved results of two summed configurations.

The proposed heuristic criteria can be divided in two classes with two different behaviors. The first class is composed of the AMCC, AMBP and AMSP heuristic criteria, and the second class is composed of the remaining five heuristic criteria. The best results for the first class are obtained for values of the parameters quite close to the deterministic configuration. Whereas the best results for the heuristic in the second class are obtained for configuration near the complete random choice, and the worst results are obtained for the deterministic criterion.

In the following we consider only the AMCC criterion because it clearly outperforms all the other heuristic criteria, as shown in Figure 3 and in Table 3. In Table 4 we report the results for the selected three configurations of the AGH algorithm, i.e. the deterministic AMCC$(1.00, 0.00)$ and the AMCC$(0.80, 0.10)$ (the

Table 4. Performance of the AMCC heuristics

Instance	Size (n, m)	Optimal value	AMCC$(1.00, 0.00)$		AMCC$(0.80, 0.10)$		AMCC$(0.80, 0.10)_{20}$	
			Solution	Error (%)	Solution	Error (%)	Solution	Error (%)
ABZ5	(10,10)	1234	1318	6.81	1346.0	9.08	1287	4.29
ABZ6	(10,10)	943	985	4.45	974.0	3.29	948	0.53
ABZ7	(15,20)	656	753	14.96	774.5	18.06	744	13.41
FT10	(10,10)	930	985	5.91	1019.5	9.62	969	4.19
ORB01	(10,10)	1059	1213	14.54	1280.0	20.87	1219	15.11
ORB02	(10,10)	888	924	4.05	928.0	4.50	916	3.15
ORB03	(10,10)	1005	1113	10.75	1147.5	14.18	1073	6.77
ORB04	(10,10)	1005	1108	10.25	1112.0	10.65	1068	6.27
ORB05	(10,10)	887	924	4.17	953.0	7.44	905	2.03
ORB06	(10,10)	1010	1107	9.60	1123.0	11.19	1069	5.84
ORB07	(10,10)	397	440	10.83	440.0	10.83	421	6.05
ORB08	(10,10)	899	950	5.67	975.0	8.45	917	2.00
ORB09	(10,10)	934	1015	8.67	1015.0	8.67	976	4.50
ORB10	(10,10)	944	1030	9.11	1028.0	8.90	958	1.48
LA01	(10,5)	666	666	0.00	666.0	0.00	666	0.00
LA02	(10,5)	655	694	5.95	691.5	5.57	656	0.15
LA03	(10,5)	597	735	23.12	690.5	15.66	617	3.35
LA04	(10,5)	590	679	15.08	625.0	5.93	599	1.53
LA05	(10,5)	593	593	0.00	593.0	0.00	593	0.00
LA06	(15,5)	926	926	0.00	932.0	0.65	926	0.00
LA07	(15,5)	890	984	10.56	911.5	2.42	890	0.00
LA08	(15,5)	836	873	1.16	899.0	7.54	863	3.23
LA09	(15,5)	951	986	3.68	951.0	0.00	951	0.00
LA10	(15,5)	958	1009	5.32	958.0	0.00	958	0.00
LA11	(20,5)	1222	1239	1.39	1253.5	2.58	1222	0.00
LA12	(20,5)	1039	1039	0.00	1059.5	1.97	1039	0.00
LA13	(20,5)	1150	1161	0.96	1166.5	1.43	1150	0.00
LA14	(20,5)	1292	1305	1.01	1292.0	0.00	1292	0.00
LA15	(20,5)	1207	1369	13.42	1384.5	14.71	1276	5.72
LA16	(10,10)	945	979	3.60	1002.5	6.08	973	2.96
LA17	(10,10)	784	800	2.04	813.0	3.70	800	2.04
LA18	(10,10)	848	916	8.02	905.5	6.78	876	3.30
LA19	(10,10)	842	846	0.48	861.0	2.26	846	0.48
LA20	(10,10)	902	930	3.10	930.0	3.10	913	1.22
LA21	(15,10)	1046	1241	18.64	1167.5	11.62	1132	8.22
LA22	(15,10)	927	1032	11.33	1024.0	10.46	985	6.26
LA23	(15,10)	1032	1131	9.59	1120.5	8.58	1066	3.29
LA24	(15,10)	935	999	6.84	1046.0	11.87	1014	8.45
LA25	(15,10)	977	1071	9.62	1087.5	11.31	1039	6.35
LA26	(20,10)	1218	1378	13.14	1363.5	11.95	1249	2.55
LA27	(20,10)	1235	1353	9.55	1441.0	16.68	1332	7.85
LA28	(20,10)	1216	1322	8.72	1377.5	13.28	1308	7.57
LA29	(20,10)	1152	1392	20.83	1403.0	21.78	1315	14.14
LA30	(20,10)	1355	1476	8.93	1473.0	8.71	1392	2.73
LA31	(30,10)	1784	1871	4.88	1939.5	8.72	1858	4.15
LA32	(30,10)	1850	1942	4.97	2006.0	8.43	1957	5.78
LA33	(30,10)	1719	1897	10.35	1862.5	8.35	1785	3.84
LA34	(30,10)	1721	1934	12.38	1922.0	11.68	1783	3.60
LA35	(30,10)	1888	2017	6.83	2020.5	7.02	1924	1.91
LA36	(15,15)	1268	1347	6.23	1381.5	8.95	1329	4.81
LA37	(15,15)	1397	1547	10.74	1535.5	9.91	1462	4.65
LA38	(15,15)	1196	1342	12.21	1395.5	16.68	1324	10.70
LA39	(15,15)	1233	1361	10.38	1417.0	14.92	1366	10.79
LA40	(15,15)	1222	1340	9.66	1343.0	9.90	1297	6.14
Median error				8.34%		8.69%		3.32%
Average error				7.86%		8.46%		4.13%
Maximum error				23.12%		21.78%		15.11%

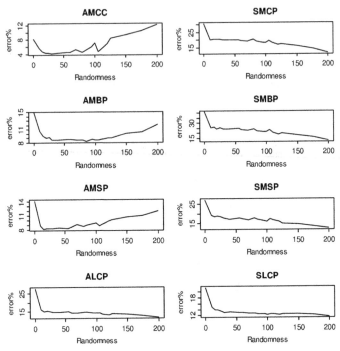

Fig. 3.   The influence of the Randomness on the performances of the AGH algorithms

median over twenty runs), and the AMCC$(0.80, 0.10)_{20}$ (the best over twenty runs). In columns 1, 2 and 3 we report the name of the instances, the size of the instance (n, m) and the values of the optimal solutions, respectively. In the subsequent columns we report, for each instance, the solutions found by the heuristics and the relative distance from the optimum (percentage).

In Table 5 we compare the results of our best proposed algorithm which is AMCC$(0.8, 0.1)_{20}$ with other constructive heuristics taken from the literature. The comparisons are made only on a subset of 13 hard instances as proposed in Vaessens et al. [26]. This subset includes a 5×10 problem (LA02), two 10×10 (FT10 and LA19), three 15×10 problems (LA21, LA24, LA25), two 20×10 (LA27, LA29), and five instances 15×15 (LA36 - LA40). In the comparisons we consider the best Shifting Bottleneck procedure (SB'88) proposed in Adams, Balas & Zawack [1], and the best Modified Shifting Bottleneck (SB'95) procedure proposed in Balas et al. [4]. Moreover we consider the Beam Search algorithm (BS) of Sabuncuoglu and Bayiz [24] that makes use of PDR rules and the INSA insertion heuristic of Nowicki and Smutnicki [18]. The proposed algorithm is able to obtain comparable results in terms of solution quality with the SB'88. It clearly outperforms the BS and the INSA heuristic, whereas the SB'95 algorithm dominates it. In any case it has to be pointed out that the SB'95 algorithm

Table 5. Comparisons on hard instances

Instance	Optimum	AMCC$(0.80, 0.10)_{20}$	SB'88	SB'95	BS	INSA
FT10	930	969	952	940	1016	994
LA02	655	656	684	667	704	722
LA19	842	846	863	878	882	971
LA21	1046	1132	1128	1071	1154	1179
LA24	935	1014	1015	976	992	1021
LA25	977	1039	1061	1012	1073	1147
LA27	1235	1332	1353	1272	1361	1466
LA29	1152	1315	1233	1227	1252	1385
LA36	1268	1329	1326	1319	1401	1445
LA37	1397	1462	1471	1425	1503	1726
LA38	1196	1324	1307	1294	1297	1307
LA39	1233	1366	1301	1278	1369	1393
LA40	1222	1297	1347	1262	1347	1387
Average error		7.04%	6.76%	3.78%	8.96%	14.58%

cannot be considered as simple greedy heuristics. Since our algorithm is much more general, thus resulting in a code which is less efficient than the code of specialized algorithms, we decide not to compare the computational times even if the time needed by our algorithm is acceptable.

6   Conclusions and Future Research

In this paper we have introduced a new class of randomized greedy heuristics for a general formulation of scheduling problems. Extensive computational results are presented and a comparison with other algorithms for the classical job shop scheduling is carried on. In particular, it turns out that our approach outperforms other simple greedy schemes but some specialized algorithms dominate it. Nevertheless the computational times required by our algorithm are acceptable. In particular in the case of ideal job shop the algorithms are always able to find a feasible solution, whereas in the other constrained cases (blocking with swap, blocking no swap and no-wait) the proposed algorithms are not often able to find a feasible solution in a single run. The problem of finding a feasible solution and improving it has to be tackled by a metaheuristic scheme. Future research deals with developing metaheuristic algorithms based on the proposed heuristics for solving more constrained scheduling problems.

Acknowledgements

Marco Pranzo would like to thank Thomas Stützle for his helpful suggestions.

References

[1] Adams, J., Balas, E., Zawack, D.: The shifting bottleneck procedure for job shop scheduling. Management Science, **34** (3) (1988) 391–401 224, 229, 233

[2] Applegate, D., Cook, W.: A computational study of the job shop scheduling problem. ORSA Journal on Computing, **3** (2) (1991) 149–156 229

[3] Balas, E.: Disjunctive programming. Annals of Discrete Mathematics, **5** (1979) 3–51 225

[4] Balas, E., Lenstra, J. K., Vazacopoulos, A.: The one-machine problem with delayed precedence constraints and its use in job shop scheduling. Management Science, **41** (1) (1995) 94–109 224, 233

[5] Blackstone, J. H., Phillips, D. T., Hogg, G. L.: A state-of-the-art survey of dispatching rules for manufacturing job shop operations. International Journal of Production Research, **21** (1982) 27–45 223

[6] Carlier, J.: The one-machine sequencing problem. European Journal of Operational Research, **11** (1982) 42–47 224

[7] Chang, Y. L., Sueyoshi, T., Sullivan, R. S.: Ranking dispatching rules by data envelopment analysis in a job-shop environment. IIE Transaction, **28** (8) (1996) 631–642 223

[8] Dell'Amico, M., Trubian, M.: Applying taboo search to the job-shop scheduling problem. Annals of Operations Research, **41** (1993) 231–252 224

[9] Dorigo, M., Gambardella, L. M.: Ant Colony System: A cooperative learning approach to the travelling salesman problem. IEEE Transaction on Evolutionary Computation, **1** (1) (1997) 53–66 227

[10] Haupt, R.: A survey of priority rule-based scheduling. OR Spektrum, **11** (1989) 3–16 223

[11] Lawrence, S.: Supplement to Resource constrained project scheduling: an experimental investigation of heuristic scheduling techniques. GSIA, Carnegie Mellon University, Pittsburgh, PA (1984) 224, 229

[12] Mascis, A., Pacciarelli, D.: Machine Scheduling via Alternative Graphs. Technical Report DIA-46-2000, Dipartimento di Informatica e Automazione, Università Roma Tre, Roma, Italy (2000) 223, 224, 226, 229

[13] Mascis, A., Pacciarelli, D.: Job shop scheduling with blocking and no-wait constraints. European Journal of Operational Research, **143** (3) (2002) 498–517 224, 226

[14] Mascis, A., Pacciarelli, D., Pranzo, M.: Train scheduling in a regional railway network. Proceedings of the 4th Triennial Symposium on Transportation Analysis (TRISTAN IV), Sao Miguel, Portugal (2001) 487–492 226

[15] Morton, T. E., Pentico, D. W.: Heuristic Scheduling Systems. John Wiley and Sons, New York (1993) 224

[16] Muth, J. F., Thompson, G. L. (eds.): Industrial scheduling. Kluwer Academic Publishers, Amsterdam (1963) 229

[17] Nawaz, M., Enscore, E. E., Ham, I.: A heuristic algorithm for the m-machines and n-jobs flow-shop sequencing problem. OMEGA The International Journal of Management Science, **11** (1) (1983) 91–95 224

[18] Nowicki, E., Smutnicki, C.: A fast taboo search algorithm for the job shop scheduling problem. Management Science, **42** (6) (1996) 797–813 224, 233

[19] Ovacik, I. M., Uzsoy, R.: Decomposition methods for complex factory scheduling problems. Prentice-Hall, Englewood Cliffs, NJ (1997) 224

[20] Panwalkar, S. S., Iskander, W.: A survey of scheduling rules. Operations Research, **25** (1) (1977) 45–61 223

[21] Pinson, E.: The job shop scheduling problem: a concise survey and some recent developments. In Chrétienne, P., Coffman, E. G., Lenstra, J. K., Liu Z. (eds.), Scheduling theory and its applications, Wiley (1997) 277–294 224

[22] Pranzo, M.: Algorithms and applications for complex job shop scheduling problems. Ph.D. Thesis, DSPSA, Università La Sapienza, Roma, Italy (2002) 226

[23] Roy, B., Sussman, R.: Les problèmes d'ordonnancement avec contraintes disjonctives. Note DS No. 9bis, SEMA, Paris (1964) 223, 226

[24] Sabuncuoglu, I., Bayiz, M.: Job shop scheduling with beam search. European Journal of Operational Research, **118** (1999) 390–412 224, 233

[25] Werner, F., Winkler, A.: Insertion techniques for the heuristic solution of the job-shop problem. Discrete Applied Mathematic, **58** (2) (1995) 191–211 224

[26] Vaessens, R. J. M., Aarts, E. H. L., Lenstra, J. K.: Job shop scheduling by local search. INFORMS Journal on Computing, **8** (3) (1996) 302–317 233

Algorithmic Techniques for Memory Energy Reduction

Mitali Singh and Viktor K. Prasanna

Department of Computer Science, University of Southern California
Los Angeles, CA-90089, USA
{mitalisi,prasanna}@usc.edu

Abstract. Energy dissipation is a critical concern for battery-powered embedded systems. Memory energy contributes significantly to overall energy in data intensive applications. Low power memory systems are being designed that support multiple power states of memory banks. In low power states, energy dissipation is reduced but time to access memory is increased. We abstract an energy model for the memory system and exploit it to develop algorithmic techniques for memory energy reduction. This is achieved by exploring the structure and data access pattern of a given algorithm to devise memory power management schedules. We illustrate our approach through two well-known embedded benchmarks - Matrix Multiplication and Fast Fourier Transform. The optimality of our schemes is discussed using information theoretic lower bounds on memory energy. Simulations demonstrate that significant energy reduction can be achieved by using our approach over state-of-the-art implementations.

1 Introduction

Due to the explosive growth of portable, wireless devices and battery-operated embedded systems, energy efficiency has become a critical concern for designers today. Design technologies at all levels of abstraction are evolving with the common goal of energy reduction. The significance of high level analysis in the design cycle cannot be underestimated. It is rapid, fairly accurate, and platform independent. Moreover, decisions made at higher levels are likely to have a larger impact on energy reduction than those at the lower levels of abstraction. We have thus been motivated to explore energy efficient design and analysis methodologies at the algorithmic level.

Until recently, majority of the research focus has been on optimizing processor energy by exploiting techniques such as dynamic voltage scaling [26] [17], precision management, and IPC management [14]. Advancements in processor technology have led to development of low power processors such as the XScale PXA250 [15] that dissipate with less than 1nJ per instruction. The next challenge lies in reduction of memory energy which accounts for as much as 90% of the overall system energy for CPU systems with peripherals [7]. For several

K. Jansen et al. (Eds.): WEA 2003, LNCS 2647, pp. 237–252, 2003.

wireless applications implemented on the pico radio [25], more than 50% of the overall energy is dissipated in the memory [27].

Advanced memory technologies such as the Mobile SDRAM [21] support several low power features such as multiple power states, bank/row specific activation, and partial array refresh (PASR). Our goal is to abstract the advanced features of the state-of-the-art memory systems and exploit them to design algorithms that reduce memory energy dissipation. This is achieved by designing algorithms optimized for reduced number of memory accesses, and implementing energy optimal memory power management schedules. We analyzed several benchmark kernels from the EDN Embedded Microprocessor Benchmark Consortium (EEMBC) [20] and the freely available Mibench Benchmarks [19], which are discussed in [34]. In this paper we present our results for Matrix Multiplication and Fast Fourier Transform(FFT). We discuss information theoretic lower bounds for energy dissipation in the memory and use them to guide our algorithm design. Simulation results (see Section 5) demonstrate that using our techniques significant energy reduction can be achieved.

Rest of this paper is organized as follows. In Section 2 we discuss related research. A high-level energy analysis is presented in Section 3 and lower bounds on memory energy dissipation are discussed. An energy efficient memory architectural design is proposed in Section 3.3. Optimal memory activation schedules for some well-known kernels are discussed in Section 4. Our simulation framework and results are presented in Section 5. Finally, we conclude in Section 6.

2   Related Work

Memory technologies such as the SDRAM reduce energy dissipation by switching to lower power states based on the time of inactivity in the current state. We call this *implicit power management* as the memory power state switching is defined only by the duration of inactivity (wait time) of the memory bank.

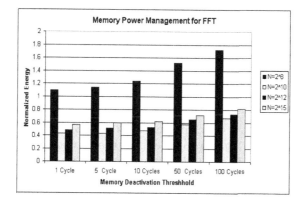

Fig. 1. Implicit Power Management

Several researchers have exploited implicit power management for memory energy reduction. Energy efficient page allocation techniques for general-purpose processors have been proposed in [16]. In [3], the authors investigate array allocation schemes to minimize memory energy dissipation. A large number of benchmark applications have been analyzed in [9] to find the optimal wait time for memory power state switching. The conclusion drawn suggests that memory should immediately transition to a lower power state when inactive. However, our simulations show that this is not the optimal policy always. Fig. 1 illustrates normalized values (w.r.t. case with no power management) for memory energy dissipation for FFT with implicit power management for various wait times. For smaller size FFT ($n = 2^8$) all the policies result in increased energy dissipation. Energy reduction is observed for larger size problems with optimal wait-time as one cycle. Thus, we propose algorithm specific power management of the memory. In this paper, we focus on design of optimal power management schemes for the memory based upon the structure and access pattern of the algorithms.

The AMRM project [1] focuses on adaptation of the memory hierarchy to reduce latency and improve power efficiency. Techniques such as off-chip memory assignment, set associativity and tiling to improve cache performance and energy efficiency have been investigated in [29]. The memory segmentation problem has been shown to be NP complete in [10]. Several researchers [24] [18] [22] [6] [4] have explored memory organization and optimization for embedded systems. Their approach has been summarized below.

Memory Architecture Customization. Memory allocation and memory bank customization problems deal with selection of memory parameters such as type, size, and number of ports. Memory building blocks and organization for application customized memory architectures are design synthesis problems and are not in the scope of this paper. We focus on algorithmic optimizations rather than hardware customizations.

Application Specific Code and Data Layout Optimizations. Application specific platform-independent code (loop) and data flow transformations have been proposed to optimize the algorithm's storage (memory, cache) and transfer requirements. This is followed by hardware customization (such as selection of a smaller cache) to reduce energy dissipation. Our approach is reverse as we optimize algorithms for a fixed architecture.

Scratch Pad Memory. : On-chip memory is partitioned into data cache and scratch pad. Split spatial-temporal caches have been proposed to improve spatial and temporal data reuse in the cache. In our analysis, we propose use of a small (of the order of cache line size) memory buffer to aid in implementation of power management schemes. A partial bank of a Mobile SDRAM that can be power controlled independently can be used as buffer. Scratch pad memory can act as a buffer but at the expense of smaller cache size.

Our approach aims at optimizing memory energy dissipation by improving the memory access pattern of the algorithms. Higher data reuse reduces the number of memory accesses. Algorithm directed power management schemes are described within the algorithm to dynamically alter the memory power states. Buffering, prefetching and blocking strategies are used to reduce energy and latency overheads for memory power management. Memory energy is analyzed using a simple, high-level, memory energy model that considers memory to be organized as multiple banks that can be power controlled independently. Since we assume, the architecture to be fixed, memory parameters (such as no of ports, banks, interconnect bandwidth) are considered constant.

3   Memory Energy

We define a high level model for memory energy analysis of algorithms and discuss lower bounds on memory energy dissipation of algorithms.

3.1   Our Energy Model

We consider our system to comprise of a computational unit with an internal memory (e.g. cache) connected to a memory unit over an interconnect (Fig. 2(a)). We abstract the low power features of the memory systems as discussed below.

– Memory has multiple power modes (states). Fig. 2(b) illustrates the the current drawn in various power modes of the SDRAM [21] memory system.

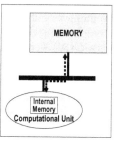

(a) System Model

$V_{DD}/V_{DDQ} = 2.5/1.8$	Current (mA)
Active Mode	150
Active Burst Mode	90
Standby Active Mode	35
Self Refresh Mode	0.100 to 0.35
Power Down Mode	0.350

(b) 128Mb Mobile SDRAM

Fig. 2.

Active: Memory can be read or written to only in the active mode. Power dissipation is the highest in this mode. The energy dissipation is higher when the memory is being accessed and lower when it is in standby active mode. We do not consider the burst access mode in our current analysis.

Refresh (idle/inactive): Power dissipation in this mode is low (zero for analysis). Data is preserved in this mode. Memory access in this mode results in higher latency as the memory must transit to the Active mode.

Power Down: The power dissipation in this mode is the least but data is not preserved. For our analysis, we do not consider transition to this mode to ensure that the data is not lost.

- Memory is organized as banks. Each bank can be placed in any power mode independent of the other banks. For example, Mobile SDRAM memory supports bank and row specific activation and deactivation (precharge) and partial-array refresh (PASR).
- Transition of memory from one power mode to another incurs energy and time overheads.
- Memory power management can be controlled through software.

The total memory energy $E(N)$ for problem size N is defined as the sum of the memory access energy $E_a(N)$, the data storage energy $E_s(N)$, and state transition overheads $E_p(N)$. The memory access energy is proportional to the memory data traffic. It also depends on parameters such as the load capacitance, frequency and voltage of the memory I/O, but we assume these to be fixed. The data storage energy is a function of the memory size and the time for which it is active. Energy dissipation in the idle state is considered to be negligible. Let K_A denote the memory access energy cost per unit of data, K_s be the storage energy cost per unit of data per unit time, and K_p represent the energy overheads for each power state transition. The memory energy is defined as follows.

$$E(N) = E_a(N) + E_s(N) = K_a \times C(N) + K_s \times S(N) \times A(N) + K_p \times P(N)$$

Here, $C(N)$ represents the total number of memory accesses and $S(N)$ is the space complexity of the algorithm. We define memory *activation complexity* $A(N)$, as the time for which memory is in the active mode. For an algorithm of time complexity $T(N)$, if memory is active for time fraction α then $A(N) = \alpha T(N)$. For conventional systems that do not support memory power management $A(N) = T(N)$. $P(N)$ denotes the number of power mode transitions of the memory banks.

3.2   Lower Bounds on Memory Energy

Algorithms have been extensively analyzed and optimized in the past using the I/O complexity model [12]. This model measures performance as a function of the number of I/O operations, and abstracts systems where the latencies involved in accessing external memory are much larger as compared to internal processing. Prior results on I/O complexity are of significant interest as they can be used to determine lower bounds for memory energy dissipation.

Every computational unit in an embedded system has an internal memory of fixed size as illustrated in Fig. 2(a). The interaction between the internal memory and the (external to computational unit) on-chip memory, can be captured by using the I/O model for analysis. We defined $C(N)$ as the number of memory accesses which is the asymptotically same as the I/O complexity $T_{I/O}(N)$ [28].

Theorem 1. *A lower bound on memory energy dissipation $E(N)$ for an algorithm with problem size N and I/O complexity $T_{I/O}(N)$ is given by $\Omega(T_{I/O}(N))$.*

Proof: We know $E(N) = K_a \times C(N) + K_s \times S(N) \times A(N) + K_p \times P(N)$ (see Section 3.1). The memory must remain active for at least the time it is accessed. Thus if the access latency is l cycles, $A(N) = \Omega(l \times C(N))$. $S(N)$ represents the space complexity of the algorithm or the size of memory required to store data. $S(N) = k \times M$, where M is the smallest segment (bank incase of multi-banked architecture) that can be power controlled independently. Each memory access requires at least one memory bank to be active. In the best case scenario memory power management overheads $P(N)$ are negligible. Thus,
$$E(N) = K_a \times C(N) + K_s \times S(N) \times A(N)$$
$$= \Omega(K_a \times C(N) + K_s \times M \times C(N)) = \Omega(C(N)) = \Omega(T_{I/O}(N)) \quad \square$$
Note that the above lower bound on memory energy dissipation is independent of how the computation is performed (on RISC, DSP, FPGA). For a given kernel mapped on a single computation unit, it depends only on the size of the input data and the size of the internal memory of the computational unit.

3.3   Memory Architecture Design

Memory energy can be optimized by introduction of a small memory buffer of size B between the computation unit and the memory unit. Data is transferred from memory to buffer before it is accessed by the processor. The buffer is active all the time, but it permits the **much larger sized** memory module to remain *idle* for a longer time. The buffer is placed near the memory modules with a simple data transfer policy to/from memory. Thus, the latency for data transfer between the memory and the buffer is much lower than a memory access from a computational unit. The latter (for example a PCI) involves complex scheduling and bandwidth allocation. We consider memory to buffer latency to be unity while the buffer to computational unit latency is l.

A simple power management scheme is described as follows. A memory bank is activated only when there is a data transfer required from/to the buffer. The memory access energy and activation overheads are $O(T_{I/O}(N))$, which is optimal. Next, consider the memory storage energy $E_s(N)$. In absence of the buffer, the entire memory must remain active while it is accessed. Hence by definition $E_s(N) = K_s \times S(N) \times A(N)$, where $A(N) = \Omega(l.C(N))$ and $A(N) = O(T(N))$. The power management scheme described above reduces memory activation time to $A(N) = 1 \times C(N) = O(T_{I/O}(N))$. Thus, $E_s(N) = K_s \times (M \times T_{I/O}(N) + B \times T(N))$. Any scheduled data transfer from buffer to computational unit need not be of size larger than the line size of the internal memory of the computational unit. Hence, it is sufficient to have $B = O(L)$, where L is the cache size. Cache size is typically in Kilobytes whereas memory size ranges from Megabytes implying $M \gg B$. Therefore, energy dissipation in the buffer can be ignored as compared to the memory, and $E_s(N) = K_s \times M \times T_{I/O}(N) = \Omega(T_{I/O}(N))$. Thus, $E(N) = O(T_{I/O}(N))$.   □

Remark 1. Since the memory access time is reduced from l to 1, memory storage energy is reduced by a factor of l by using the buffer even in the best case scenario when $A(N) = O(l.C(N))$ in absence of buffer.

Remark 2. The memory buffer increases memory to processor data access latency from l to $l+1$ as data is fetched to buffer before it is transferred to (from) memory. However, data prefetch can be utilized to schedule transfer of data into the buffer before it is accessed by the computational unit.

4   Memory Energy Optimization

Memory energy optimization involves designing algorithms that reduce memory-processor data transfers and implement energy optimal memory activation schedules. Our approach can be summarized as follows:

- Understanding the memory access behavior of the kernel algorithm.
- Minimizing the number of memory accesses required by optimizing the cache complexity of the algorithm. For example, data layout can be altered.
- Designing power management schedule based on the memory access pattern.
- Reducing the power management overheads.

It is important to note that our analysis holds for all computational units with an internal memory. For our simulations, we consider a RISC computational unit. The data cache (internal memory for RISC) is of size $O(L)$.

4.1   Matrix Multiplication

Matrix Multiplication is an embedded automotive/industrial benchmark [20].

Baseline (MMS). Consider multiplication of two $N \times N$ matrices A and B to produce matrix D. Computation complexity of this algorithm is $O(N^3)$. Computation of each element of D requires the corresponding row from A and column from B to be fetched into the cache. There is no data reuse. The number of memory to cache transfers $C(N)$ is at least $2N \times N^2 = O(N^3)$. $3N^2$ elements need to be stored in memory. The memory is active all the time. The storage energy is $O(A(N) \times S(N))$, which is $O(N^3) \times O(N^2) = O(N^5)$. Thus, memory energy is given by $O(C(N)) + O(A(N) \times S(N)) = O(N^3) + O(N^5) = O(N^5)$.

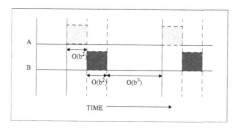

Fig. 3. Memory Access Schedule

Blocked (MMB). We investigate an alternative implementation using blocked (tiled) data layout with block size b. The block should be able to fit into the cache and thus, $b^2 = O(L)$. The arithmetic complexity for this implementation remains $O(N^3)$. Data is fetched and operated upon as blocks, resulting in higher data reuse in the cache. To compute b^2 elements of D, we require $2N \times b$ transfers. $C(N)$ for this algorithm is reduced to $O(2N^3/b)$, which is an improvement by a factor of b. Since the memory modules remain active for the entire duration, the memory energy is given by $O(C(N)) + O(A(N) \times S(N)) = O(N^3/b) + O(N^3 \times 3N^2) = O(N^5)$. The memory access energy is decreased with reduced data traffic, but the storage energy remains same.

Memory Power Management (MMBPM). As the next level of optimization, we reduce the activation time for the memory. This is achieved by explicitly scheduling the memory power state transitions (see Fig. 5) based on the memory access pattern of the algorithm. The data access timing diagram is illustrated in Fig. 3. After each successive fetch of $O(b^2)$ from matrix A and B computation of $O(b^3)$ takes place. Energy can be saved by deactivating the memory modules for this duration. Thus, memory activation time $A(N)$ is reduced to $O(N^3/b)$. The storage energy is reduced to $O(N^3/b) \times O(N^2) = O(N^5/b)$.

Memory Buffer (MMBBUF). Using a memory buffer and blocking, the energy of the system can be decreased to $O(N^3/b)$ as described in Section 3.3. Since we can predict which block will be required for computation next, data prefetch [2] can be used to hide buffer to memory data transfer latencies.

The I/O complexity as analyzed by Hong and Kung [12] is given by :

Lemma 1. *For matrix multiplication of two $N \times N$ matrices on a processor with L words of memory, the I/O complexity bound is given by $\Omega(N^3/\sqrt{L})$.*

Using Theorem 1 and Lemma 1 it follows,

Theorem 2. *Memory energy dissipation for Matrix Multiplication of two $N \times N$ matrices is $\Omega(N^3/\sqrt{L})$, where L is the cache size.*

Remark 3. We discussed an energy optimal implementation of Matrix multiplication (MMBBUF). For $b = \sqrt{L}$, where L represents the cache size, the memory energy are reduced to $\Omega(N^3/\sqrt{L})$. This is the optimal as shown by Theorem 2.

4.2   Fast Fourier Transform

Fourier Transforms are used in digital signal processing in several embedded applications such as the Asynchronous Digital Subscriber Line (ADSL) where data is converted from the time domain to the frequency domain to reduce error rate. The problem size in Mibench is 2^{15} and we use the same for analysis. We examine the Cooley-Tukey algorithm, which involves recursive decomposition of a larger FFT into smaller sub-problems that can be solved efficiently. For computation of an N-point FFT, the computation complexity of this algorithm is $O(N \log_2 N)$. Consider computation of an $N1 \times N2$-point FFT. The data layout in the memory can be analyzed in terms of data *stride*, which is defined as the distance between data blocks that are accessed successively. The stride of data for the $N2$-point computation is determined by the $N1$-point FFT computation. The cache performance is poor if the stride is too large. Dynamic Data Layouts [23] [31] can be used to improve the cache performance. These have been exploited to improve the time performance [23].

Baseline (FFT). As our baseline case, we chose an optimal implementation from the FFTW library and analyze its cache behavior as a function of the data stride. For small strides, there is a large spatial locality in the data. Therefore, there are only compulsory cache misses, and for an N-point FFT, $O(N/b)$ misses are encountered. Here b is the cache block size. However, if the stride is very large, cache performance is very poor due to increased number of conflict misses. A cache miss could incur for every data access required for the FFT computation, which is $O(N \log_2 N)$.

Dynamic Data Layouts (FFTD). Data reorganization could be performed prior to every FFT computation to reduce the data stride. However, this itself dissipates a lot of energy as it could involve $O(N)$ memory accesses. Therefore, prior to each FFT computation, a tradeoff is performed between the data reorganization overheads involved and the acceptable data stride. The FFT decomposition strategy is selected taking this tradeoff into consideration. Significant energy reduction is achieved by improving the cache performance. Details of the algorithm are discussed in [34].

Memory Power Management (FFTDPM). Conventional computation of FFT follows the following sequence of operations. Two elements are fetched from memory, operated on and the result is placed back in the memory. This approach requires the memory to be active all the time. By increasing the interval between successive data fetches, we can exploit memory power management for energy reduction. This is achieved by block computation. FFT is computed over blocks of size $h = O(L)$, where L is the cache size. h elements are fetched, FFT is computed over this block and the result is placed back. This involves only $O(N \log_h N)$ data fetches. Moreover, it permits us to schedule the memory power state switching. Each time h elements of data are fetched and computed upon for $O(h \log h)$ time before the next data transfer is scheduled. Memory is activated and deactivated based on this schedule. The memory activation time is reduced from $O(N \log_2 N)$ to $O(N \log_2 N)/(\log_2 h)$, which results in improvement of storage energy by $O(\log_2 h)$.

Memory Buffer (FFTDBUF). Next, we utilize the memory buffer along with data prefetch. The storage energy in all the previously discussed implementations is $A(N) \times S(N)$, which is $O(N^2 \log_h N)$. Using a memory buffer, we can reduce this to $O(N \log_h N)$ (see Section 3.3).

Hong and Kung [12] have shown the following result:

Lemma 2. *The I/O time for computing an N-point FFT on a processor with L words of memory is at least $\Omega(N \log_2 N / \log_2 L)$.*

From Theorem 1 and Lemma 2, it follows,

Theorem 3. *For computing an N-point Fast Fourier Transform, the lower bound for memory energy is $\Omega(N \log_2 N / \log_2 L)$, where L is the cache size.*

Remark 4. The energy reduction techniques discussed above achieve an energy optimal implementation of FFT. The optimal bound for memory energy for our implementation is $\Omega(N \log_2 N / \log_2 L)$.

5   Simulations

Energy estimation for the algorithms described above is challenging as currently there is no hardware or middle-ware support, or simulators that support algorithm directed power management. Thus, we designed a high level energy estimation framework for fast and yet fairly accurate energy estimation of algorithms.

5.1   Simulation Framework

The simulator is based upon instruction level analysis, where cost of each instruction depends on the power state of the system. The choice of this level of abstraction is based on several reasons. The foremost being the speed without much loss in accuracy. Moreover, it is a suitable abstraction from an algorithm designer's perspective. An algorithm description augmented with the power management schedule (see Fig. 5) is supplied as input to the simulator. The SimpleScalar Toolset [5] is configured for the chosen architecture and modified to provide an instruction execution trace with an embedded power schedule. The trace consists of a limited set of instructions (7 currently). The classification is based on their energy costs.

- *State:* The State instruction permits the designer to change the power state of the system by changing the power mode of any component. For example the memory can be placed in a low power inactive mode.
- *Compute:* This represents all processor only instructions that do not require any cache or memory access. Measurements [30] have demonstrated that there is little power variation among these instructions.
- *ReadCache and WriteCache:* Data is accessed from the cache.
- *ReadMem and WriteMem:* A cache miss occurs resulting in a memory access.
- *Prefetch:* Data is prefetched into the buffer. Energy costs are similar to *ReadMem* but there is no latency in fetching the data.

Each instruction has an associated power and time cost depending on the power state of the system. The *State* instruction denotes change in the power state of the memory and accounts for the state transition overheads. A profile is maintained for power dissipation in each architecture component to identify hot spots. We have modeled the Itsy Pocket Computer for which detailed measurements are available [35]. Currently we model only power modes of the memory and the switching clock enabled/disable mode of the processor. The Itsy measurements do not profile state transitions. The LART [26] measurements are used to obtain overheads for changing the processor mode. The memory activation/deactivation latency is considered to be one cycle.

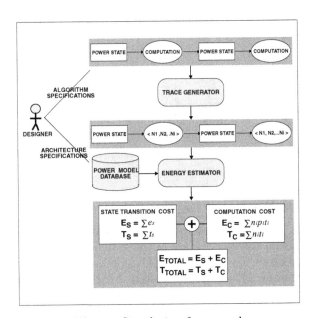

Fig. 4. Simulation framework

```
TransformNtoB(A, Ab, k, m, kblk, mblk)
Tr_NtoB(B, Bb, n, k, nblk, kblk);
TransformNtoB(C, Cb, n, m, nblk, mblk);
    for(jb=jj=0;jb<nblk;jb++){
        jbs = ((jj+=BS) < n) ? BS: n-jj+BS;
        for(lb=ll=0;lb<kblk;lb++){
            lbs = ((ll+=BS) < k) ? BS: k-ll+BS;
            bpt = &(Bb(jb,lb));
            for(ib=ii=0;ib<mblk;ib++){
                ibs = ((ii+=BS) < m) ? BS: m-ii+BS;
                a = &(Ab(ib,lb)); t = &(Cb(ib,jb));
        POWERMG(STATE MEMORY MODE 0)
                for(i=0;i<ibs;i++){ b = bpt;
                for(j=0;j<jbs;j++){ temp_c=0;
                    for(l=0;l<lbs;l++){
                    temp_c += a[l] * b[l];}
                    t[j] += temp_c; b += BS;}
                a += BS;  t += BS;}
        POWERMG(STATE MEMORY MODE 1)}}}
    TransformBtoN(Cb, C, n, m, nblk, mblk);
```

Fig. 5. Sample Code

The goal of this simulator is to guide algorithm design by identifying the trend. It cannot guarantee high accuracy due to several reasons. It does not simulate the dynamic effects of changing the power state. For example, consider a scenario where memory is inactive. The trace shows a cache miss. There is a delay associated with memory activation. This could have resulted in more cache misses in a running system. Since our trace is pre-computed, such effects are not accounted.

The accuracy of the simulator depends on the power models incorporated. Some of the values have been approximated when measurements are not available. For example, the power during a state transition is approximated as the average power between previous and next state. Variation in input data may change the switching activity. We assume an average power cost for each instruction. We have used benchmark data when available, and randomly generated data otherwise. The framework is fast, modular and sufficiently accurate for algorithmic analysis. Once some algorithm designs are identified to be energy efficient they can be tested on lower level simulators or implementation boards for higher accuracy.

5.2   Simulation Results

Matrix Multiplication. We simulated matrix multiplication for $n = 32$ and $n = 64$ and the results are illustrated in Fig. 6. Note that to improve the clarity of the figure we have scaled down the results for $n = 64$ by a factor of 8. The energy reduction is incremental with problem size. We consider $n = 64$. Our algorithm using blocking (MMB) with block size 16 reduced energy dissipation by 93.6% keeping memory active all the time. This reduction was achieved due to improved cache behavior and thus the execution time also decreased. We examined the effect of implicit power management (MMSP) on the two algorithms. We assumed memory becomes idle if not accessed for 1 cycle. The energy of the conventional algorithm was reduced by 57.4%, but the execution time was increased due to memory activation overheads. The same policy applied to our blocked algorithm (MMBP) reduced energy by 96.4% with no increase in time. Explicit power management (see Fig. 5)) (MMBPM) reduced the number of accesses to the memory when it was inactive. Memory energy was reduced by 96.43%. Memory power management in presence of a memory buffer (MMB-BUF) reduced memory energy by 37% over MMBPM. Thus, an overall memory energy reduction by 97.7% was achieved with no increase in execution time.

Fast Fourier Transform. Simulation results for FFT are illustrated in Fig. 6. We have scaled down the results for large n to improve clarity. Data reorganization (FFTD) and implicit power management (FFTP) are only beneficial for large size problems. We observe an increase in energy for both these techniques for $n = 2^8$ due to high overheads. For larger problems, energy is reduced and the improvement is proportional to problem size. Implicit power management

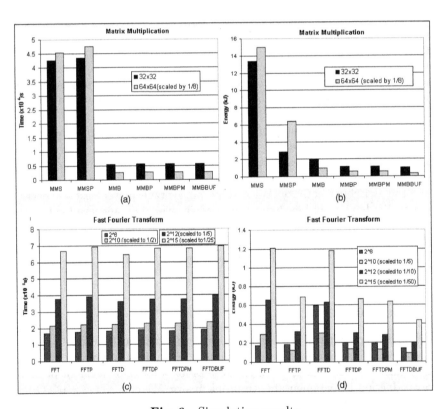

Fig. 6. Simulation results

(FFTP) reduces energy by 57%. For $n = 2^{12}$, we observe energy reduction 5% by using our algorithm (FFTD) without power management, 59.7% using implicit power management (FFTDP), 60% using memory power management (FFT-DPM) and 70% using memory buffer and power management (FFTDBUF).

6   Conclusion

In this paper, we presented algorithmic techniques for memory energy reduction by reducing data traffic and design of efficient power management schedules. Note that reduction in data traffic also decreases energy dissipation over the (high capacitance) interconnect. Currently, we do not exploit all the features of the memory such as lower energy dissipation by accessing data in a burst (see Fig. 2(b)), which will be investigated in our future work. We will also integrate other power management schemes in our analysis such as the DVS for the processor to understand the interactions between the processor and the memory. For example, slowing the processor may reduce processor energy but increase memory latency and energy.

Acknowledgment

This work is supported by the DARPA Power Aware Computing and Communication Program under contract no. F33615-02-2-4005.

References

[1] The AMRM project, http://www1.ics.uci.edu/ amrm/. 239
[2] T. Alexander and G. Kedem, "Distributed Prefetch-buffer/Cache Design for High Performance Memory systems," Symposium on High-Performance Computer Architecture (HPCA), February 1996. 244
[3] R. Athavale, N. Vijaykrishnan, M. Kandemir, and M. J. Irwin, "Influence of Array Allocation Mechanisms on Memory System Energy," International Parallel and Distributed Processing Symposium (IPDPS), April 2001. 239
[4] L. Benini, L. Macchiarulo, A. Macii, and M. Poncino, "Layout-Driven Memory Synthesis for Embedded Systems-on-Chip," IEEE Transactions on VLSI Systems, Vol. 10(2), April 2002. 239
[5] D. Burger, T. M. Austin, and S. Bennett, "The SimpleScalar Tool Set, Version 2.0," Technical Report, UW-Madison, 1997. 246
[6] F. Catthoor, K. Danckaert, S. Wuytack, and N. D. Dutt, "Code Transformations for Data Transfer and Storage Exploration Preprocessing in Multimedia Processors," IEEE Design & Test of Computers, Vol. 18(3), pp 70-82, May/June 2001. 239
[7] R. Y. Chen and M. J. Irwin, "Architectural-Level Power Estimation and Design Experiments," ACM Transactions on Design Automation of Electronic Systems (TODAES), Vol. 6(1), pp 50-66, January 2001. 237
[8] N. D. Dutt, "Memory Organization and Exploration for Embedded Systems-on-Silicon," International Conference on VLSI and CAD (ICVC), October 1997.

[9] X. Fan, C. Ellis, and A. R. Lebeck, "Memory Controller Policies for DRAM Power Management," International Symposium on Low Power Electronics and Design (ISLPED), August 2001. 239

[10] A. H. Farrahi, G. E. Tellez, and M. Sarrafzadeh, "Memory Segmentation to Exploit Sleep Mode Operation," Design Automation Conference (DAC), June 1995. 239

[11] FFTW, http://www.fft.org.

[12] J. W. Hong and H. T. Kung, "I/O Complexity: The Red-Blue Pebble Game," Symposium on Theory of Computing (STOC), May 1981. 241, 244, 246

[13] P. Kirschenhofer, P. H. Prodinger, and W. Szpankowski, "On the balance property of patricia tries: External path length view," Theoretical Computer Science, Vol. 68, pp 1-17, 1989.

[14] P. M. Kogge, V. W. Freeh, K. Ghose, N. Toomarian, and N. Aranki, "Morph: Adding an Energy Gear to a High Performance Microarchitecture for Embedded Applications," Kool Chips Workshop, MICRO-33, December 2000. 237

[15] Intel PXA250 Processor, http://www.intel.com/design/pca/prodbref/298620.htm. 237

[16] A. R. Lebeck, X. Fan, H. Zeng, and C. Ellis, "Power Aware Page Allocation," International Conference on Architectural Support for Programming Languages and Operating Systems (ASPLOS), November 2000. 239

[17] J. Luo and N. K. Jha, "Static and dynamic variable voltage scheduling algorithms for real-time heterogeneous distributed embedded systems," International Conference on VLSI Design, Jan 2002. 237

[18] T. V. Meeuwen, A. V. Zelst, F. Catthoor, "System-level Interconnect Architecture Exploration for Custom Memory Organzations," International Symposium on Systems Synthesis (ISSS), October 2001. 239

[19] Mibench version 1.0, http://eecs.umich.edu/ jringenb/mibench. 238

[20] EEMBC-Embedded Microprocessor Benchmarking Consortium, http://www.eembc.org. 238, 243

[21] "Mobile SDRAM Power Saving Features," Technical Note TN-48-10, MICRON, http://www.micron.com. 238, 240

[22] L. Nachtergaele, F. Catthoor, and C. Kulkarni, "Random-Access Data Storage Components in Customized Architectures," IEEE Design & Test of Computers, Vol. 18(3), pp 70-82, May/June 2001. 239

[23] N. Park and V. K. Prasanna, "Cache Conscious Walsh-Hadamard Transform," International Conference on Acoustics, Speech, and Signal Processing (ICASSP), May 2001. 245

[24] P. R. Panda, N. D. Dutt, A. Nicolau, F. Catthoor, A. Vandecapplelle, E. Brockmeyer, C. Kulkarni, and E. D. Greef, "Data Memory Organization and Optimizations in Application-Specific Systems," IEEE Design & Test of Computers, Vol. 18(3), pp 56-68, May/June 2001. 239

[25] PICO Radio, http://bwrc.eecs.berkeley.edu/Research/. 238

[26] J. Pouwelse, K. Langendoen, and H. Sips, "Dynamic Voltage Scaling on a Low-Power Microprocessor," UbiCom-Tech. Report, 2000. 237, 247

[27] Jan Rabaey, "Piconodes for Sensor Networks," DARPA PACC PI Meeting, 2001. 238

[28] S. Sen and S. Chatterjee, "Towards a Theory of Cache-Efficient Algorithms," Symposium on Discrete Algorithms (SODA), January 2000. 242

[29] W. T. Shiue and C. Chakrabarti, "Memory Exploration for Low Power Embedded Systems," Design Automation Conference (DAC), October 1999. 239

[30] A. Sinha and A. P. Chandrakasan, "JouleTrack - A Web Based Tool for Software Energy Profiling," Design Automation Conference (DAC), April 2001. 247

[31] The SPIRAL Project, http://www.ece.cmu.edu/~ spiral/. 245

[32] W. Tang, A. V. Veidenbaum, and R. Gupta, "Architectural Adaptation for Power and Performance," International Conference on ASIC, October 2001.

[33] K. V. Palem, R. M. rabbah, V. J. Mooney III, P. Korlmaz, and K. Puttaswamy, "Power Optimization of Embedded Memory Systems via Data Remapping," CREST Technical report, Georgia Institute of Technology, February 2002.

[34] M. Singh and V. K. Prasanna, "Application Directed Power Management for Optimizing Memory Energy Dissipation", Technical Report, EEB-Systems, University of Southern California, 2003. 238, 245

[35] M. A. Viredaz and D. A. Wallach, "Power Evaluation of a Handheld Computer: A Case Study," COMPAQ WRL Research Report, January 2001. 247

A Framework for Designing Approximation Algorithms for Scheduling Problems

Roberto Solis-Oba*

Department of Computer Science
The University of Western Ontario, London, Canada
solis@csd.uwo.ca

Abstract. Scheduling problems have attracted the attention of the algorithms community for several decades. A large number of scheduling problems have been proposed and studied, and many different techniques have been devised for solving them. Among the reasons why scheduling problems are so fascinating are their rich variety, both in form and in complexity. Furthermore, there is a sea of applications for scheduling problems, which arise from an equally varied number of areas.

This note briefly surveys a technique that has been successfully used to solve (or approximately solve) a large number of scheduling problems with minimax objective function.

1 Introduction

Scheduling problems have attracted the attention of the algorithms community for several decades. The importance of scheduling problems stems, partly, from their large and varied set of applications, including manufacturing, transportation and planning, operating systems, and communication networks, among others.

Many artificial and real-life scheduling problems have been studied, yielding as a result a rich and powerful scheduling theory. A variety of techniques for solving these problems have been devised, along with methods for determining their complexity. Many of these results have strongly influenced research in other areas.

Most scheduling problems can be described very concisely, and all of them involve a a set J of jobs that needs to be processed by a group M of machines. The processing of every job j_i requires a certain amount of time, which might or not depend on the machine(s) selected to process it. Additionally, jobs might have some properties that need to be considered when producing a schedule for them: there might be some constraints in the order in which jobs might be processed; a job might require more than one machine for its processing; a job might be suspended and resumed at a later time; a job might have a due date before which it must be completed; and so on.

---

* Author partially supported by the Natural Sciences and Engineering Research Council of Canada grant R3050A01.

Machines might be identical, or they might have different speeds. Machines might be available only during some times, or a machine might need to collaborate with other machines to process a job.

When scheduling a set of jobs, it is normally desired to minimize or maximize some objective function. Typical objective functions are minimize the maximum completion time of the jobs, minimize their average completion time, and minimize the maximum tardiness (completion time beyond the due date) of the jobs.

Equally varied is the set of techniques and methods that have been proposed to deal with scheduling problems. For an overview of these techniques the reader is referred to [5, 12, 13]. In this paper we are interested in studying polynomial time algorithms that provide exact or approximate solutions for scheduling problems. Furthermore, we are mainly interested in scheduling problems that have a min-max objective function: minimize the maximum *completion time, lateness,* or *tardiness* of the jobs. The completion time of a job is the time when the job completes its processing. If jobs have due dates, then the lateness of a job is defined as the difference between its completion time and its due date (note that the lateness of a job is negative if the job completes before its due date). The tardiness of a job is the maximum between zero and its lateness.

We describe a framework for designing *polynomial time approximation schemes* for a large number of scheduling problems with min-max objective function. This framework has been successfully used on a very large number of problems [1, 2, 4, 6, 7, 8, 9, 11, 10, 14]. A polynomial time approximation scheme (PTAS) is an algorithm that produces a solution of value at most a factor $1 + \varepsilon$ times larger than the value of an optimum solution for any precision value $\varepsilon > 0$. The running time of a PTAS is polynomial in the size of the input, but it might be super-polynomial in the inverse $1/\varepsilon$ of the precision.

2   The Framework

We assume that we know lower LB and upper UB bounds for the value OPT of the optimum solution and, furthermore, we assume that the bounds are "tight" in the sense that $UB/LB \leq \alpha$, for some constant value α. This framework might be used on problems for which there is an algorithm $\mathcal{A}$ that computes solutions of value $LB + O(P_{\max})$, where $P_{\max}$ is the maximum contribution of a job to the value of the objective function. We will specify in the next section the meaning of the "contribution of a job to the value of the objective function". For example, for the problem of scheduling jobs on identical machines to minimize the maximum completion time, $P_{\max}$ is the maximum job processing time. Note that algorithm $\mathcal{A}$ produces a near-optimal solution if all jobs are "small" in the sense of the magnitude of their individual contributions to the total value of the objective function. Many scheduling problems have the property that they can be solved almost optimally when all jobs are "small".

The set of jobs is split into 3 groups, commonly called the *large, medium,* and *small* jobs. Let P_i be the contribution of job j_i to the value of the solution.

Medium jobs j_i are chosen so that each one of them has small P_i value and the sum of P_i values of all the medium jobs is also small. More specifically, for some small constant value $\varepsilon > 0$, and constant positive integer value κ, medium job j_i has value $\varepsilon^{2\kappa} LB \leq P_i < \varepsilon^{2(\kappa-1)} LB$, and the sum of P_i values of the medium jobs is at most $\varepsilon m(LB)$. Note that $\kappa < \lceil 1/\varepsilon \rceil$. The set of medium jobs constitutes a smaller instance of the original problem formed by small jobs only. If a lower bound LB_M for this instance is small compared to LB, then algorithm $\mathcal{A}$ above can be used to find a small value solution for it. If it is possible to combine this solution with a near-optimal schedule for the large and small jobs, then we obtain a near-optimal solution for the original problem.

The interesting property about this partitioning is that the instance composed of large and small jobs consists only of jobs (large) with value $P_i \geq \varepsilon^{2(\kappa-1)} LB$ and jobs (small) with value $P_i < \varepsilon^{2\kappa} LB$. This gap between the values of small and large jobs makes it possible to deal with large and small jobs "almost" independently. We deal with the large jobs by first rounding their values up so they are multiples of $\varepsilon^{2\kappa-1} LB$. This rounding increases the value of a large job by at most a factor $1 + \varepsilon$ and, thus, it increases the value of an optimum schedule for the large and small jobs by at most the same factor.

Since each large job has a value that is only a constant factor smaller than the value of an optimum solution, then only a constant number of them can be processed by a particular machine. Hence, we can use dynamic programming (or enumeration if there is a constant number of large jobs) to find a short feasible schedule for them. Divide the time interval $[0, UB]$ into sub-intervals of size $\varepsilon^{2\kappa-1} LB$. Consider a machine and a feasible schedule for the large jobs on that machine. This schedule can be described by a $\alpha \varepsilon^{1-2\kappa}$ dimensional binary vector, called a *configuration*. The number of different configurations is $\rho = 2^{\alpha \varepsilon^{1-2\kappa}}$. Note that ρ is a huge, but constant number. The schedule for the entire set of machines is described by a ρ-dimensional vector, whose i-th component indicates the number of machines with the i-th configuration. We can find feasible short schedules for the large jobs by taking every large job and trying to place it in each one of the available configurations. The running time of this dynamic program is $O(m^\rho)$.

For each feasible schedule for the large jobs, the small jobs are assigned to the empty gaps left by them. This is usually done by using a linear program or a greedy algorithm. When assigning small jobs to gaps we must ensure that the instance defined by the small jobs on a gap of size d has optimum value at most $d + O(P'_{|rmmax})$, where P'_m is the value of the largest small job in the gap. Finally, algorithm $\mathcal{A}$ is used on each one of the gaps containing small jobs to find a feasible schedule for them. The size of each gap is increased by $O(P'_{max})$ to accommodate the schedule produced by $\mathcal{A}$. Since $\mathcal{A}$ schedules only small jobs, this increase is of value $O(\varepsilon^{2\kappa}) LB$. As there are at most $\alpha \varepsilon^{1-2\kappa}$ gaps, the overall increase caused by $\mathcal{A}$ is $O(\varepsilon) LB$.

Among all the solutions computed by the above algorithm, the one with smallest value is finally selected.

3 Some Problems

Let us apply the framework on a few scheduling problems in which the objective is to minimize the maximum completion time or *makespan*. The first four problems deal with identical machines.

Identical Machines. For each job P_i is its processing time. It is well known that valid bounds for the optimum solution are $LB = \max\{\frac{1}{m}\sum_i P_i, P_{\max}\}$, and $UB = \frac{1}{m}\sum_i P_i + P_{\max}$; $UB/LB \leq 2$. Moreover, the list scheduling algorithm behaves exactly as described for algorithm $\mathcal{A}$. For the instance of the problem formed by the medium jobs, $\mathcal{A}$ finds a schedule of length at most $\varepsilon LB + \varepsilon^{2(\kappa-1)}LB < 2\varepsilon OPT$. The schedule for the medium jobs is appended to the schedule for the large and small jobs. A greedy algorithm is used to assign small jobs to empty gaps so that a gap of size d with m' idle machines gets small jobs of total size at most $m'd$.

Release Dates. For each job j_i we choose $P_i = p_i + r_i$, where p_i is its processing time and r_i is its release date. Tight bounds for the optimum solution are $LB = \max\{\frac{1}{m}\sum_i p_i, p_{\max}, r_{\max}\}$ and $UB = \frac{1}{m}\sum_i p_i + p_{\max} + r_{\max} \leq 3LB$, where $p_{\max}$ and $r_{\max}$ are the largest processing time and release date, respectively. Note that if a feasible schedule for J is shifted by $\varepsilon^{2\kappa-1}LB$ units of time, then we can disregard release dates for small jobs (recall that small jobs have small P_i value and not just small processing times). Also, if medium jobs are scheduled after time $r_{\max}$, we can disregard their release dates and, thus, the list scheduling algorithm can be used as algorithm $\mathcal{A}$. Observe also that after rounding, we might assume that processing times and release dates of large jobs are integer multiples of $\varepsilon^{2\kappa-1}LB$.

Delivery Times. Assume that besides its processing time, every job j_i also has a delivery time q_i. Given a feasible schedule, the *delivery completion time* of a job j_i is defined as $C_i + q_i$, where C_i is the time when job j_i finishes its processing. The problem is to minimize the maximum delivery completion time. This problem is equivalent to that of minimizing the maximum lateness of the jobs.

For every job we choose $P_i = p_i$. If $q_{\max}$ is the maximum delivery time, then we can choose $LB = \max\{\frac{1}{m}\sum_i p_i, p_{\max}, q_{\max}\}$ and $UB = \frac{1}{m}\sum_i p_i + p_{\max} + q_{\max}$. To account for the delivery times we introduce a non-bottleneck machine M_0 which can "process" the delivery times. Each configuration is augmented to accommodate M_0. In each augmented configuration we only need to keep track of the largest delivery completion time in M_0. If two augmented configurations are identical except for the delivery completion time on M_0, we only keep the smaller one.

Once more, we can use the list scheduling algorithm as algorithm $\mathcal{A}$. However, when we assign small jobs to gaps, we consider the small jobs in non-increasing order of delivery time. Thus, small jobs with large delivery times are placed early in the schedule, and small jobs with small delivery times are placed near the end of the schedule.

Medium jobs are scheduled with the list scheduling algorithm, but this time the schedule for the small and large jobs is appended to the schedule for the medium jobs.

Chain Precedence Constraints. Consider that there are chain precedence constraints restricting the processing order of the jobs. Assume that each precedence chain has fixed size μ.

We consider every chain S_i as a single job and define $P_i = \sum_{j_k \in S_i} p_i$. Let $S_{\max}$ be the total processing time of the largest chain. Then, we can choose $LB = \max\{\frac{1}{m} \sum_i p_i, S_{\max}\}$ and $UB = \frac{1}{m} \sum_i p_i + S_{\max}$. The upper bound is achieved by the list scheduling algorithm.

In [9] it is shown that the value ε can be chosen so that all large chains S_i consist of only jobs with length at least $\varepsilon^{2\kappa-2} LB$ (smaller jobs in long chains can be disregarded by increasing the length of the schedule by $O(\varepsilon)LB$). By considering each chain as a single job, the list scheduling algorithm can be used as algorithm $\mathcal{A}$.

Medium and small chains are scheduled as in the case without precedence constraints.

Uniform Machines. Assume that each machine j has a speed s_j, so the time needed to process job j_i on machine j is p_i/s_j. Assume that the minimum and maximum machine speeds are 1 and $s_{\max}$, respectively. Furthermore, consider that $s_{\max}$ is constant.

For each job j_i we choose $P_i = p_i$. Tight bounds for the value of the optimum solution are $LB = \max\{\sum_i p_i / \sum_i s_i, p_{\max}/s\}$ and $UB = \sum_i p_i / \sum_i s_i + p_{\max} \leq 2LB$. The upper bound is achieved by an algorithm that considers the jobs in non-increasing order of processing time and schedules a job on the machine that minimizes its completion time [3]. This is algorithm $\mathcal{A}$.

The small jobs are greedily assigned to empty gaps, so that a gap of length d in which the subset S of machines is idle gets small jobs of total processing time $d \sum_{j \in S} s_j$.

3.1   Fixed Number of Machines

If the number m of machines is fixed, then the number of large jobs is also fixed. Hence, we can use enumeration instead of dynamic programming to schedule the large jobs. The main advantage of using enumeration is that we can ensure that one of the schedules constructed for the large jobs is identical to the schedule for the large jobs in an optimum solution. This is a powerful observation, that allows us to use the framework to design approximation algorithms for very complex problems [1, 2, 4, 7, 8, 10, 11, 14]. When considering such complex problems it is usually not easy to assign the small jobs to the empty gaps left by the large jobs.

Thus, linear programming is used instead of a greedy algorithm. A disadvantage of using linear programming is that it might produce fractional assignments of small jobs to gaps.

To illustrate the use of the framework when the number of machines is fixed, let us consider a few problems involving minimizing the makespan.

Unrelated Machines. When machines are unrelated, the processing time of a job depends on the machine that processes it. The processing time of job j_i on machine k is denoted as p_{ik}. Let $d_i = \min_k \; p_{ik}$. We choose $LB = \frac{1}{m}\sum_i d_i$ and $UB = \sum_i d_i \leq m(LB)$. The upper bound is achieved if the jobs are scheduled sequentially in their fastest machines. We choose $P_i = d_i$.

The total processing time of the medium jobs is $\varepsilon m(LB) = O(\varepsilon)LB$. Hence, the medium jobs are scheduled sequentially in their fastest machines, and this schedule is appended to the schedule for the large and small jobs.

Fix a feasible schedule for the large jobs. Let t be the maximum completion time of the largest jobs in this schedule. The time interval $[0, t]$ is divided in sub-intervals of size $\varepsilon^{2\kappa-1}LB$ as above, and we consider an additional sub-interval that starts at time t and ends at time T (T is a variable). For each job j_i we introduce variables x_{ik}^ℓ to denote the fraction of job j_i that is scheduled on machine k during sub-interval ℓ. For each sub-interval ℓ, let S_ℓ be the set of machines available for processing small jobs and let L_ℓ be the length of the sub-interval. Then, we assign small jobs to the empty gaps left by the large jobs by using the following linear program.

$$\text{Minimize } T$$
$$\begin{aligned}
\text{s.t.} \quad & \textstyle\sum_{k\ell} x_{ik}^\ell = 1 && \text{for all small jobs } j_i \\
& \textstyle\sum_{ik} x_{ik}^\ell p_{ik} \leq L_\ell |S_\ell| && \text{for all subintervals} \\
& x_{ik}^\ell = 0 && \text{for all } k \notin S_\ell \\
& x_{ik}^\ell \geq 0 && \text{for all } k \in S_\ell \\
& T \geq 0
\end{aligned}$$

This linear program has one constraint for each small job, plus a constant number, $m\varepsilon^{1-2\kappa}$, of additional constraints for the subintervals. Hence, a basic feasible solution of the linear program has only $m\varepsilon^{1-2\kappa}$ fractional values on the variables x_{ik}^ℓ. Small jobs with fractional assignments are simply moved to the end of the schedule, where they are placed sequentially in their fastest machines. This increases the length of the schedule by only $m\varepsilon(LB) = O(\varepsilon)LB$.

Finally, the other small jobs are simply scheduled according to the solution of the linear program.

Unrelated Machines with Precedence Constraints. We assume now that there are precedence constraints restricting the order in which jobs can be processed. The precedence constraints partition the set of jobs into dependent groups S_i, such that two jobs from the same group S_i need to be processed in an order consistent with the precedence constraints, but jobs belonging to different groups are independent. Here we assume that each one of the groups has fixed size, and the largest group has μ jobs.

For a group S_i let $D_i = \sum_{j_k \in S_i} d_k$. We re-define our set of jobs by considering each group as a single job, and choosing $P_i = D_i$. The jobs in each large group are rounded so that each one of them has size at least $\varepsilon^{2\kappa-1}LB$. Note that there is a constant number of different relative orders for the jobs belonging to the same group and, hence, there is a constant number of different schedules for the jobs belonging to large groups.

Fix a schedule fro the jobs in the large groups. To allocate small jobs to empty gaps we use a slightly more complicated linear program than above. This time for each small group S_i we define variables x_{ik}^{ℓ}, where k and ℓ are μ-dimensional vectors indicating the machines and intervals where the jobs from S_i are to be placed.

After solving the linear program, the small and medium jobs are scheduled as above.

References

[1] A. K. Amoura, E. Bampis, C. Kenyon, and Y. Manoussakis, Scheduling independent multiprocessor tasks, Proceedings of the 5th Annual European Symposium on Algorithms (1997), 1-12. 254, 257

[2] J. Chen and A. Miranda, A polynomial time approximation scheme for general multiprocessor job scheduling, Proceedings of the 31st Annual ACM Symposium on the Theory of Computing (1999), 418-427. 254, 257

[3] T. Gonzalez, O. Ibarra, and S. Sahni, Bounds for LPT schedules on uniform processors, *SIAM Journal on Computing* **6** (1977), 155-166. 257

[4] L. Hall, Approximability of flow shop scheduling, *Mathematical Programming* **82** (1998), 175-190. 254, 257

[5] L. Hall, Approximation algorithms for scheduling, in *Approximation algorithms for NP-hard problems* edited by D. S. Hochbaum, 1-45. 254

[6] D. Hochbaum and D. B. Shmoys, Using dual approximation algorithms for scheduling problems: practical and theoretical results, *Journal of the Association for Computing Machinery* **34** (1987), 144-162. 254

[7] K. Jansen, M. Mastrolilli, and R. Solis-Oba, Job shop scheduling problems with controllable processing times, Proceedings of the 7th Italian Conference on Theoretical Computer Science (2001), 107-122. 254, 257

[8] K. Jansen, L. Porkolab, Polynomial time approximation schemes for general multiprocessor job shop scheduling, 27th International Colloquium on Automata, Languages and Programming, ICALP 2000, LNCS 1853, 878-889, (2000). 254, 257

[9] K. Jansen and M. Sviridenko, Polynomial time approximation schemes for the multiprocessor open and flow shop scheduling problem, 17th Symposium on Theoretical Aspects of Computer Science, STACS 2000, LNCS 1770, 455-565, (2000). 254, 257

[10] K. Jansen, R. Solis-Oba, and M. Sviridenko, Makespan minimization in job shops: a polynomial time approximation scheme, SIAM Journal on Discrete Mathematics, to appear. 254, 257

[11] K. Jansen and R. Solis-Oba, Scheduling jobs with chain precedence constraints, manuscript. 254, 257

[12] D. Karger, C. Stein, J. Wein, Scheduling algorithms, in *CRC Algorithms and Theory of Computation Handbook* edited by M. J. Atallah, Chapter 35. 254

[13] E. L. Lawler, J. K. Lenstra, A. H. G. Rinooy Kan, and D. B. Shmoys, Sequencing and scheduling: algorithms and complexity, in *Handbooks in Operations Research and Management Science* **4** (1993), 445-522. 254

[14] S. V. Sevastianov and G. J. Woeginger, Makespan minimization in open shops: a polynomial time approximation scheme, *Mathematical Programming* **82** (1998), 191-198. 254, 257

Analysis and Visualization of Social Networks*

Dorothea Wagner

University of Karlsruhe
Department of Computer Science, Germany

Abstract. Social network analysis is a subdiscipline of the social sciences using graph-theoretic concepts to understand and explain social structure. We describe the main issues in social network analysis. General principles are laid out for visualizing network data in a way that conveys structural information relevant to specific research questions. Based on these innovative graph drawing techniques integrating the analysis and visualization of social networks are introduced.

1 Introduction

Social Network Analysis is a subdiscipline of the social sciences using graph-theoretic concepts to describe, understand and explain, sometimes even predict or design, social structure. It is focused on uncovering the patterning of people's interaction and based on the intuitive notion that these patterns are important features of the lives of the individuals who display them. Starting from social sciences the study of social networks became an interdisciplinary field. On one hand, it is guided by formal theory organized in mathematical terms, on the other hand grounded in the systematic analysis of empirical data. Network analysis has found important applications in organizational behavior, inter-organizational relations, the spread of contagious diseases, mental health, social support, the diffusion of information and animal social organization.

Since the 1980s, a yearly international conference on social network analysis, called *SUNBELT* is organized by the *International Network for Social Network Analysis, INSNA* [1]. A comprehensive, though non-visual, tool for social network analysis is UCINET [2]. For a comprehensive summary of social network analysis, its levels of analysis and its methodological tools see [24].

Also applications such as the analysis of Web graphs, bibliographic data, or financial data, often use similar or identical methods like in social network analysis. Recently, there is growing interest to understand the structure, dynamics and evolution of the Internet and WWW, and this way social network analysis has been rediscovered in other fields. Especially physicists in the complex systems community are interested in the statistical mechanics of complex networks. The very general questions in complex systems are how networks emerge, what

---

* The author gratefully acknowledges financial support from DFG under grant WA 654/13-1 and from the European Commission within FET Open Project COSIN (IST-2001-33555). © Vis◦n◦ logos by Christiane Nöstlinger and Ulrik Brandes.

they look like, and how they evolve. This includes networks from such diverse areas as physics, biology, economics, ecology, and computer science. Modeling networks as dynamical systems, network morphogenesis and self-organization, as well as new graph theoretical aspects and network reconstruction from experimental data are considered, [3], [4] and [5]. It seems that because of this new emerging interest in networks at all graph theory and graph algorithms attract increasing attention from other sciences [18], [19].

In 1996, we began a cooperation with researchers from political science, aimed at providing the methodology of social network analysis with tailor-made means of automated visualization. Given the importance of visualizations for scientific development, it is astonishing how little attention the subject had received so far in the analysis of social networks. One of the rare exceptions is [21]. Even though a fair amount of software has been available to facilitate graphical editing, and even automatic layout of networks, the State of the Art that time seemed too heuristic to be satisfactory for supporting network analysis.

One of the first outcomes of our interdisciplinary cooperation was a survey of visualization methods in use at that time [11]. In that paper, general principles are laid out for visualizing network data in a way that conveys structural information relevant to specific research questions. These general principles resulted in innovative uses of graph drawing methods for social network visualization, and prototypical implementations thereof. With the growing demand for access to these methods, we started implementing an integrated tool for public use, the tool $vis\circ n_e$ [15]. The main application area of $vis\circ n_e$ is a methodological approach in the social sciences. Its usage is focused on graphs of small to medium size. As an alternative especially for large graphs, we recommend to try *Pajek* [8].

2   Social Networks

A *social network* consists of entities such as persons, organizations, or things, that are linked by binary relations such as social relations, dependencies, or exchange. These relations may be directed or undirected, weighted or unweighted, and weights, if present, may be interpreted as increasing or decreasing the tie between the two entities. Since data is often gathered by means of questionnaires, not even the existence of an edge is a sure thing. The two respondents corresponding to the end-vertices of a potential edge may have different perceptions regarding the presence of a specific type of tie between them. It is a long-standing debate whether unconfirmed edges should be included in an analysis. Typically, researchers decide to either treat unconfirmed edges like confirmed edges, or to exclude them completely. A crucial feature in many studies is the interrelation between the structure of a social network and the attributes of its elements.

We define a *social network* to be a labeled directed graph $G = (V, E = E_C \cup E_U; \delta, \omega)$, where E_C and E_U are disjoint sets of *confirmed* and *unconfirmed* edges, $\delta : E \rightarrow \mathbb{R}_{\geq 0}$ is a non-negative edge *length*, and $\omega : E \rightarrow \mathbb{R}_{\geq 0}$ a non-negative edge *strength*. A vertex or edge *attribute* is a (partial) function assigning *nominal* or *numerical* values to vertices or edges.

Although we cannot put any restrictions on the class of graphs, typical examples from social science projects are sparse, locally dense, and exhibit a small average distance between pairs of vertices.

3   Analysis

The purpose of *social network analysis* is to identify important vertices, crucial relationships, subgroups, roles, network characteristics, and so on, to answer substantive questions about structures. There are three main levels of interest: the element, group, and network level. On the element level, one is interested in properties (both absolute and relative) of single actors, links, or incidences. Examples for this type of analysis are bottleneck identification and structural ranking of network items. On the group level, one is interested in classifying the elements of a network and properties of subnetworks. Examples are actor equivalence classes and cluster identification. Finally, on the network level, one is interested in properties of the overall network such as connectivity or balance.

While we have an intuitive understanding what makes a vertex important or central, there is no universally accepted definition of importance. *Centrality of a vertex* may for example be measured according to the degree of that vertex, its distance to all other vertices or the number of shortest paths between two other vertices that contain the vertex itself. Similarly, there are different notions of importance or status in a directed graph. We refer to [15] for an unification and overview of such indices. Similarly, mathematical terms that capture to what extend networks tend to build clusters, like the *clustering coefficient*, or how networks evolve, like the *degree distribution*, are of interest [4]. Questions regarding the overall structure ask for example to what extend the network exhibits the *small-world phenomenon* [25].

Algorithmic aspects in network analysis concern the fast computation of such indices. Vertex indices are often easily computable in polynomial time. However, more efficient algorithms that are applicable also for large graphs as the fast algorithm for betweenness centrality presented in [9], are of increasing interest in this context.

4   Visualization

In *graph drawing* algorithms are designed that try to produce what is often termed an "aesthetic" visualization of a graph. In *network analysis* the demand that visualizations are not misleading is maybe even more important. Hence there are two obvious criteria for the quality of social network visualizations:

1. Is the information manifest in the network represented accurately?
2. Is this information conveyed efficiently?

With these criteria in mind, the following three aspects should be carefully thought through when creating network visualizations [11]:

- the *substantive aspect* the viewer is interested in,
- the *design* (i.e. the mapping of data to graphical variables), and
- the *algorithm* employed to realize the design (artifacts, efficiency, etc.).

Depending on the context, actors of high structural importance are interpreted as a being *central* or as having *high status*. With this substantive aspect in mind, we designed visualizations that represent vertex indices by constraining vertex positions to fixed distances from the center or from the bottom of the drawing, in either case depending linearly on the vertex index. These ideas have been further developed and applied in the following projects. We also refer to [15] for a more detailed description and figures illustrating the results.

Drug Policy. This project [20] studies the presence of HIV-preventive measures for IV-drug users in nine selected German municipalities. The substantive question underlying this research is, why municipalities with comparable problem pressure differ significantly in the provision of HIV-preventive measures such as methadone substitution or needle exchange. The policy networks under scrutiny comprise all local organizations directly or indirectly involved in the provision of such measures. The actors included in the study were queried about relations such as strategic collaboration, common activities, or informal communication with other organizations in the same municipality. None of the networks has more than 120 edges of the same type, and typically more than 50% of them are unconfirmed. In [12] a three-stage force-directed method for centrality layouts is presented, and in [6] a simple, purely combinatorial algorithm is developed.

Industry Privatization. The second study [23] deals with networks of public, societal and private organizations that developed during the privatization of industrial conglomerates in East Germany as part of the economic transformation after German unification in 1990. Their privatization is understood as political bargaining processes between actors that are connected by ties such as exchange of resources, command, or consideration of interest. The privatization was foreseen to be carried out by the Treuhandanstalt, a public agency of the federal government. Due to its institutional position and its ownership of all companies, it was generally assumed to be one of the most powerful actors in the transformation of East Germany. As part of the analysis, status indices are used as indicators for the power or influence of actors. In [14] a layered layout algorithm is outlined that visually supports status analyses of networks. A refinement of this algorithm uses the linear-time algorithm of [13] for coordinate assignment.

Topic Identification. Our third example illustrates the use of methods from social network analysis in another domain, namely topic identification in texts by centering resonance analysis [16]. The structure of texts is represented by graphs that have a vertex for each word occurring in a noun phrase and an edge for each pair of words that appear together in the same noun phrase or consecutively in the same sentence. It is argued that words corresponding to nodes with high betweenness centrality in such a graph are important for the structure of the text and thus a proxy for its topic. This method was applied to Reuters news dealing with the terrorist attacks of September 11, 2001 [7] to identify, among other things, the main topics, topic changes, side stories, etc. in the news. Centrality visualizations can then be used to show for example the main topics identified for the very first day of media coverage.

5   Visone

The Visone software [15] is implemented in C++ using LEDA, the *Library of Efficient Data Types and Algorithms* [22]. While the user interface is a customized version of LEDA's GraphWin class, all graph generation, analysis, and layout algorithms (except for LEDA's force-directed layout routine) have been implemented from scratch.

Starting with version 1.1, the main data format used in Visone will be the XML sublanguage GraphML (Graph Markup Language) [10]. GraphML support is implemented in a LEDA extension package which will be made available for public use. It will hence be possible to administer project files with several social networks and any number of attributes. Data attributes can be mapped freely to graphical attributes like color, shape, and so on.

Acknowledgments

First of all, the author indebted to Ulrik Brandes who is a leading part of the project.The author thanks Sabine Cornelsen, Patrick Kenis, Jörg Raab, and Volker Schneider for many years of fruitful cooperation, the participants of POL-NET summer schools for their feedback and suggestions, and Michael Baur, Marc Benkert, Marco Gaertler, Boris Köpf, and Jürgen Lerner for their implementation efforts in developing the software tool Visone.

References

[1] International Network for Social Network Analysis, INSNA. See http://www.sfu.ca/ insna/ 261

[2] Analytic Technologies. *UINET V* . Network analysis software. See http://www.analytictech.com/ 261

[3] A.-L. Barabási and R. Albert. Emergence of scaling in random networks. *Science*, 286:509–512, 1999. 262

[4] A.-L. Barabási and R. Albert. Statistical mechanics of complex networks *Review of Modern Physics*, Vol. 74, January 2002, 47–97. 262, 263

[5] A.-L. Barabasi. LINKED: The New Science of Networks. Perseus Books, 2002. 262

[6] M. Baur and U. Brandes. An improved heuristic for crossing minimization in circular layouts. Working Paper, 2003. 264

[7] V. Batagelj, U. Brandes, J. C. Johnson, S. Kobourov, L. Krempel, A. Mrvar, and D. Wagner. Analysis and visualization of network data. Special Session during *Sunblt Social Netark Conference N*   , New Orleans, February 2002. 265

[8] Pajek. See `http://vlado.fmf.uni-lj.si/pub/networks/pajek/`. 262

[9] U. Brandes. A faster algorithm for betweenness centrality. *Journal of Mathematical Sociology*, 25(2):163–177, 2001. 263

[10] U. Brandes, M. Eiglsperger, I. Herman, M. Himsolt, and M. S. Marshall. GraphML progress report: Structural layer proposal. In: P. Mutzel, M. Jünger, and S. Leipert (eds.) *Proceedings 9th International Symposium on Gaph Drawing (G '01)* , Springer Lecture Notes in Computer Science 2265:501–512, 2002. For up-to-date information see `http://graphml.graphdrawing.org/`. 265

[11] U. Brandes, P. Kenis, J. Raab, V. Schneider, and D. Wagner. Explorations into the visualization of policy networks. *Journal of Theoretical Politics*, 11(1):75–106, 1999. 262, 263

[12] U. Brandes, P. Kenis, and D. Wagner. Communicating centrality in policy network drawings. *IEEE Transactions on Visualiation and Computer Gaphics*   , 9(2), 2003. To appear. 264

[13] U. Brandes and B. Köpf. Fast and simple horizontal coordinate assignment. In: P. Mutzel, M. Jünger, and S. Leipert (eds.) *Proceedings 9th International Symposium on Gaph Draing (G '01)*   , Springer Lecture Notes in Computer Science 2265:31–44, 2002. 264

[14] U. Brandes, J. Raab, and D. Wagner. Exploratory network visualization: Simultaneous display of actor status and connections. *Journal of Social Structure*, 2(4), 2001. 264

[15] U. Brandes and D. Wagner. Vis ne Analysis and visualization of social networks. In: P. Mutzel and M. Jünger (eds.) *Special issue on Gaph Draing Softare*   , Mathematics and Visualization. To appear. For up-to-date information on Vis ne see `http://www.visone.de/`. 262, 263, 264, 265

[16] S. R. Corman, T. Kuhn, R. D. McPhee, and K. J. Dooley. Studying complex discursive systems: Centering resonance analysis of communication. *Human Communication Research*, 28(2):157–206, 2002. 264

[17] L. Freeman Visualizing Social Networks. *Journal of Social Structure*, 1, 2000, (1).

[18] B. Hayes Computing Science: Graph Theory in Practice: Part I. *American Scientist*, Vol. 88, No. 1, January-February 2000, 9–13. 262

[19] B. Hayes Computing Science: Graph Theory in Practice: Part II. *American Scientist*, Vol. 88, No. 2, March-April 2000, 104–109. 262

[20] P. Kenis. An analysis of cooperation structures in local drug policy in germany, 1998. Unpublished Report. 264

[21] A. S. Klovdahl. A note on images of networks. *Social Netarks* 3: 197–214. 262

[22] K. Mehlhorn and S. Näher. *The LEDA Platform of Combinatorial and Gometric Computing*. Cambridge University Press, 1999. 265

[23] J. Raab. *Steuerung von Privatisierung*. Westdeutscher Verlag, 2002. 264

[24] S. Wasserman and K. Faust. Social Network Analysis. Cambridge: Cambridge University Press. 1994. 261

[25] D. J. Watts and S. H. Strogatz. Collective dynamics of "small-world" networks. *Nature*, 393:440–442, 1998. 263

Author Index